T0297452

# Advances in Computer Vision and Pattern Recognition

More information about this series at http://www.springer.com/series/4205

Radu Tudor Ionescu · Marius Popescu

# Knowledge Transfer between Computer Vision and Text Mining

## Similarity-based Learning Approaches

Radu Tudor Ionescu
Department of Computer Science
University of Bucharest
Bucharest
Romania

Marius Popescu
Department of Computer Science
University of Bucharest
Bucharest
Romania

ISSN 2191-6586 ISSN 2191-6594 (electronic)
Advances in Computer Vision and Pattern Recognition
ISBN 978-3-319-30365-9 ISBN 978-3-319-30367-3 (eBook)
DOI 10.1007/978-3-319-30367-3

Library of Congress Control Number: 2016932522



Printed on acid-free paper

This Springer imprint is published by Springer Nature
The registered company is Springer International Publishing AG Switzerland

*To our dear families and friends*

# Foreword

The present book basically studies similarity-based learning approaches for two different fields: computer vision and string processing. However, the discussed text goes far beyond the goal of a general or even of a comprehensive presentation. From the very beginning, the reader is faced with a genuine scientific challenge: accepting the authors' view according to which image and text can and should be treated in a similar fashion.

Computer vision and string processing seem and are traditionally considered two unrelated fields of study. A question which naturally arises is whether this classical view of the two fields can or should be modified.

While learning from data is a central scientific issue nowadays, no one should claim to be a data analyst without having performed string processing. Information retrieval and extraction ultimately depend on string manipulation. From a different angle, computer vision is concerned with the theory behind artificial systems that extract information from images. One is finally concerned with a goal of the same nature: information acquisition. From this perspective, the approach proposed by the authors seems more natural and indeed scientifically justified.

The authors consider treating images as text and improving text processing techniques with knowledge coming from computer vision. Indeed, corresponding concepts like word and visual word do exist, while the existing literature regards, for instance, modeling object recognition as machine translation. The authors present improved methods that exploit such concepts as well as novel approaches, while broadening the meaning of classical concepts like string processing by taking into account tasks ranging from phylogenetic analysis and DNA sequence alignment to native language identification and text categorization by topic. All in all, the authors gradually build a strong case in favor of the theory they are promoting:

knowledge transfer from one of the studied domains to the other being extremely productive. The very topic of this book represents a scientific challenge, one that the authors master with a great precision and that offers interesting perspectives for future scientific research.

Florentina Hristea

# Preface

Machine learning is currently a vast area of research with applications in a broad range of fields such as computer vision, bioinformatics, information retrieval, natural language processing, audio processing, data mining, and many others. Among the variety of state-of-the-art machine learning approaches for such applications are the similarity-based learning methods. Learning based on similarity refers to the process of learning based on pairwise similarities between the training samples. The similarity-based learning process can be both supervised and unsupervised, and the pairwise relationship can be either a similarity, a dissimilarity, or a distance function.

This book studies several similarity-based learning approaches, such as nearest neighbor models, local learning, kernel methods, and clustering algorithms. A nearest neighbor model based on a novel dissimilarity for images is presented in this book. It is used for handwritten digit recognition and achieves impressive results. Kernel methods are used in several tasks investigated in this book. First, a novel kernel for visual word histograms is presented. It achieves state-of-the-art performance for object recognition in images. Several kernels based on a pyramid representation are presented next. They are used for facial expression recognition from static images. The same pyramid representation is successfully used for text categorization by topic. Moreover, an approach based on string kernels for native language identification is also presented in this work. The approach achieves state-of-the-art performance levels, while being language independent and theory neutral. An interesting pattern can already be observed, namely that the machine learning tasks approached in this book can be divided into two different areas: computer vision and string processing.

Despite the fact that computer vision and string processing seem to be unrelated fields of study, image analysis and string processing are in some ways similar. As will be shown by the end of this book, the concept of treating image and text in a similar fashion has proven to be very fertile for specific applications in computer vision. In fact, one of the state-of-the-art methods for image categorization is inspired by the *bag of words* representation, which is very popular in information retrieval and natural language processing. Indeed, the *bag of visual words* model, which builds a vocabulary of visual words by clustering local image descriptors extracted from images, has demonstrated impressive levels of performance for image categorization and image retrieval. By adapting string processing techniques for image analysis or the other way around, knowledge from one domain can be transferred to the other. In fact, many breakthrough discoveries have been made by transferring knowledge between different domains. This book follows this line of research and presents novel approaches or improved methods that rely on this concept. First of all, a dissimilarity measure for images is presented. The dissimilarity measure is inspired by the rank distance measure for strings. The main concern is to extend rank distance from one-dimensional input (strings) to two-dimensional input (digital images). While rank distance is a highly accurate measure for strings, the empirical results presented in this book suggest that the proposed extension of rank distance to images is very accurate for handwritten digit recognition and texture analysis. Second of all, a kernel that stems from the same idea is also presented in this book. The kernel is designed to encode the spatial information in an efficient way and it shows performance improvements in object class recognition and text categorization by topic. Third of all, some improvements to the popular bag of visual words model are proposed in the present book. As mentioned before, this model is inspired by the bag of words model from natural language processing and information retrieval. A new distance measure for strings is introduced in this work. It is inspired by the image dissimilarity measure based on patches that is also described in the present book. Designed to conform to more general principles and adapted to DNA strings, it comes to improve several state-of-the-art methods for DNA sequence analysis. Furthermore, another application of this novel distance measure for strings is discussed. More precisely, a kernel based on this distance measure is used for native language identification. To summarize, all the contributions presented in this book come to support the concept of treating image and text in a similar manner.

It is worth mentioning that the studied methods exhibit state-of-the-art performance levels in the approached tasks. A few arguments come to support this claim. First of all, an improved bag of visual words model described in this work obtained the fourth place at the Facial Expression Recognition (FER) Challenge of the ICML 2013 Workshop in Challenges in Representation Learning (WREPL). Second of all, the system based on string kernels presented in this book ranked on third place in the closed Native Language Identification Shared Task of the BEA-8 Workshop of NAACL 2013. Third of all, the PQ kernel for visual word histograms described in this work received the Caianiello Best Young Paper Award at ICIAP 2013. Together, these achievements reflect the significance of the methods described in the present book.

# Contents

1 **Motivation and Overview** . . . . . 1
1.1 Introduction . . . . . 1
1.2 Knowledge Transfer between Image and Text . . . . . 2
1.3 Overview and Organization . . . . . 7
References . . . . . 11

2 **Learning based on Similarity** . . . . . 15
2.1 Introduction . . . . . 15
2.2 Nearest Neighbor Approach . . . . . 17
2.3 Local Learning . . . . . 20
2.4 Kernel Methods . . . . . 22
2.4.1 Mathematical Preliminaries . . . . . 22
2.4.2 Overview of Kernel Classifiers . . . . . 24
2.4.3 Kernel Functions . . . . . 26
2.4.4 Kernel Normalization . . . . . 28
2.4.5 Generic Kernel Algorithm . . . . . 29
2.4.6 Multiple Kernel Learning . . . . . 29
2.5 Cluster Analysis . . . . . 30
2.5.1 K-Means Clustering . . . . . 32
2.5.2 Hierarchical Clustering . . . . . 33
References . . . . . 35

**Part I Knowledge Transfer from Text Mining to Computer Vision**

3 **State-of-the-Art Approaches for Image Classification** . . . . . 41
3.1 Introduction . . . . . 41
3.2 Image Distance Measures . . . . . 42
3.2.1 Color Image Distances . . . . . 42
3.2.2 Grayscale Image Distances . . . . . 44
3.2.3 Earth Mover's Distance . . . . . 44
3.2.4 Tangent Distance . . . . . 44
3.2.5 Shape Match Distance . . . . . 45

3.3 Patch-based Techniques . . . . 45
3.4 Image Descriptors . . . . 46
3.5 Bag of Visual Words . . . . 47
3.5.1 Encoding Spatial Information . . . . 48
3.6 Deep Learning . . . . 49
References . . . . 50

**4 Local Displacement Estimation of Image Patches and Textons** . . . . 53
4.1 Introduction . . . . 53
4.2 Local Patch Dissimilarity . . . . 54
4.2.1 Extending Rank Distance to Images . . . . 54
4.2.2 Local Patch Dissimilarity Algorithm . . . . 56
4.2.3 LPD Algorithm Optimization . . . . 58
4.3 Properties of Local Patch Dissimilarity . . . . 60
4.4 Experiments and Results . . . . 61
4.4.1 Data Sets Description . . . . 61
4.4.2 Learning Methods . . . . 62
4.4.3 Parameter Tuning . . . . 64
4.4.4 Baseline Experiment . . . . 67
4.4.5 Kernel Experiment . . . . 72
4.4.6 Difficult Experiment . . . . 74
4.4.7 Filter-based Nearest Neighbor Experiment . . . . 75
4.4.8 Local Learning Experiment . . . . 78
4.4.9 Birds Experiment . . . . 79
4.5 Local Texton Dissimilarity . . . . 81
4.5.1 Texton-based Methods . . . . 81
4.5.2 Texture Features . . . . 82
4.5.3 Local Texton Dissimilarity Algorithm . . . . 83
4.6 Texture Experiments and Results . . . . 85
4.6.1 Data Sets Description . . . . 86
4.6.2 Learning Methods . . . . 88
4.6.3 Brodatz Experiment . . . . 89
4.6.4 UIUCTex Experiment . . . . 91
4.6.5 Biomass Experiment . . . . 95
4.7 Discussion . . . . 96
References . . . . 96

**5 Object Recognition with the Bag of Visual Words Model** . . . . 99
5.1 Introduction . . . . 99
5.2 Bag of Visual Words Model . . . . 101
5.3 PQ Kernel for Visual Word Histograms . . . . 103
5.4 Spatial Non-Alignment Kernel . . . . 107
5.4.1 Translation and Size Invariance . . . . 109
5.5 Object Recognition Experiments . . . . 110
5.5.1 Data Sets Description . . . . 112
5.5.2 Implementation and Evaluation Procedure . . . . 113

5.5.3 PQ Kernel Results on Pascal VOC Experiment . . . . . . . 115
5.5.4 PQ Kernel Results on Birds Experiment . . . . . . . . . . . . 118
5.5.5 SNAK Parameter Tuning . . . . . . . . . . . . . . . . . . . . . . 119
5.5.6 SNAK Results on Pascal VOC Experiment . . . . . . . . . . 120
5.5.7 SNAK Results on Birds Experiment . . . . . . . . . . . . . . . 121
5.6 Bag of Visual Words for Facial Expression Recognition . . . . . . 122
5.7 Local Learning . . . . . . . . . . . . . . . . . . . . . . . . . . . . . 125
5.8 Facial Expression Recognition Experiments . . . . . . . . . . . . . . 125
5.8.1 Data Set Description . . . . . . . . . . . . . . . . . . . . . . . 125
5.8.2 Implementation . . . . . . . . . . . . . . . . . . . . . . . . . . . 127
5.8.3 Parameter Tuning and Results . . . . . . . . . . . . . . . . . . 127
5.9 Discussion . . . . . . . . . . . . . . . . . . . . . . . . . . . . . . . 129
References . . . . . . . . . . . . . . . . . . . . . . . . . . . . . . . . . 130

**Part II Knowledge Transfer from Computer Vision to Text Mining**

**6 State-of-the-Art Approaches for String and Text Analysis** . . . . . . . 135
6.1 Introduction . . . . . . . . . . . . . . . . . . . . . . . . . . . . . . . 135
6.2 String Distance Measures . . . . . . . . . . . . . . . . . . . . . . . 136
6.2.1 Hamming Distance . . . . . . . . . . . . . . . . . . . . . . . . . 136
6.2.2 Edit Distance . . . . . . . . . . . . . . . . . . . . . . . . . . . . 137
6.2.3 Rank Distance . . . . . . . . . . . . . . . . . . . . . . . . . . . . 137
6.3 Computational Biology . . . . . . . . . . . . . . . . . . . . . . . . . 139
6.3.1 Sequencing and Comparing DNA . . . . . . . . . . . . . . . . 139
6.3.2 Phylogenetic Analysis . . . . . . . . . . . . . . . . . . . . . . . 140
6.4 Text Mining . . . . . . . . . . . . . . . . . . . . . . . . . . . . . . . 141
6.4.1 String Kernels . . . . . . . . . . . . . . . . . . . . . . . . . . . . 142
References . . . . . . . . . . . . . . . . . . . . . . . . . . . . . . . . . 144

**7 Local Rank Distance** . . . . . . . . . . . . . . . . . . . . . . . . . . . 149
7.1 Introduction . . . . . . . . . . . . . . . . . . . . . . . . . . . . . . . 149
7.2 Approach . . . . . . . . . . . . . . . . . . . . . . . . . . . . . . . . . 151
7.3 Local Rank Distance Definition . . . . . . . . . . . . . . . . . . . . 153
7.4 Local Rank Distance Algorithm . . . . . . . . . . . . . . . . . . . . 155
7.5 Properties of Local Rank Distance . . . . . . . . . . . . . . . . . . 158
7.6 Local Rank Distance Sequence Aligners . . . . . . . . . . . . . . . 161
7.6.1 Indexing Strategies and Efficiency Improvements . . . . . . 163
7.7 Experiments and Results . . . . . . . . . . . . . . . . . . . . . . . . 165
7.7.1 Data Sets Description . . . . . . . . . . . . . . . . . . . . . . . 165
7.7.2 Phylogenetic Analysis . . . . . . . . . . . . . . . . . . . . . . . 167
7.7.3 DNA Comparison . . . . . . . . . . . . . . . . . . . . . . . . . . 171
7.7.4 Alignment in the Presence of Contaminated Reads . . . . . 172
7.7.5 Clustering an Unknown Organism . . . . . . . . . . . . . . . 180
7.7.6 Time Evaluation of Sequence Aligners . . . . . . . . . . . . . 184
7.7.7 Experiment on Vibrio Species . . . . . . . . . . . . . . . . . . 185

7.8 Discussion. . . . . . . . . . . . . . . . . . . . . . . . . . . . . . . . . . . . . . . 187
References . . . . . . . . . . . . . . . . . . . . . . . . . . . . . . . . . . . . . . . . 189

**8 Native Language Identification with String Kernels** . . . . . . . . . . . . 193
8.1 Introduction. . . . . . . . . . . . . . . . . . . . . . . . . . . . . . . . . . . . . . 193
8.2 Related Work . . . . . . . . . . . . . . . . . . . . . . . . . . . . . . . . . . . . . 195
8.2.1 Native Language Identification . . . . . . . . . . . . . . . . . . 195
8.2.2 Methods that Work at the Character Level . . . . . . . . . . 196
8.3 Similarity Measures for Strings . . . . . . . . . . . . . . . . . . . . . . . . 197
8.3.1 String Kernels . . . . . . . . . . . . . . . . . . . . . . . . . . . . . . . 197
8.3.2 Kernel based on Local Rank Distance . . . . . . . . . . . . . 200
8.4 Learning Methods . . . . . . . . . . . . . . . . . . . . . . . . . . . . . . . . . . 200
8.5 Experiments . . . . . . . . . . . . . . . . . . . . . . . . . . . . . . . . . . . . . . 203
8.5.1 Data Sets Description . . . . . . . . . . . . . . . . . . . . . . . . . 203
8.5.2 Parameter Tuning and Implementation Choices . . . . . . . 205
8.5.3 Experiment on TOEFL11 Corpus. . . . . . . . . . . . . . . . . . 207
8.5.4 Experiment on ICLE Corpus . . . . . . . . . . . . . . . . . . . . 210
8.5.5 Experiment on TOEFL11-Big Corpus. . . . . . . . . . . . . . 211
8.5.6 Cross-Corpus Experiment . . . . . . . . . . . . . . . . . . . . . . 213
8.5.7 Experiment on ALC Subset Corpus . . . . . . . . . . . . . . . 214
8.5.8 Experiment on ASK Corpus . . . . . . . . . . . . . . . . . . . . 217
8.6 Language Transfer Analysis. . . . . . . . . . . . . . . . . . . . . . . . . . 220
8.7 Discussion. . . . . . . . . . . . . . . . . . . . . . . . . . . . . . . . . . . . . . . 224
References . . . . . . . . . . . . . . . . . . . . . . . . . . . . . . . . . . . . . . . . 225

**9 Spatial Information in Text Categorization** . . . . . . . . . . . . . . . . . . 229
9.1 Introduction. . . . . . . . . . . . . . . . . . . . . . . . . . . . . . . . . . . . . . 229
9.2 Related Work . . . . . . . . . . . . . . . . . . . . . . . . . . . . . . . . . . . . . 231
9.3 Methods to Encode Spatial Information. . . . . . . . . . . . . . . . . . 232
9.3.1 Spatial Pyramid for Text. . . . . . . . . . . . . . . . . . . . . . . . 233
9.3.2 Spatial Non-Alignment Kernel for Text . . . . . . . . . . . . 234
9.4 Experiments . . . . . . . . . . . . . . . . . . . . . . . . . . . . . . . . . . . . . . 236
9.4.1 Data Set Description . . . . . . . . . . . . . . . . . . . . . . . . . . 236
9.4.2 Implementation Choices . . . . . . . . . . . . . . . . . . . . . . . . 236
9.4.3 Evaluation Procedure . . . . . . . . . . . . . . . . . . . . . . . . . . 237
9.4.4 Experiment on Reuters-21578 Corpus. . . . . . . . . . . . . . 238
9.5 Discussion. . . . . . . . . . . . . . . . . . . . . . . . . . . . . . . . . . . . . . . 239
References . . . . . . . . . . . . . . . . . . . . . . . . . . . . . . . . . . . . . . . . 240

**10 Conclusions** . . . . . . . . . . . . . . . . . . . . . . . . . . . . . . . . . . . . . . 243
10.1 Discussion and Conclusions. . . . . . . . . . . . . . . . . . . . . . . . . . 243
References . . . . . . . . . . . . . . . . . . . . . . . . . . . . . . . . . . . . . . . . 245

**Index** . . . . . . . . . . . . . . . . . . . . . . . . . . . . . . . . . . . . . . . . . . . . . 247

# List of Figures


Figure 1.1 An example in which the context helps to disambiguate an object (kitchen glove), which can easily be mistaken for something else if the rest of the image is not seen. The image belongs to the Pascal VOC 2007 data set. **a** A picture of a kitchen glove. **b** A picture of the same glove with context . . . . . . . . . . . . . . . . . . . . . . . . . . . . . 4

Figure 1.2 An object that can be described by multiple categories such as toy, bear, or both.. . . . . . . . . . . . . . . . . . . . . . . . . 5

Figure 2.1 A 3-NN model for handwritten digit recognition. For visual interpretation, digits are represented in a two-dimensional feature space. The figure shows 30 digits sampled from the popular MNIST data set. When the new digit $x$ needs to be recognized, the 3-NN model selects the nearest 3 neighbors and assigns label 4 based on a majority vote . . . . . . . . . . . . . . . . . . . . . . . . . . . . . . . . . . . . 18

Figure 2.2 A 1-NN model for handwritten digit recognition. The figure shows 30 digits sampled from the popular MNIST data set. The decision boundary of the 1-NN model generates a Voronoi partition of the digits. . . . . . . . . . . . . . 18

Figure 2.3 Two classification models are used to solve the same binary classification problem. The two test samples depicted in *red* are misclassified by a global linear classifier (*left-hand* side). The local learning framework produces a nonlinear decision boundary that fixes this problem (*right-hand* side). **a** A global linear classifier misclassifies the test samples depicted in *red*. **b** A local learning model based on an underlying linear classifier is able to correctly classify the test samples depicted in *red*. . . . . . . . . . . . . . . . . . . . . . . . . . . . . . . . . . . 21

Figure 2.4 The function $\phi$ embeds the data into a feature space where the nonlinear relations now appear linear. Machine learning methods can easily detect such linear relations .... 24

Figure 4.1 Two images that are compared with LPD. **a** For every position $(x_1; y_1)$ in the first image, LPD tries to find a similar patch in the second image. First, it looks at the same position $(x_1; y_1)$ in the second image. The patches are not similar. **b** LPD gradually looks around position $(x_1; y_1)$ in the second image to find a similar patch. **c** LPD sum up the spatial offset between the similar patches at $(x_1; y_1)$ from the first image and $(x_2; y_2)$ from the second image .... 57

Figure 4.2 A random sample of 15 handwritten digits from the MNIST data set.............................. 62

Figure 4.3 A random sample of 12 images from the Birds data set. There are two images per class. Images from the same class sit next to each other in this figure............... 63

Figure 4.4 Average accuracy rates of the 3-NN based on LPD model with patches of $1 \times 1$ pixels at the *top* and $2 \times 2$ pixels at the *bottom*. Experiment performed on the MNIST subset of 100 images. **a** Accuracy rates with patches of $1 \times 1$ pixels. **b** Accuracy rates with patches of $2 \times 2$ pixels...... 65

Figure 4.5 Average accuracy rates of the 3-NN based on LPD model with patches of $3 \times 3$ pixels at the *top* and $4 \times 4$ pixels at the *bottom*. Experiment performed on the MNIST subset of 100 images. **a** Accuracy rates with patches of $3 \times 3$ pixels. **b** Accuracy rates with patches of $4 \times 4$ pixels...... 66

Figure 4.6 Average accuracy rates of the 3-NN based on LPD model with patches of $5 \times 5$ pixels at the *top* and $6 \times 6$ pixels at the *bottom*. Experiment performed on the MNIST subset of 100 images. **a** Accuracy rates with patches of $5 \times 5$ pixels. **b** Accuracy rates with patches of $6 \times 6$ pixels...... 67

Figure 4.7 Average accuracy rates of the 3-NN based on LPD model with patches of $7 \times 7$ pixels at the *top* and $8 \times 8$ pixels at the *bottom*. Experiment performed on the MNIST subset of 100 images. **a** Accuracy rates with patches of $7 \times 7$ pixels. **b** Accuracy rates with patches of $8 \times 8$ pixels...... 68

Figure 4.8 Average accuracy rates of the 3-NN based on LPD model with patches of $9 \times 9$ pixels at the *top* and $10 \times 10$ pixels at the *bottom*. Experiment performed on the MNIST subset of 100 images. **a** Accuracy rates with patches of $9 \times 9$ pixels. **b** Accuracy rates with patches of $10 \times 10$ pixels .... 69

Figure 4.9 Average accuracy rates of the 3-NN based on LPD model with patches ranging from $2 \times 2$ pixels to $9 \times 9$ pixels. Experiment performed on the MNIST subset of 300 images . . . . . 70
Figure 4.10 Similarity matrix based on LPD with patches of $4 \times 4$ pixels and a similarity threshold of 0.12, obtained by computing pairwise dissimilarities between the samples of the MNIST subset of 1000 images . . . . . 72
Figure 4.11 Euclidean distance matrix based on $L_2$-norm, obtained by computing pairwise distances between the samples of the MNIST subset of 1000 images . . . . . 73
Figure 4.12 Error rate drops as K increases for 3-NN (○) and 6-NN (◇) classifiers based on LPD with filtering . . . . . 77
Figure 4.13 Sample images from three classes of the Brodatz data set . . . . . 87
Figure 4.14 Sample images from four classes of the UIUCTex data set. Each image is showing a textured surface viewed under different poses. **a** Bark. **b** Pebbles. **c** Brick. **d** Plaid . . . . . 88
Figure 4.15 Sample images from the biomass texture data set. **a** Wheat. **b** Waste. **c** Corn . . . . . 89
Figure 4.16 Similarity matrix based on LTD with patches of $32 \times 32$ pixels and a similarity threshold of 0.02, obtained by computing pairwise dissimilarities between the texture samples of the Brodatz data set . . . . . 92
Figure 4.17 Similarity matrix based on LTD with patches of $64 \times 64$ pixels and a similarity threshold of 0.02, obtained by computing pairwise dissimilarities between the texture samples of the UIUCTex data set . . . . . 94
Figure 5.1 The BOVW learning model for object class recognition. The feature vector consists of SIFT features computed on a regular grid across the image (dense SIFT) and vector quantized into visual words. The frequency of each visual word is then recorded in a histogram. The histograms enter the training stage. Learning is done by a kernel method . . . . . 102
Figure 5.2 The spatial similarity of two images computed with the SNAK framework. First, the center of mass is computed according to the objectness map. The average position and the standard deviation of the spatial distribution of each visual word are computed next. The images are aligned according to their centers, and the SNAK kernel is computed by summing the distances between the average positions and the standard deviations of each visual word in the two images . . . . . 111

Figure 5.3 A random sample of 12 images from the Pascal VOC data set. Some of the images contain objects of more than one class. For example, the image at the *top left* shows a dog sitting on a couch, and the image at the *top right* shows a person and a horse. Dog, couch, person, and horse are among the 20 classes of this data set . . . . . . . . . . . . . . . . 112
Figure 5.4 A random sample of 12 images from the Birds data set. There are two images per class. Images from the same class sit next to each other in this figure . . . . . . . . . . . . . . . 113
Figure 5.5 The BOVW learning model for facial expression recognition. The feature vector consists of SIFT features computed on a regular grid across the image (dense SIFT) and vector quantized into visual words. The presence of each visual word is then recorded in a presence vector. Normalized presence vectors enter the training stage. Learning is done by a local kernel method . . . . . . . . . . . . . 124
Figure 5.6 An example of SIFT features extracted from two images representing distinct emotions: fear (*left*) and disgust (*right*) . . . . . . . . . . . . . . . . . . . . . . . . . . . . . . . . . . . . . . . 125
Figure 5.7 The six nearest neighbors selected with the presence kernel from the vicinity of the test image are visually more similar than the other six images randomly selected from the training set. Despite of this fact, the nearest neighbors do not adequately indicate the test label (disgust). Thus, a learning method needs to be trained on the selected neighbors to accurately predict the label of the test image . . . . . . . . . . . . . . . . . . . . . . . . . . . . . . . . . . . . . . 126
Figure 7.1 Phylogenetic tree obtained for 22 mammalian mtDNA sequences using LRD based on 2-mers . . . . . . . . . . 168
Figure 7.2 Phylogenetic tree obtained for 22 mammalian mtDNA sequences using LRD based on 4-mers . . . . . . . . . . . . . . . . 168
Figure 7.3 Phylogenetic tree obtained for 22 mammalian mtDNA sequences using LRD based on 6-mers . . . . . . . . . . 169
Figure 7.4 Phylogenetic tree obtained for 22 mammalian mtDNA sequences using LRD based on 8-mers . . . . . . . . . . . . . . . . 169
Figure 7.5 Phylogenetic tree obtained for 22 mammalian mtDNA sequences using LRD based on 10-mers . . . . . . . . . 170
Figure 7.6 Phylogenetic tree obtained for 22 mammalian mtDNA sequences using LRD based on sum of $k$-mers . . . . . . . . . . 170
Figure 7.7 Phylogenetic tree obtained for 27 mammalian mtDNA sequences using LRD based on 18-mers . . . . . . . . . 171

Figure 7.8 The distance evolution of the best chromosome at each generation for the rat–mouse–cow experiment. The *green line* represents the rat–house mouse (RH) distance, the *blue line* represents the rat–fat dormouse (RF) distance, and the *red line* represents the rat–cow (RC) distance. . . . . . 173

Figure 7.9 The precision–recall curves of the state-of-the-art aligners versus the precision–recall curves of the two LRD aligners, when 10,000 contaminated reads of length 100 from the orangutan are included. The two variants of the BOWTIE aligner are based on local and global alignment, respectively. The LRD aligner based on hash tables is a fast approximate version of the original LRD aligner . . . . . . 175

Figure 7.10 The precision–recall curves of the state-of-the-art aligners versus the precision–recall curves of the two LRD aligners, when 50,000 contaminated reads of length 100 from 5 mammals are included. The two variants of the BOWTIE aligner are based on local and global alignment, respectively. The LRD aligner based on hash tables is a fast approximate version of the original LRD aligner . . . . . . 178

Figure 7.11 Local Rank Distance computed in the presence of different types of DNA changes such as point mutations, indels, and inversions. In the first three cases **a–c**, a single type of DNA polymorphism is included in the second (*bottom*) string. The last case **d** shows how LRD measures the differences between the two DNA strings when all the types of DNA changes occur in the second string. The nucleotides affected by changes are marked with bold. To compare the results for the different types of DNA changes, the first string is always the same in all the four cases. Note that in all the four examples, LRD is based on 1-mers. In each case, $\Delta_{LRD} = \Delta_{\text{left}} + \Delta_{\text{right}}$. **a** Measuring LRD with point mutations. The *T* at index 7 is substituted with *C*. **b** Measuring LRD with indels. The substring *GT* is deleted. **c** Measuring LRD with inversions. The substring *AGTT* is inverted. **d** Measuring LRD with point mutations, indels, and invensions . . . . . . . . . . . . . . . . . . . . . . . . . . . 188

Figure 8.1 An example with three classes that illustrates the masking problem. Class *A* is masked by classes *B* and *C*. . . . . . . . . . 203

# List of Tables


Table 4.1 Results of the experiment performed on the MNIST subset of 300 images, using the 3-NN based on LPD model with patches ranging from $2 \times 2$ pixels to $9 \times 9$ pixels . . . . . . . . . 70
Table 4.2 Results of the experiment performed on the MNIST subset of 300 images, using various maximum offsets, patches of $4 \times 4$ pixels, and a similarity threshold of 0.12 . . . . . . . . . . . 71
Table 4.3 Baseline 3-NN versus 3-NN based on LPD . . . . . . . . . . . . . 71
Table 4.4 Accuracy rates of several classifiers based on LPD versus the accuracy rates of the standard SVM and KRR . . . . . . . . . 73
Table 4.5 Comparison of several classifiers (some based on LPD) . . . . . 74
Table 4.6 Error and time of the 3-NN classifier based on LPD with filtering, for various K values . . . . . . . . . . . . . . . . . . . . . . . 76
Table 4.7 Confusion matrix of the 3-NN based on LPD with filtering using $K = 50$ . . . . . . . . . . . . . . . . . . . . . . . . . . . 78
Table 4.8 Error rates on the entire MNIST data set for baseline 3-NN, $k$-NN based on Tangent distance, and $k$-NN based on LPD with filtering . . . . . . . . . . . . . . . . . . . . . . . . . . . . . 78
Table 4.9 Error rates of different $k$-NN models on Birds data set . . . . . 80
Table 4.10 Error on Birds data set for texton learning methods of Lazebnik et al. (2005a) and kernel methods based on LPD . . . . . . . . . . . . . . . . . . . . . . . . . . . . . . . . . . 80
Table 4.11 Accuracy rates on the Brodatz data set using 3 random samples per class for training . . . . . . . . . . . . . . . . . . . . . . . 90
Table 4.12 Accuracy rates of several MKL approaches that include LTD compared with state-of-the-art methods on the Brodatz data set . . . . . . . . . . . . . . . . . . . . . . . . . . . . . 90
Table 4.13 Accuracy rates on the UIUCTex data set using 20 random samples per class for training . . . . . . . . . . . . . . . . . . . . . . . 93
Table 4.14 Accuracy rates of several MKL approaches that include LTD compared with state-of-the-art methods on the UIUCTex data set . . . . . . . . . . . . . . . . . . . . . . . . . . . . . 93

Table 4.15 Accuracy rates on the Biomass Texture data set using 20, 30 and 40 random samples per class for training and 70, 60 and 50 for testing, respectively . . . . . . . . . . . . . . 95
Table 5.1 Mean AP on Pascal VOC 2007 data set for SVM based on different kernels . . . . . . . . . . . . . . . . . . . . . . . . . . . . . . . . 115
Table 5.2 Mean AP on the 20 classes of the Pascal VOC 2007 data set for the SVM classifier based on 3000 visual words using the spatial pyramid representation and different kernels . . . . . . . . . . . . . . . . . . . . . . . . . . . . . . . . . . . . . . 117
Table 5.3 Running time required by each kernel to compute the two kernel matrices for training and testing, respectively . . . . . . . 117
Table 5.4 Classification accuracy on the Birds data set for SVM based on different kernels . . . . . . . . . . . . . . . . . . . . . . . . . . 118
Table 5.5 Mean AP on Pascal VOC 2007 data set for different representations that encode spatial information into the BOVW model . . . . . . . . . . . . . . . . . . . . . . . . . . . . . . . . . 120
Table 5.6 Classification accuracy on the Birds data set for different representations that encode spatial information into the BOVW model . . . . . . . . . . . . . . . . . . . . . . . . . . . . . . . . . 122
Table 5.7 Accuracy levels for several BOVW models obtained on the FER validation, test, and private test sets . . . . . . . . . . . . . . . 128
Table 7.1 The 27 mammals from the EMBL database used in the phylogenetic experiments . . . . . . . . . . . . . . . . . . . . . . . . . . . 166
Table 7.2 The genomic sequence information of three vibrio pathogens consisting of two circular chromosomes . . . . . . . . 167
Table 7.3 The number of misclustered mammals for different clustering techniques on the 22 mammals data set . . . . . . . . . 171
Table 7.4 Closest string results for the genetic algorithm based on LRD with 3-mers . . . . . . . . . . . . . . . . . . . . . . . . . . . . . . . . 172
Table 7.5 Several statistics of the state-of-the-art aligners versus the LRD aligner, when 10,000 contaminated reads of length 100 sampled from the orangutan genome are included . . . . . . 176
Table 7.6 Metrics of the human reads mapped to the human mitochondrial genome (true positives) by the hash LRD aligner versus the human reads that are not mapped to the genome (false negatives) . . . . . . . . . . . . . . . . . . . . . . . . . . 176
Table 7.7 Several statistics of the state-of-the-art aligners versus the LRD aligner, when 50,000 contaminated reads of length 100 sampled from the genomes of five mammals are included . . . . . . . . . . . . . . . . . . . . . . . . . . . . . . . . . . . . . . 178
Table 7.8 The recall at best precision of the state-of-the-art aligners versus the LRD aligner, when 10,000 contaminated reads of length 100 sampled from the orangutan genome are included . . . . . . . . . . . . . . . . . . . . . . . . . . . . . . . . . . . . . . 179

Table 7.9 The recall at best precision of the state-of-the-art aligners versus the LRD aligner, when 40,000 contaminated reads of length 100 sampled from the blue whale, the harbor seal, the donkey, and the house mouse genomes are included, respectively . . . . 179
Table 7.10 The results for the real-world setting experiment on mammals . . . . 182
Table 7.11 The results for the hard setting experiment on mammals . . . . 183
Table 7.12 The running times of the BWA aligner, the BLAST aligner, the BOWTIE aligner, and the LRD aligner . . . . 184
Table 7.13 The results of the rank-based aligner on vibrio species . . . . 186
Table 8.1 Summary of corpora used in the experiments . . . . 203
Table 8.2 Distribution of the documents per native language in the ALC subset . . . . 204
Table 8.3 Average word length and optimal $p$-gram range for the TOEFL11 corpus (English L2), the ALC subset (Arabic L2), and the ASK corpus (Norwegian L2) . . . . 206
Table 8.4 Accuracy rates on TOEFL11 corpus (English L2) of various classification systems based on string kernels compared with other state-of-the-art approaches . . . . 208
Table 8.5 Accuracy rates on the raw text documents of the TOEFL11 corpus (English L2) of various classification systems based on string kernels . . . . 209
Table 8.6 Accuracy rates on ICLEv2 corpus (English L2) of various classification systems based on string kernels compared with a state-of-the-art approach . . . . 211
Table 8.7 Accuracy rates on TOEFL11-Big (English L2) corpus of various classification systems based on string kernels compared with a state-of-the-art approach . . . . 212
Table 8.8 Accuracy rates on TOEFL11-Big corpus (English L2) of various classification systems based on string kernels compared with a state-of-the-art approach . . . . 213
Table 8.9 Accuracy rates on ALC subset (Arabic L2) of various classification systems based on string kernels compared with a state-of-the-art approach . . . . 215
Table 8.10 Accuracy rates on three subsets of five languages of the ASK corpus (Norwegian L2) of various classification systems based on string kernels compared with a state-of-the-art approach . . . . 218
Table 8.11 Accuracy rates on the ASK corpus (Norwegian L2) of various classification systems based on string kernels . . . . 220

Table 8.12 Examples of discriminant overused character sequences with their ranks (left) according to the KRR model based on blended spectrum presence bits kernel extracted from the TOEFL11 corpus (English L2) . . . . . . . . . . . . . . . . . . . . 222
Table 8.13 Examples of discriminant underused character sequences (ranks omitted for readability) according to the KRR model based on blended spectrum presence bits kernel extracted from the TOEFL11 corpus (English L2) . . . . . . . . . . . . . . . 223
Table 9.1 Confusion matrix (also known as contingency table) of a binary classifier with labels $+1$ or $-1$ . . . . . . . . . . . . . . . . 237
Table 9.2 Empirical results on the Reuters-21578 corpus obtained by the standard bag of words versus two methods that encode spatial information, namely spatial pyramids and SNAK . . . . 238

# Chapter 1
# Motivation and Overview

## 1.1 Introduction

Machine learning is a branch of artificial intelligence that studies computer systems that can learn from data. In this context, learning is about recognizing complex patterns and making intelligent decisions based on data. In the early years of artificial intelligence, the idea that human thinking could be rendered logically in a numerical computing machine emerged, but it was unclear if such a machine could model the complex human brain, until Alan Turing proposed a test to measure its performance in 1950. The Turing test states that a machine exhibits human-level intelligence if a human judge engages in a natural language conversation with the machine and cannot distinguish it from another human. Despite the fact that intelligent machines that can pass the Turing test have not been developed yet, many interesting and useful systems that can learn from data have been proposed since then.

Several learning paradigms have been proposed in the context of machine learning. The two most popular ones are supervised and unsupervised learning. *Supervised learning* refers to the task of building a classifier using labeled training data. The most studied approaches in machine learning are supervised and they include Support Vector Machines (Cortes and Vapnik 1995), Naïve Bayes classifiers (Manning et al. 2008), neural networks (Bishop 1995; Krizhevsky et al. 2012; LeCun et al. 2015), Random Forests (Breiman 2001), and many others (Caruana and Niculescu-Mizil 2006). *Unsupervised learning* refers to the task of finding hidden structure in unlabeled data. The best known form of unsupervised learning is *cluster analysis*, which aims at clustering objects into groups based on their similarity. Among the other learning paradigms are *semi-supervised learning*, which combines both labeled and unlabeled data, and *reinforcement learning*, which learns to take actions in an environment in order to maximize a long-term reward. Depending on the desired outcome of the machine learning algorithm or on the type of training input available for an application, a particular learning paradigm may be more suitable than the others.



R.T. Ionescu and M. Popescu, *Knowledge Transfer between Computer Vision and Text Mining*, Advances in Computer Vision and Pattern Recognition,
DOI 10.1007/978-3-319-30367-3_1

Machine learning is currently a vast area of research with applications in a broad range of fields, such as computer vision (Fei-Fei and Perona 2005; Forsyth and Ponce 2002; Sebastiani 2002; Zhang et al. 2007), bioinformatics (Dinu and Ionescu 2013; Inza et al. 2010; Leslie et al. 2002) information retrieval (Chifu and Ionescu 2012; Ionescu et al. 2015b; Manning et al. 2008), natural language processing (Lodhi et al. 2002; Popescu and Grozea 2012; Sebastiani 2002), and many others (Ionescu et al. 2015a). Among the variety of state-of-the-art machine learning approaches for such applications are the similarity-based learning methods (Chen et al. 2009).

This book studies similarity-based learning approaches such as nearest neighbor models, kernel methods (Shawe-Taylor and Cristianini 2004), and clustering algorithms. The studied approaches have interesting applications and exhibit state-of-the-art performance levels in two different areas: computer vision and string processing. It is important to note that, in this book, *string processing* refers to any task that needs to process string data such as text documents, DNA sequences, and so on. This work investigates string processing tasks ranging from phylogenetic analysis (Ionescu 2013) and sequence alignment (Dinu et al. 2014) to native language identification (Ionescu et al. 2014b; Popescu and Ionescu 2013) and text categorization by topic, from a machine learning perspective. These tasks belong to one of two separate fields, namely text mining or computational biology, but they are gathered under one umbrella called string processing. On the other hand, a broad variety of computer vision tasks are also investigated in this book, including object recognition (Ionescu and Popescu 2013b, 2015a, b), facial expression recognition (Ionescu et al. 2013), optical character recognition (Dinu et al. 2012; Ionescu and Popescu 2013a) and texture classification (Ionescu et al. 2014a). While all the topics enumerated so far seem to be unrelated, each and every one of them includes at least a concept that is borrowed from the other fields of study covered by this book. Further details about this transfer of knowledge between domains are given in the following section. Before going into the next section, it is worth mentioning that the core part of this book is mostly based on recently published works by the authors, yet, it also includes (previously) unpublished work and results.

## 1.2 Knowledge Transfer between Image and Text

In recent years, computer science specialists are faced with the challenge of processing massive amounts of data. The largest part of this data is actually unstructured and semi-structured data, available in the form of text documents, images, audio files, video files, and so on. Researchers have developed methods and tools that extract relevant information and support efficient access to unstructured and semi-structured content. Such methods that aim at providing access to information are mainly studied by researchers in machine learning and related fields. In fact, a tremendous amount of effort has been dedicated to this line of research (Agarwal and Roth 2002; Lazebnik et al. 2005, 2006; Leung and Malik 2001; Manning et al. 2008). In the context of machine learning, the aim is to obtain a good representation of the data that can later be used to build an efficient classifier. In computer vision, image representations are obtained by *feature detection* and *feature extraction*. Most of the feature extraction methods are handcrafted by researchers that have a good understanding of the application and a vast experience. This is the case of the bag of visual words model (Leung and Malik 2001; Sivic et al. 2005) in computer vision. A different approach is *representation learning*, which aims at discovering a better representation of the data provided during training. This is the case of deep learning algorithms (Bengio 2009; LeCun et al. 2015; Montavon et al. 2012) that aim at discovering multiple levels of representation, or a hierarchy of features. Deep algorithms learn to transform one representation into another, by better disentangling the factors of variation that explain the observed data.

Whether the representation of the data is obtained through a handcrafted method or learned by a fully automatic process, common concepts of treating different kinds of unstructured and semi-structured data, such as image and text, naturally arise. Despite the fact that computer vision and string processing seem to be unrelated fields of study, the concept of treating image and text in a similar fashion has proven to be very fertile for several applications. Furthermore, by adapting string processing techniques to image analysis or the other way around, knowledge from one domain can be transferred to the other.

An example of similarity between text and image is discussed next. It refers to word sense disambiguation and object recognition in images. *Word sense disambiguation* (WSD) is a core research problem in computational linguistics and natural language processing, which was recognized since the beginning of the scientific interest in machine translation, and in artificial intelligence, in general. WSD is about determining the meaning of a word in a specific context. Actually all the WSD methods use the context to determine the meaning of an ambiguous word, because the entire information about the word sense is contained in the context (Agirre and Edmonds 2006). The basic concept is to extract features from the context that could help the WSD process. In a similar fashion, an object in an image can be recognized using the entire image as a context. For example, a method that could detect the presence of a kitchen glove in the image would have to look for distinctive features such as the texture of the material, the shape, and perhaps even the color. However, there could be other objects that have similar shape or color, and in more difficult situations, such as illustrated in Fig. 1.1a, it may be almost impossible to distinguish the glove. Thus, a better approach could be to look for other distinctive features in the image provided by the context. For instance, a human can easily figure out that a glove is hanging by a kitchen cabinet knob in the scene illustrated in Fig. 1.1b. It is easier to understand the entire scene as a whole than taking the glove out of context. In conclusion, the idea of using the context can help to avoid any confusion. Not surprisingly, this intuitive idea has already been studied in the computer vision literature (Galleguillos and Belongie 2010; Rabinovich et al. 2007). In (Rabinovich et al. 2007), the semantic context is incorporated into object categorization to reduce ambiguity in objects' visual appearance and to improve accuracy. The paper of (Galleguillos and Belongie 2010) goes even further and makes a distinction between three types of context, namely semantic context, spatial context, and scale context.

**Fig. 1.1** An example in which the context helps to disambiguate an object (kitchen glove), which can easily be mistaken for something else if the rest of the image is not seen. The image belongs to the Pascal VOC 2007 data set. **a** A picture of a kitchen glove. **b** A picture of the same glove with context

Another example of treating image and text in a similar manner is a state-of-the-art method for image categorization and image retrieval inspired by the *bag of words* representation, which is very popular in information retrieval and natural language processing. The bag of words model represents a text as an unordered collection of words, completely disregarding grammar, word order, and syntactic groups. The bag of words model has many applications from information retrieval (Manning et al. 2008) to natural language processing (Manning and Schütze 1999) and word sense disambiguation (Agirre and Edmonds 2006; Chifu and Ionescu 2012). In the context of image analysis, the concept of *word* needs to be somehow defined. Computer vision researchers have introduced the concept of *visual word*. Local image descriptors, such as SIFT (Lowe 1999), are vector quantized to obtain a vocabulary of visual words. The vector quantization process can be done, for example, by k-means clustering (Leung and Malik 2001) or by probabilistic Latent Semantic Analysis (Sivic et al. 2005). The frequency of each visual word is then recorded in a histogram which represents the final feature vector for the image. This histogram is the equivalent of the bag of words representation for text. The idea of representing images as *bag of visual words* has demonstrated very good performance for image categorization (Zhang et al. 2007) and image retrieval (Philbin et al. 2007).

One of the most important problems in computer vision is object recognition. Machine learning methods represent the state-of-the-art approach for the object recognition problem. A common approach is to make some assumptions in order to treat object recognition as a classification problem. First, object categories are considered to be fixed and known. Second, each instance belongs to a single category. However, some researchers argue that these assumptions do not adequately describe the reality. The following example shows that these assumptions are indeed wrong. The object presented in Fig. 1.2 can be described either as a toy, a bear, or both. It is clear that the object does not belong to a single category. Furthermore, the category of the object might be irrelevant for particular applications. Another

**Fig. 1.2** An object that can be described by multiple categories such as toy, bear, or both

drawback of this approach is that it misses out some of the subtle aspects of object recognition. For example, an object classification system does not understand the properties of an object and it cannot deal with unfamiliar objects. In other words, it fails to extract aspects of meaning. Thus, some computer vision researchers have proposed different approaches for the object recognition task. One alternative approach, proposed by Duygulu et al. (2002), is to model object recognition as machine translation. The model is based on the observation that object recognition is a little like translation, in that a picture (or text in a source language) goes in, and a description (or text in a target language) comes out. In this model, object recognition becomes a process of annotating image regions with words. First, images are segmented into regions, which are then classified into region types. Next, a mapping between region types and keywords provided with the images is learned. This process is similar to learning a *lexicon* from data, a standard problem in the machine translation literature (Jurafsky and Martin 2000; Manning and Schütze 1999). This approach has been proven productive for this interpretation of object recognition. Research in this area has led to the development of other systems, such as the one described by Farhadi et al. (2010), which generates sentences from images. The system computes a score linking an image to a sentence. This score can be used to attach a descriptive sentence to a given image, or to obtain images that illustrate a given sentence. To take this even further, the work of Sadeghi and Farhadi (2011) suggests that it is easier and more effective to generate descriptions of images in terms of chunks of meaning, such as "a person riding a horse", rather than individual components, such as "person" or "horse". In this approach, categories are replaced with visual phrases for recognition.

The examples described so far are successful cases of treating image as text. However, research that studies ways to improve text processing techniques with the knowledge from computer vision has also been conducted. A good example is the method introduced by Barnard and Johnson (2005), which proposes the use of images for WSD, either alone, or in conjunction with traditional text-based methods. To integrate image information with text data, the authors exploit previous work on linking images and words (Barnard et al. 2003; Duygulu et al. 2002). The empirical results strongly suggest that images can help disambiguate senses of words.

In the recent years, deep neural networks have demonstrated impressive levels of performance in various computer vision tasks (Krizhevsky et al. 2012; LeCun et al. 2015; Simonyan and Zisserman 2014). After their success in computer vision, researchers have tried to adapt the deep learning algorithms for text data (Johnson and Zhang 2015; Mikolov et al. 2013; Sutskever et al. 2014). One of the most popular approaches is to use neural networks in order to build word embeddings (Mikolov et al. 2013) by mapping words from a vocabulary to vectors of real-value numbers in a low dimensional space. Although deep learning algorithms are now widely used in the NLP community, others have suggested that equally good word embeddings can be produced without the help of deep models, for example, by using Hellinger Principal Component Analysis (Lebret et al. 2013). Nevertheless, deep learning models, which essentially promote the idea of *end-to-end* learning, are now extremely popular in both computer vision and natural language processing, which only comes to support the main argument behind the present book, namely that knowledge transfer between these domains is fruitful.

The concept of treating image and text in a similar manner is exploited in one way or another in the previous examples. The knowledge transfer from one domain to another has proven to be very fertile in the case of computer vision and natural language processing. This book follows the same line of research and presents novel approaches or improved methods that rely on this cornerstone concept. One of the concepts recurring several times throughout this book refers to measuring the local non-alignment among two objects, and it stems from the rank distance measure (Dinu and Manea 2006). Remarkably, this idea shows its uses in various tasks when used for different kinds of objects, including strings and images.

First, a dissimilarity measure for images termed Local Patch Dissimilarity is presented in Chap. 4. The dissimilarity measure is inspired by the rank distance measure (Dinu and Manea 2006). The main concern is to extend rank distance from one-dimensional input (strings) to two-dimensional input (digital images). While rank distance is a highly accurate measure for strings, the experiments presented in Chap. 4 suggest that the proposed extension of rank distance to images is very accurate for handwritten digit recognition and, with some redesign work, for texture analysis.

Second, some improvements to the popular bag of visual words model are proposed in Chap. 5. As mentioned before, this model is inspired by the bag of words model from text mining and information retrieval. Not only that the bag of visual words itself is transferred from text analysis, but its improvements described in this book are also inspired from the same domain. Indeed, the PQ kernel is a viable kernel for visual word histograms that was previously used as a stylistic measure for authorship identification by Dinu and Popescu (2009). The Spatial Non-Alignment Kernel is another kernel that improves the object recognition performance of the bag of visual words model by encoding spatial information in the similarity of two images. Compared to the popular spatial pyramid representation (Lazebnik et al. 2006), it improves performance while consuming less space and time. Just as Local Patch Dissimilarity, the Spatial Non-Alignment Kernel is inspired by rank distance.

Third, a new distance measure for strings is introduced in Chap. 7. It is inspired by the image dissimilarity measure presented in Chap. 4. Designed to conform to more general principles, while being better adapted for specific data types, such as DNA strings or text, it shows interesting results in DNA sequence analysis. Furthermore, another application of this novel distance measure for strings is presented in Chap. 8. More precisely, a kernel based on this distance measure is used for native language identification. Furthermore, the (histogram) intersection kernel is used for the first time as a kernel for strings, in the context of native language identification. Notably, the intersection kernel has successfully been used in computer vision for object class recognition from images (Maji et al. 2008; Vedaldi and Zisserman 2010).

Computer vision researchers have demonstrated that the object recognition performance can be improved by including spatial information into the bag of visual words model. A state-of-the-art approach is the spatial pyramid representation (Lazebnik et al. 2006), which divides the image into spatial bins. The Spatial Non-Alignment Kernel is another general approach to encode spatial information. In Chap. 9, these two approaches are used to improve performance of the bag of words model in the context of text categorization by topic, showing that spatial information can also be useful for text analysis.

To summarize, all the contributions presented in this book are based on the cornerstone concept of treating image and text in a similar manner. Moreover, there are several other contributions (Barnard and Johnson 2005; Duygulu et al. 2002; LeCun et al. 2015) that transfer knowledge between computer vision and text mining, but altogether, this concept is far from saturated. When a breakthrough discovery is made in one domain, researchers can always consider adapting and using the respective discovery to another domain, even though their attempt may not prove to be successful in the end.

## 1.3 Overview and Organization

The rest of this book is organized as follows. All the machine learning methods that are employed to obtain results for different applications in computer vision and string processing are described in Chap. 2. The chapter gives an overview of the main concepts of learning based on similarity. Specific machine learning methods based on these concepts are then presented. First, nearest neighbor models are discussed. A nonstandard learning formulation based on the notions of similarity and nearest neighbors, known as *local learning*, is then presented. An overview of *kernel methods* is also given, since the state-of-the-art methods consistently used in the supervised learning tasks presented throughout this book are kernel methods. Chapter 2 ends with a discussion about cluster analysis. Clustering techniques are used throughout this book in various contexts, from building vocabularies of visual words to phylogenetic analysis.

The main content of this book is organized in two parts. Part I presents machine learning methods and applications in computer vision that are based on knowledge and concepts borrowed from text mining. Part II presents machine learning methods and applications in text and string processing, or more precisely, in computational biology and text mining. These are based on concepts transferred from computer vision. Chapters 3–5 belong to Part I, while Chaps. 6–9 belong to Part II. Finally, the conclusions are drawn in Chap. 10.

The content of each chapter is briefly discussed next. Chapter 3 discusses the state-of-the-art methods in computer vision for several tasks such as object recognition, texture analysis, and optical character recognition. Most of the state-of-the-art approaches are based on patches or image descriptors, but other approaches, such as those based on deep learning, have shown impressive levels of performance. Image descriptors, which form another class of local image features besides patches, are also presented in Chap. 3. Since this book is focused on learning based on similarity, an entire section is dedicated to image distance measures.

The first contribution of this book is discussed in Chap. 4. The chapter presents a novel dissimilarity measure for images, called Local Patch Dissimilarity (LPD), that was introduced by Dinu et al. (2012). This new distance measure is inspired by rank distance, which is a distance measure for strings. Hence, it shows the concept of treating image and text in a similar way, in practice. An algorithm to compute LPD and theoretical properties of this dissimilarity are also given. Chapter 4 describes several ways of improving LPD in terms of efficiency, such as using a hash table to store precomputed patch distances or skipping the comparison of overlapping patches. Another way to avoid the problem of the higher computational time on large sets of images is to turn to local learning methods. All these efficiency improvements were published by Ionescu and Popescu (2013a). Several experiments are conducted on two data sets using both standard machine learning methods and local learning methods. The obtained results come to support the fact that LPD is a very good dissimilarity measure for images with applications in handwritten digit recognition and image classification. A variant of LPD introduced by Ionescu et al. (2014a), called Local Texton Dissimilarity (LTD), is also presented in Chap. 4. Local Texton Dissimilarity aims at classifying texture images. It is based on textons, which are represented as a set of features extracted from image patches. Textons provide a lighter representation of patches, allowing for a faster computational time and a better accuracy when used for texture analysis. The performance level of the machine learning methods based on LTD is comparable to the state-of-the-art methods for texture classification.

Chapter 5 presents some improvements of the bag of visual words model for two applications, namely object recognition and facial expression recognition. For the bag of visual words approach, images are represented as histograms of visual words from a codebook that is usually obtained with a simple clustering method. Next, kernel methods are used to compare such histograms. Chapter 5 introduces a novel kernel for histograms of visual words, namely the PQ kernel. The PQ kernel was initially presented by Ionescu and Popescu (2013b) and extensively studied by Ionescu and Popescu (2015a). A proof that PQ is indeed a kernel is also given in Chap. 5. The proof is based on building its feature map. Object recognition experiments are conducted to compare the PQ kernel with other state-of-the-art kernels on two benchmark data sets. The PQ kernel has the best performance on both data sets. Researchers have demonstrated that the object recognition performance with the bag of visual words can be improved by including spatial information. A state-of-the-art approach is the spatial pyramid representation, which divides the image into spatial bins. In Chap. 5, another general approach that encodes the spatial information in a much better and efficient way is described. The approach is to embed the spatial information into a kernel function termed the Spatial Non-Alignment Kernel (SNAK) (Ionescu and Popescu 2015b). For each visual word, the average position and the standard deviation is computed based on all the occurrences of the visual word in the image. These are computed with respect to the center of the object, which is determined with the help of the objectness measure (Alexe et al. 2010, 2012). The pairwise similarity of two images is then computed by taking into account the difference between the average positions and the difference between the standard deviations of each visual word in the two images. In all the experiments, the SNAK framework shows a better recognition accuracy than the spatial pyramid. Finally, the chapter presents a bag of visual words model based on local multiple kernel learning for facial expression recognition.

Chapter 6 presents the state-of-the-art methods for several problems that involve string processing. The string processing problems studied by this work belong to two major scientific fields, namely computational biology and text mining. Consequently, the chapter is divided into two separate sections corresponding to the two fields of study. First of all, clustering methods used for phylogenetic analysis and other methods for sequencing and comparing DNA are discussed. Second of all, an overview of state-of-the-art text processing techniques is given. These include the bag of words model and approaches that work at the character level, namely string kernels.

In Chap. 7, a novel distance measure, called Local Rank Distance (LRD), inspired from the image dissimilarity measure presented in Chap. 4, is introduced. LRD was initially presented by Ionescu (2013). The novel distance measure is designed to comprise more general principles than rank distance, but it is also developed having a practical motivation in mind, specifically to be more suitable for DNA strings or text. Chapter 7 shows three applications of LRD. The first application is the phylogenetic analysis of mammals. Experiments show that phylogenetic trees produced by LRD are better or at least similar to those reported in the literature. The second application is to find the closest string for a set of DNA strings, using a genetic algorithm based on LRD. The results produced by LRD come to support the fact that the concepts that make a good dissimilarity measure for images can be transferred with success in comparing and analyzing strings. The third application is about assigning a set of short DNA reads to a reference genome. As such, a genome sequence aligner based on LRD (Dinu et al. 2014) is described in Chap. 7. The LRD aligner presented in this book aims to improve correctness over speed. However, some indexing strategies to speed up the aligner are also described.

Chapter 8 presents an application of machine learning methods that work at the character level. More precisely, several string kernels, one of which is based on Local Rank Distance, are combined to obtain state-of-the-art results for native language identification (NLI). A broad set of NLI experiments are conducted to compare the string kernels approach with other state-of-the-art methods on English, Arabic, and Norwegian corpora. In all the experiments, strings kernels obtain results better than the state-of-the-art methods, sometimes by a very large margin. For instance, there is a 32.3 % improvement in accuracy over the state-of-the-art system of (Tetreault et al. 2012), when the systems based on string kernels are trained on the TOEFL11 corpus and tested on the TOEFL11-Big corpus. The results are even more impressive considering that the proposed approach is language independent and linguistic theory neutral. To gain additional insights about the string kernels approach, the features selected by the classifier as being more discriminant are analyzed in Chap. 8. The analysis also offers information about localized language transfer effects, since the features used by the proposed model are $p$-grams of various lengths. It is worth noting that the string kernels approach for NLI was described in a series of papers (Ionescu et al. 2014b; Popescu and Ionescu 2013).

In Chap. 9, two approaches to encode spatial information in the bag of visual words model are transferred to text analysis. These are the spatial pyramid (Lazebnik et al. 2006) and the Spatial Non-Alignment Kernel (Ionescu and Popescu 2015b). In the context of object recognition from images, the spatial information helps to significantly improve the performance (Ionescu and Popescu 2015b; Lazebnik et al. 2006). The empirical results presented in Chap. 9 indicate that spatial information can also be useful for text categorization by topic. The spatial pyramid for text divides the text into sections (or parts) using multiple levels of granularity and extracts features from each of these sections. The final representation, obtained by concatenating all the features, roughly indicates what features appear in a certain section of a text document, such as the introduction or the conclusion. The Spatial Non-Alignment Kernel for text replaces visual words from images with words from text documents, providing a soft assignment alternative to the spatial pyramid. As in the case of object recognition from images, the Spatial Non-Alignment Kernel seems to be a better approach in terms of performance, probably because it represents the spatial information (location of words in text) in a more accurate way than the spatial pyramid.

The conclusions presented in Chap. 10 point to the fact that the concept of treating image and text in a similar way is indeed fertile. Some advice on future work and new directions that could arise from this concept is also provided in the final chapter.

## References

Agarwal S, Roth D (2002) Learning a sparse representation for object detection. In: Proceedings of ECCV, pp 113–127

Agirre E, Edmonds PG (2006) Word Sense Disambiguation: Algorithms and applications. Springer

Alexe B, Deselaers T, Ferrari V (2010) What is an object? In: Proceedings of CVPR, pp 73–80

Alexe B, Deselaers T, Ferrari V (2012) Measuring the objectness of image windows. IEEE Trans Pattern Anal Mach Intell 34(11):2189–2202

Barnard K, Duygulu P, Forsyth D, De Freitas N, Blei DM, Jordan MI (2003) Matching words and pictures. J Mach Learn Res 3:1107–1135

Barnard K, Johnson M (2005) Word sense disambiguation with pictures. Artif Intell 167(1–2):13–30

Bengio Y (2009) Learning deep architectures for AI. Found Trends Mach Learn 2(1):1–127

Bishop CM (1995) Neural networks for pattern recognition. Oxford University Press Inc, New York, USA

Breiman L (2001) Random forests. Mach Learn 45(1):5–32

Caruana R, Niculescu-Mizil A (2006). An empirical comparison of supervised learning algorithms. In: Proceedings of ICML, pp 161–168

Chen Y, Garcia EK, Gupta MR, Rahimi A, Cazzanti L (2009) Similarity-based classification: concepts and algorithms. J Mach Learn Res 10:747–776

Chifu AG, Ionescu RT (2012) Word sense disambiguation to improve precision for ambiguous queries. Cent Eur J Comput Sci 2(4):398–411

Cortes C, Vapnik V (1995) Support-vector networks. Mach Learn 20(3):273–297

Dinu LP, Ionescu RT, Popescu M (2012) Local Patch Dissimilarity for images. In: Proceedings of ICONIP 7663:117–126

Dinu LP, Ionescu RT (2013) Clustering based on median and closest string via rank distance with applications on DNA. Neural Comput Appl 24(1):77–84

Dinu LP, Ionescu RT, Tomescu AI (2014) A rank-based sequence aligner with applications in phylogenetic analysis. PLoS ONE, 9(8):e104006. doi:10.1371/journal.pone.0104006

Dinu LP, Manea F (2006) An efficient approach for the rank aggregation problem. Theoret Comput Sci 359(1–3):455–461

Dinu LP, Popescu M (2009) Comparing statistical similarity measures for stylistic multivariate analysis. In: Proceedings of RANLP

Duygulu P, Barnard K, De Freitas JFG, Forsyth DA (2002) Object recognition as machine translation: learning a lexicon for a fixed image vocabulary. In: Proceedings of ECCV 97–112

Farhadi A, Hejrati M, Sadeghi MA, Young P, Rashtchian C, Hockenmaier J, Forsyth D (2010) Every picture tells a story: generating sentences from images. In: Proceedings of ECCV 15–29

Fei-Fei L, Perona P (2005) A bayesian hierarchical model for learning natural scene categories. In: Proceedings of CVPR 2:524–531

Forsyth DA, Ponce J (2002) Computer vision: a modern approach. Prentice Hall Professional Technical Reference

Galleguillos C, Belongie S (2010) Context based object categorization: a critical survey. Comput Vis Image Underst 114:712–722

Inza I, Calvo B, Armañanzas R, Bengoetxea E, Larrañaga P, Lozano JA (2010) Machine learning: an indispensable tool in bioinformatics. Meth Mol Biol (Clifton, N.J.) 593:25–48

Ionescu RT, Chifu A-G, Mothe J (2015b) DeShaTo: describing the shape of cumulative topic distributions to rank retrieval systems without relevance judgments. In: Proceedings of SPIRE 9309:75–82

Ionescu RT, Popescu AL, Popescu M, Popescu D (2015a) BiomassID: a biomass type identification system for mobile devices. Comput Electron Agric 113:244–253

Ionescu RT, Popescu AL, Popescu D, Popescu M (2014a) Local Texton Dissimilarity with applications on biomass classification. In: Proceedings of VISAPP

Ionescu RT, Popescu M (2013a) Speeding up Local Patch Dissimilarity. In: Proceedings of ICIAP 8156:1–10

Ionescu RT, Popescu M (2013b) Kernels for visual words histograms. In: Proceedings of ICIAP 8156:81–90
Ionescu RT, Popescu M (2015a) PQ kernel: a rank correlation kernel for visual word histograms. Pattern Recogn Lett 55:51–57
Ionescu RT, Popescu M (2015b) Have a SNAK. Encoding spatial information with the Spatial Non-Alignment Kernel. In: Proceedings of ICIAP 9279:97–108
Ionescu RT, Popescu M, Cahill A (2014b) Can characters reveal your native language? A language-independent approach to native language identification. In: Proceedings of EMNLP, pp 1363–1373
Ionescu RT, Popescu M, Grozea C (2013) Local learning to improve bag of visual words model for facial expression recognition. In: Workshop on Challenges in Representation Learning, ICML, 2013
Ionescu RT (2013) Local Rank Distance. In: Proceedings of SYNASC 221–228
Johnson R, Zhang T (2015) Effective use of word order for text categorization with convolutional neural networks. In: Proceedings of NAACL, pp 103–112
Jurafsky D, Martin JH (2000) Speech and language processing: an introduction to natural language processing, computational linguistics, and speech recognition, 1st edn. Prentice Hall PTR, Upper Saddle River, NJ, USA
Krizhevsky A, Sutskever I, Hinton GE (2012) ImageNet classification with deep convolutional neural networks. In: Proceedings of NIPS 1106–1114
Lazebnik S, Schmid C, Ponce J (2005) A sparse texture representation using local affine regions. IEEE Trans Pattern Anal Mach Intell 27(8):1265–1278
Lazebnik S, Schmid C, Ponce J (2006) Beyond bags of features: spatial pyramid matching for recognizing natural scene categories. In: Proceedings of CVPR 2:2169–2178
Lebret R, Legrand J, Collobert R (2013) Is deep learning really necessary for word embeddings? Deep learning workshop NIPS
LeCun Y, Bengio Y, Hinton G (2015) Deep learning. Nature 521(7553):436–444
Leslie CS, Eskin E, Noble WS (2002) The spectrum kernel: a string kernel for svm protein classification. In: Proceedings of Pacific Symposium on Biocomputing, pp 566–575
Leung T, Malik J (2001) Representing and recognizing the visual appearance of materials using three-dimensional textons. Int J Comput Vis 43(1):29–44
Lodhi H, Saunders C, Shawe-Taylor J, Cristianini N, Watkins CJCH (2002) Text classification using string kernels. J Mach Learn Res 2:419–444
Lowe DG (1999) Object recognition from local scale-invariant features. In: Proceedings of ICCV 2:1150–1157
Maji S, Berg AC, Malik J (2008) Classification using intersection kernel support vector machines is efficient. In: Proceedings of CVPR
Manning CD, Raghavan P, Schütze H (2008) Introduction to information retrieval. Cambridge University Press, New York, USA
Manning CD, Schütze H (1999) Foundations of statistical natural language processing. MIT Press, Cambridge, MA, USA
Mikolov T, Sutskever I, Chen K, Corrado GS, Dean J (2013) Distributed representations of words and phrases and their compositionality. In: Proceedings of NIPS, pp 3111–3119
Montavon G, Orr GB, Müller K-R (eds) (2012) Neural Networks: Tricks of the Trade. In: Lecture notes in computer science (LNCS), vol 7700, 2nd edn. Springer
Philbin J, Chum O, Isard M, Sivic J, Zisserman A (2007) Object retrieval with large vocabularies and fast spatial matching. In: Proceedings of CVPR, pp 1–8
Popescu M, Grozea C (2012) Kernel methods and string kernels for authorship analysis. CLEF (Online Working Notes/Labs/Workshop)
Popescu M, Ionescu RT (2013) The story of the characters, the DNA and the native language. In: Proceedings of the Eighth Workshop on Innovative Use of NLP for Building Educational Applications, pp 270–278

Rabinovich A, Vedaldi A, Galleguillos C, Wiewiora E, Belongie S (2007) Objects in context. In: Proceedings of ICCV
Sadeghi MA, Farhadi A (2011) Recognition using visual phrases. In: Proceedings of CVPR, pp 1745–1752
Sebastiani F (2002) Machine learning in automated text categorization. ACM Comput Surv 34(1):1–47
Shawe-Taylor J, Cristianini N (2004) Kernel methods for pattern analysis. Cambridge University Press
Simonyan K, Zisserman A (2014) Very deep convolutional networks for large-scale image recognition. CoRR, abs/1409.1556
Sivic J, Russell BC, Efros AA, Zisserman A, Freeman WT (2005) Discovering objects and their localization in images. In: Proceedings of ICCV, pp 370–377
Sutskever I, Vinyals O, Le QV (2014) Sequence to sequence learning with neural networks. In: Proceedings of NIPS, pp 3104–3112
Tetreault J, Blanchard D, Cahill A, Chodorow M (2012) Native tongues, lost and found: resources and empirical evaluations in native language identification. In: Proceedings of COLING 2012:2585–2602
Vedaldi A, Zisserman A (2010) Efficient additive kernels via explicit feature maps. In: Proceedings of CVPR, pp 3539–3546
Zhang J, Marszalek M, Lazebnik S, Schmid C (2007) Local features and kernels for classification of texture and object categories: a comprehensive study. International J Comput Vis 73(2):213–238

# Chapter 2
# Learning based on Similarity

## 2.1 Introduction

Learning based on similarity refers to the process of learning based on pairwise similarities between the training samples. The similarity-based learning process can be both supervised and unsupervised, and the pairwise relationship can be either a similarity, a dissimilarity, or a distance function. Similarity functions may be asymmetric and even fail to satisfy other mathematical properties required for metrics or inner products, for example. When the learning process is supervised, the similarity-based method aims at estimating the class label of a test sample using both the pairwise similarities between the labeled training samples, and the similarities between the test sample and the set of training samples. When the learning process is unsupervised, the similarity-based method aims at finding hidden structure in unlabeled training samples, using the pairwise similarities between samples. An advantage of similarity-based learning is that it does not require direct access to the features, as long as the similarity function is well defined and can be computed for any pair of samples. Thus, the feature space is not required to be a euclidean space.

Similarity-based learning has a long history starting with $k$-nearest neighbors (Fix and Hodges 1951), which is one of the oldest machine learning algorithms, and stretching to the state-of-the-art kernel methods (Shawe-Taylor and Cristianini 2004). Similarity-based learning methods have been widely used in several domains such as computer vision, natural language processing, computational biology, and information retrieval. Computer vision researchers proposed several methods based on computing similarity between images for object recognition and image retrieval. Such methods range from distance measures such as the Tangent distance (Simard et al. 1996), the Earth Mover's distance (Rubner et al. 2000), or the shape matching distance (Belongie et al. 2002), to kernel methods such as the pyramid match kernel (Grauman and Darrell 2005; Lazebnik et al. 2006) or the PQ kernel (Ionescu and Popescu 2013). Most of the state-of-the-art techniques in computational biology, such as those that obtain phylogenetic trees or those that compare DNA sequences,



R.T. Ionescu and M. Popescu, *Knowledge Transfer between Computer Vision and Text Mining*, Advances in Computer Vision and Pattern Recognition,
DOI 10.1007/978-3-319-30367-3_2

are based on distance measures for strings. Popular choices for recent techniques are the Hamming distance (Chimani et al. 2011; Vezzi et al. 2012), edit distance (Shapira and Storer 2003), Kendall's tau distance (Popov 2007) or rank distance (Dinu and Ionescu 2012a, b). Other popular similarity-based tools from computational biology are the FASTA algorithm (Lipman and Pearson 1985) and the BLAST algorithm (Altschul et al. 1990). These tools compute the similarity between different amino acid sequences for protein classification. The cosine similarity between term frequency-inverse document frequency (TF-IDF) vectors is widely used in information retrieval and text mining for document classification (Manning et al. 2008). More recently, the string kernel (Shawe-Taylor and Cristianini 2004), which computes the similarity between strings by counting common character $n$-grams, has demonstrated impressive levels of performance for text categorization (by topic) (Lodhi et al. 2002), authorship identification (Popescu and Dinu 2007; Popescu and Grozea 2012; Sanderson and Guenter 2006), and native language identification (Ionescu et al. 2014; Popescu and Ionescu 2013).

The similarity-based learning paradigm consists of a wide variety of algorithms and approaches. Among the variety of similarity-based learning methods, only four of them are discussed in dedicated sections of this chapter, namely the nearest neighbor approach, the local learning methods, the kernel methods, and the cluster analysis techniques. These four approaches are used in different applications presented in this work. Since many of them are widely known and studied in literature, this chapter is rather aimed at giving an overview of the approaches used throughout this book. Other similarity-based learning methods, such as treating similarities as features, or generative classifiers, are briefly mentioned next. By treating the similarities between a sample and training samples as features, similarity-based classification problems can be regarded as standard classification problems (Chen et al. 2009; Graepel et al. 1998, 1999; Liao and Noble 2003; Pekalska and Duin 2002). In other words, each sample is represented by a feature vector obtained by computing the similarity with a set of training samples. Generative classifiers provide a structured probabilistic model of the data. Training data is used for estimating the parameters of the generative model. Given the pairwise similarity of $n$ samples, one approach to generative classification is using the similarities as features. Then, the parameters of a standard generative model can be estimated from an $n$-dimensional feature space. Recently, another generative framework for similarity-based classification, termed similarity discriminant analysis, has been proposed in (Cazzanti et al. 2008). It models the class-conditional distributions of similarity statistics. Other approaches designed to reduce bias are a local variant proposed in (Cazzanti and Gupta 2007) and a mixture model variant discussed in (Chen et al. 2009).

The rest of this chapter is organized as follows. Nearest neighbor models are discussed in Sect. 2.2. Section 2.3 presents local learning methods, which are based on the notions of similarity and nearest neighbors. An overview of kernel methods is given in Sect. 2.4. The chapter ends with Sect. 2.5, which gives an overview of clustering methods based on similarity.

**Algorithm 1**: Nearest Neighbor Algorithm

1 **Input**:
2 $\mathcal{S} = \{(x_i, t_i) \mid x_i \in \mathbb{R}^m, t_i \in \mathbb{N}, i \in \{1, 2, \ldots, n\}\}$ – the set of $n$ training samples and labels;
3 $Z = \{z_i \mid z_i \in \mathbb{R}^m, i \in \{1, 2, \ldots, l\}\}$ – the set of $l$ test samples;
4 $k$ – the number of neighbors;
5 $\Delta$ – a distance measure.

6 **Initialization**:
7 $Y \leftarrow \emptyset$;

8 **Computation**:
9 **for** $z_i \in Z$ **do**
10 $\quad \mathcal{N} \leftarrow$ the nearest $k$ neighbors to $z_i$ from $\mathcal{S}$ according to $\Delta$;
11 $\quad y \leftarrow$ the majority label obtained through a voting scheme on $\mathcal{N}$;
12 $\quad Y \leftarrow Y \cup \{y\}$;

13 **Output**:
14 $Y = \{y_i \mid y_i \in \mathbb{N}, i \in \{1, 2, \ldots, l\}\}$ – the set of predicted labels for the test samples in $Z$.

## 2.2 Nearest Neighbor Approach

Since the $k$-nearest neighbors algorithm ($k$-NN) was introduced in (Fix and Hodges 1951), it has been studied by many researchers and it is still an active topic in machine learning. The $k$-nearest neighbors algorithm is one of the simplest of all the machine learning algorithms, proving that simple models are always attractive for researchers. The nearest neighbor model is described in Algorithm 1.

The $k$-nearest neighbors classification rule employed in Step 11 of Algorithm 1 works as follows: an object is assigned to the most common class of its $k$ nearest neighbors, where $k$ is a positive integer value. If $k = 1$, then the object is simply assigned to the class of its nearest neighbor. When $k > 1$, the decision is based on a majority vote. It is convenient to let $k$ be odd, to avoid voting ties. However, if voting ties do occur, the object can be assigned to the class of its 1-nearest neighbor, or one of the tied classes can be randomly chosen to be the class assigned to the object. The output of Algorithm 1 is a set of labels associated to the test samples.

The example about handwritten digit recognition presented in Fig. 2.1 gives some insights of how the $k$-NN model works in practice. In this example, digits are represented in a two-dimensional feature space. When a new sample $x$ comes in, the algorithm selects the nearest 3 neighbors and assigns the majority class to $x$. In Fig. 2.1, the majority label among the nearest 3 neighbors of $x$ is 4. Thus, label 4 is assigned to $x$. This model can be referred to as a 3-NN model. To better understand how the decision of the $k$-NN model is taken in general, it is worth considering a 1-NN model. For this model, the decision at every point is to assign the label of the closest data point. This process generates a Voronoi partition of the training samples, as seen in Fig. 2.2. Each training data point corresponds to a Voronoi cell. When a new data point comes in, it is assigned to the class associated to the Voronoi cell that the respective data point falls in.

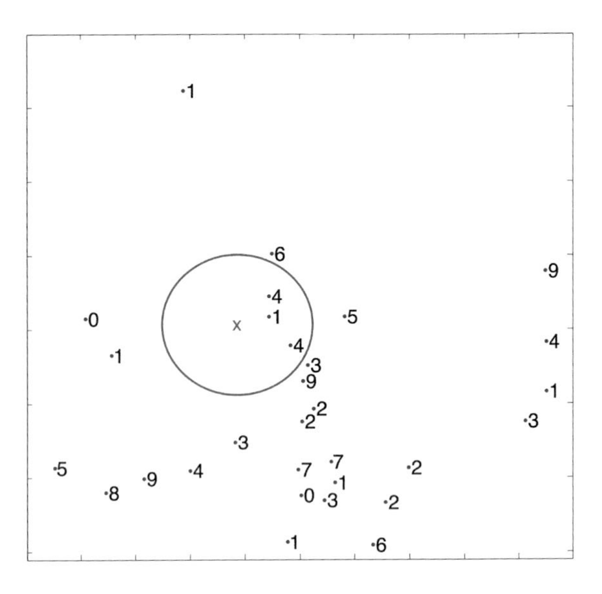


**Fig. 2.1** A 3-NN model for handwritten digit recognition. For visual interpretation, digits are represented in a two-dimensional feature space. The figure shows 30 digits sampled from the popular MNIST data set. When the new digit $x$ needs to be recognized, the 3-NN model selects the nearest 3 neighbors and assigns label 4 based on a majority vote

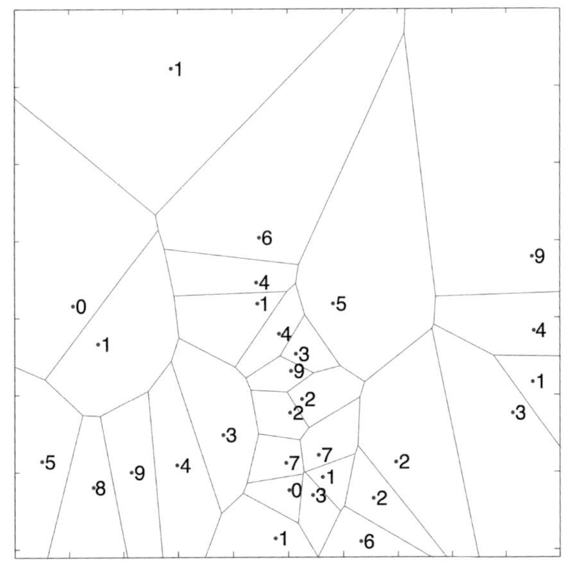


**Fig. 2.2** A 1-NN model for handwritten digit recognition. The figure shows 30 digits sampled from the popular MNIST data set. The decision boundary of the 1-NN model generates a Voronoi partition of the digits

The $k$-NN algorithm is a non-parametric method for classification. In other words, no parameters have to be learned. In fact, the $k$-NN model does not require training at all. The decision of the classifier is only based on the nearest $k$ neighbors of an object with respect to a similarity or distance function. The euclidean distance measure is a very common choice, but other similarity measures can also be used instead. Actually, the performance of the $k$-NN classifier depends on the strength and the discriminatory power of the distance measure used. It is worth mentioning that a good

**Algorithm 2**: Local Learning Algorithm

1 **Input**:
2 $\mathcal{S} = \{(x_i, t_i) \mid x_i \in \mathbb{R}^m, t_i \in \mathbb{N}, i \in \{1, 2, \ldots, n\}\}$ – the set of $n$ training samples and labels;
3 $Z = \{z_i \mid z_i \in \mathbb{R}^m, i \in \{1, 2, \ldots, l\}\}$ – the set of $l$ test samples;
4 $k$ – the number of neighbors;
5 $\Delta$ – a distance measure;
6 $\mathcal{C}$ – a classification model.
7 **Initialization**:
8 $Y \leftarrow \emptyset$;
9 **Computation**:
10 **for** $z_i \in Z$ **do**
11 $\quad \mathcal{N} \leftarrow$ the nearest $k$ neighbors to $z_i$ from $\mathcal{S}$ according to $\Delta$;
12 $\quad f \leftarrow$ the discriminant function of $\mathcal{C}$ trained on $\mathcal{N}$;
13 $\quad y \leftarrow$ the label predicted by applying $f$ on $z_i$;
14 $\quad Y \leftarrow Y \cup \{y\}$;
15 **Output**:
16 $Y = \{y_i \mid y_i \in \mathbb{N}, i \in \{1, 2, \ldots, l\}\}$ – the set of predicted labels for the test samples in $Z$.

choice of the distance metric can help to achieve invariance with respect to a certain family of transformations. For example, a distance metric that is invariant to scale, rotation, luminosity, and contrast changes is a suitable choice for computer vision tasks. Researchers continue to study and develop new similarity or dissimilarity measures for a broad variety of applications in different domains. Yet, when it comes to testing the similarity measure in machine learning tasks, the method of choice is the $k$-NN model, because it deeply reflects the strength of the similarity measure. Good examples of this fact are the Tangent distance (Simard et al. 1996) and the shape matching distance (Belongie et al. 2002), which are both used for handwritten digit recognition. For the same reason, the $k$-NN model is used to assess the performance of the new dissimilarity measure for images presented in Chap. 4 of this work.

It is interesting to mention that the $k$-NN model is one of the first classifiers for which an upper bound of its error rate has been demonstrated. More precisely, a theoretical result demonstrated in (Cover and Hart 1967) states that the nearest neighbor rule is asymptotically at most twice as bad as the Bayes rule. Furthermore, if $k$ is allowed to grow with $n$ such that $k/n \to 0$, the nearest neighbor rule is universally consistent. More consistency results and other theoretical aspects of the $k$-NN model are discussed in (Devroye et al. 1996).

The $k$-NN model defers all the computations to the test phase. This represents a great disadvantage when the computational time is taken into consideration. Searching for the $k$ nearest neighbors among $n$ training samples may take time proportional to $O(n \cdot k \cdot d)$ using a naive approach, where $d$ represents the computational cost of the distance function. Different approaches based on multidimensional search trees that partition the space and guide the search have been proposed to reduce the time complexity (Dasarathy 1991). Other fast $k$-NN approaches are proposed in (Faragó et al. 1993) and (Zhang and Srihari 2004).

## 2.3 Local Learning

Local learning belongs to the category of unconventional learning paradigms. The development of unconventional (or nonstandard) learning formulations and non-inductive types of inference was studied in (Vapnik 2006). The author argues in favor of introducing and developing unconventional learning methods, as an alternative to algorithmic improvements of existing learning methods. This view is consistent with the main principle of VC theory (Vapnik and Chervonenkis 1971), suggesting that one should always use direct learning formulations for finite sample estimation problems, rather than more general settings (such as density estimation).

Local learning methods attempt to locally adjust the performance of the training system to the properties of the training set in each area of the input space. The local learning paradigm is formally described in Algorithm 2. The local learning algorithm essentially works by first selecting a few training samples located in the vicinity of a given test sample (Step 11), then by training a classifier with only these few examples (Step 12) and finally, by applying the classifier to predict the class label of the test sample (Step 13). It is interesting to note that the $k$-NN model and the Radial Basis Function (RBF) network can be included in the family of local learning algorithms. Actually, the $k$-NN model is the simplest formulation of local learning, since the discriminant function is constant (there is no learning involved). Moreover, almost any other classifier can be employed in the local learning paradigm. Nonetheless, it is important to mention that besides the classifier, a similarity or distance measure is required to determine the neighbors located in the vicinity of a test sample.

An interesting remark is that a linear classifier such as SVM put in the local learning framework becomes nonlinear. In the standard approach, a single linear classifier trained at the global level (on the entire train set) produces a linear discriminant function. On the other hand, the discriminant function for a set of test samples is no longer linear in the local learning framework, since each prediction is given by a different linear classifier which is specifically trained for a single test sample. Moreover, the discriminant function cannot be determined without having the test samples beforehand, yet the local learning paradigm is able to rectify some limitations of linear classifiers, as illustrated in Fig. 2.3.

Local learning has a few advantages over standard learning methods. First, it divides a hard classification problem into more simple subproblems. Second, it reduces the variety of samples in the training set, by selecting samples that are most similar to the test one. Third, it improves accuracy for data sets affected by labeling noise. Considering these advantages, the local learning paradigm is suitable for classification problems with large training data.

In (Bottou and Vapnik 1992) the idea of local algorithms for pattern recognition was used. The approach is based on local linear rules instead of local constant rules, and VC bounds (Vapnik and Chervonenkis 1971) instead of the distance to the $k$-th nearest neighbor. The local linear rules demonstrated an improvement in accuracy on the popular MNIST data set (the error rate dropped from 4.1 to 3.2 %).

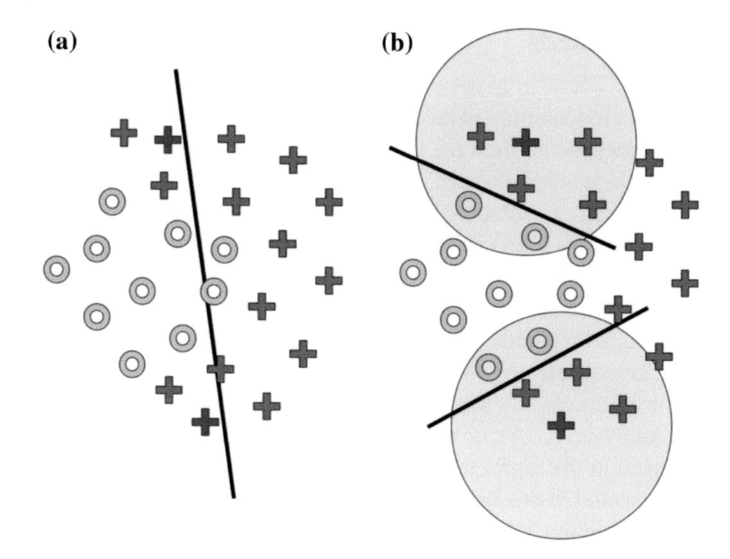


**Fig. 2.3** Two classification models are used to solve the same binary classification problem. The two test samples depicted in *red* are misclassified by a global linear classifier (*left-hand* side). The local learning framework produces a nonlinear decision boundary that fixes this problem (*right-hand* side). **a** A global linear classifier misclassifies the test samples depicted in *red*. **b** A local learning model based on an underlying linear classifier is able to correctly classify the test samples depicted in *red*

For the regression estimation problem, a similar approach was used in the Nadaraya–Watson estimator (Nadaraya 1964) with a slightly different concept of locality. Nadaraya and Watson suggested considering "soft locality" by using a kernel as a weighting function for estimating a value of interest.

In Chap. 4, a filter-based $k$-NN approach, which is a pure local learning algorithm, is used. In the filter-based $k$-NN approach, the filtering step consists of selecting the nearest $K$ neighbors (where $K$ is larger than $k$) using a distance measure that is much faster and easy to compute. Instead of training a classifier, the next step is to select the nearest $k$ neighbors from those filtered $K$ examples using another distance measure that is able to capture more subtle differences. This latter distance measure is allowed to consume more time in order to determine a better similarity (or dissimilarity) between training samples. This approach is appropriate when it is unreasonable, from the time perspective, to compute the latter distance for all training examples. This two-step selection (or filtering) process is much faster to compute than a standard $k$-NN based only on the computationally heavy distance measure, but it also provides more accurate results than using the less computationally expensive distance measure.

In Chap. 5, the learning phase of the framework used for facial expression recognition is based on a local learning algorithm. The algorithm uses a kernel based on visual word occurrences to select nearest neighbors in the vicinity of a test image. Then, it trains an SVM classifier only on the selected neighbors to predict the class label of the test image.

## 2.4 Kernel Methods

In the similarity-based learning paradigm, a popular approach is to treat the pairwise similarities as inner products in some Hilbert space or to treat pairwise dissimilarities as distances in some euclidean space. This can be achieved in roughly two ways. One is to explicitly embed the samples in a euclidean space, according to the pairwise similarities (or dissimilarities) using multidimensional scaling (Borg and Groenen 2005). Another way is to modify the similarities into kernels and apply kernel methods. This section is focused on the latter approach and it covers the following topics: an overview of kernel methods, methods of combining kernels, such as kernel alignment, multiple kernel learning (MKL), and state-of-the-art kernel methods such as Support Vector Machines (SVM), Kernel Ridge Regression (KRR), Kernel Linear Discriminant Analysis (KDA), or Kernel Partial Least Squares (KPLS). Special consideration is given to the topics that discuss kernel approaches used throughout the experiments presented in this book.

Kernel-based learning algorithms work by embedding the data into a Hilbert space, and searching for linear relations in that space using a learning algorithm. The embedding is performed implicitly, that is by specifying the inner product between each pair of points rather than by giving their coordinates explicitly. The power of kernel methods lies in the implicit use of a Reproducing Kernel Hilbert Space induced by a positive semi-definite kernel function. Despite the fact that the mathematical meaning of a kernel is the inner product in a Hilbert space, another interpretation of a kernel is the pairwise similarity between samples.

The kernel function offers to the kernel methods thc power to naturally handle input data that is not in the form of numerical vectors, such as strings, images, or even video and audio files. The kernel function captures the intuitive notion of similarity between objects in a specific domain and can be any function defined on the respective domain that is symmetric and positive definite. For strings, many such kernel functions exist with various applications in computational biology and computational linguistics (Shawe-Taylor and Cristianini 2004). For images, a state-of-the-art approach is the pyramid match kernel (Grauman and Darrell 2005; Lazebnik et al. 2006).

### *2.4.1 Mathematical Preliminaries*

This section follows the theoretical presentation given in (Shawe-Taylor and Cristianini 2004). Therefore, most of the definitions, propositions, and theorems are reproduced from (Shawe-Taylor and Cristianini 2004) for the sake of completeness of this chapter.

A definition of an inner product space is given next.

**Definition 1** A vector space $X$ over the set of real numbers $\mathbb{R}$ is an inner product space, if there exists a real-valued symmetric bilinear (linear in each argument) map

$\langle \cdot, \cdot \rangle$, that satisfies $\langle x, x \rangle \geq 0$, for all $x \in X$. The bilinear map is known as the inner product, dot product or scalar product.

An inner product space is sometimes referred to as a Hilbert space, although most researchers agree that additional properties of completeness and separability are required. Formally, a *Hilbert space* can be defined as follows.

**Definition 2** A Hilbert Space $\mathcal{H}$ is an inner product space with the additional properties of completeness and separability. A space $\mathcal{H}$ is complete if every Cauchy sequence $\{h_n\}_{n \geq 1}$ of elements of $\mathcal{H}$ converges to a element $h \in \mathcal{H}$, where a Cauchy sequence is one that satisfies the property that

$$\sup_{m>n} \|h_n - h_m\| \rightarrow 0, as\ n \rightarrow \infty.$$

A space $\mathcal{H}$ is separable if for any $\epsilon > 0$ there is a finite set of elements $\{h_1, \ldots, h_N\}$ of $\mathcal{H}$ such that for all $h \in \mathcal{H}$

$$\min_i \|h_i - h\| < \epsilon.$$

Note that $\mathbb{R}^m$ is a Hilbert space.

A kernel method performs a mapping into an embedding or feature space. An *embedding map* (or *feature map*) is a function

$$\phi : x \in \mathbb{R}^m \longmapsto \phi(x) \in F \subseteq \mathcal{H}.$$

A *kernel function* is defined as follows.

**Definition 3** A kernel is a function $k$ that for all $x, z \in X$ satisfies

$$k(x, z) = \langle \phi(x), \phi(z) \rangle,$$

where $\phi$ is a mapping from $X$ to an inner product feature space $F$

$$\phi : x \longmapsto \phi(x) \in F.$$

The choice of the map $\phi$ aims to convert the nonlinear relations from $X$ into linear relations in the embedding space $F$. An example of feature embedding where nonlinear patterns are converted into linear ones is given in Fig. 2.4.

Given a set of vectors in $X$, the pairwise kernels between these vectors generate a *kernel matrix*. The kernel matrix is defined next.

**Definition 4** Given a set of vectors $\{x_1, \ldots, x_n\}$ and a kernel function $k$ employed to evaluate the inner products in a feature space with feature map $\phi$. The kernel matrix is defined as the $n \times n$ matrix $K$ with entries given by

$$K_{ij} = \langle \phi(x_i), \phi(x_j) \rangle = k(x_i, x_j).$$

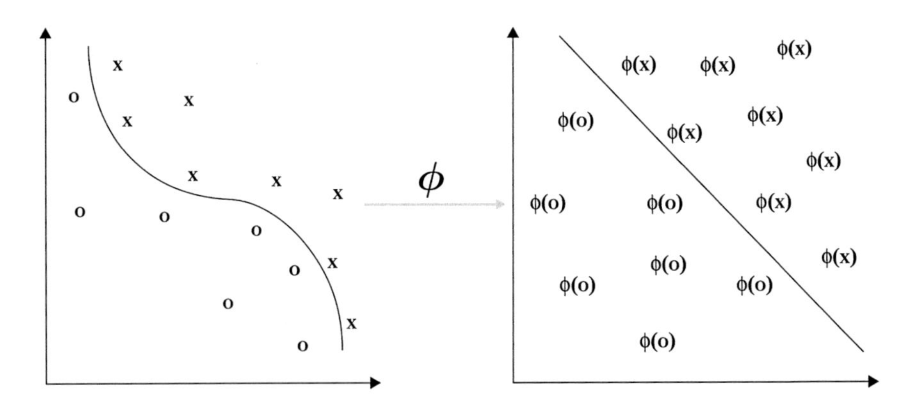


**Fig. 2.4** The function $\phi$ embeds the data into a feature space where the nonlinear relations now appear linear. Machine learning methods can easily detect such linear relations

Given a square matrix $A$, the real number $\lambda$ and the non-zero vector $x$ are an eigenvalue and the corresponding eigenvector of $A$ if $Ax = \lambda x$. A square matrix A is symmetric if $A' = A$, where $A'$ represents the transpose of $A$. A symmetric matrix is positive semi-definite, if its eigenvalues are all non-negative.

**Proposition 1** *Kernel matrices are positive semi-definite.*

Finitely positive semi-definite functions are defined next.

**Definition 5** A function $k : X \times X \longmapsto \mathbb{R}$ satisfies the finitely positive semi-definite property if it is a symmetric function for which the matrices formed by restriction to any finite subset of the space $X$ are positive semi-definite.

The following theorem gives the characterization of kernels.

**Theorem 1** *A function $k : X \times X \longmapsto \mathbb{R}$ which is either continuous or has a finite domain, can be decomposed into a feature map $\phi$ into a Hilbert space $F$ applied to both its arguments followed by the evaluation of the inner product in $F$ as follows:*

$$k(x, z) = \langle \phi(x), \phi(z) \rangle,$$

*if and only if it satisfies the finitely positive semi-definite property.*

Given a function $k$ that satisfies the finitely positive semi-definite property, its corresponding space $F_k$ can be referred to as its *Reproducing Kernel Hilbert Space* (RKHS).

### 2.4.2 *Overview of Kernel Classifiers*

In the case of binary classification problems, kernel-based learning algorithms look for a discriminant function, a function that assigns $+1$ to examples that belong to one class and $-1$ to examples that belong to the other class. This function will be a linear function in the space $\mathcal{F}$, which means it will have the following form:

$$f(x) = \operatorname{sign}(\langle w, \phi(x) \rangle + b),$$

for some weight vector $w$. The kernel can be employed whenever the weight vector can be expressed as a linear combination of the training points, $\sum_{i=1}^{n} \alpha_i \phi(x_i)$, implying that $f$ can be expressed as follows:

$$f(x) = \operatorname{sign}\left( \sum_{i=1}^{n} \alpha_i k(x_i, x) + b \right).$$

Various kernel methods differ in the way they find the vector $w$ (or equivalently the vector $\alpha$). Support Vector Machines (Cortes and Vapnik 1995) try to find the vector $w$ that defines the hyperplane that maximally separates the images in $\mathcal{F}$ of the training examples belonging to the two classes. Mathematically, the SVM classifier chooses the $w$ and the $b$ that satisfy the following optimization criterion:

$$\min_{w,b} \frac{1}{n} \sum_{i=1}^{n} [1 - y_i(\langle w, \phi(x_i) \rangle + b)]_+ + \nu ||w||^2,$$

where $y_i$ is the label (+1/−1) of the training example $x_i$, $\nu$ is a regularization parameter and $[x]_+ = \max\{x, 0\}$.

Kernel Ridge Regression (KRR) selects the vector $w$ that simultaneously has small empirical error and small norm in the RKHS generated by the kernel $k$. The resulting minimization problem is

$$\min_{w} \frac{1}{n} \sum_{i=1}^{n} (y_i - \langle w, \phi(x_i) \rangle)^2 + \lambda ||w||^2,$$

where again $y_i$ is the label (+1/−1) of the training example $x_i$, and $\lambda$ is a regularization parameter.

The *Linear Discriminant Analysis* (LDA) method, also known as *Fisher Discriminant Analysis*, maximizes the ratio of between-class variance to the within-class variance in order to guarantee maximal separability for a particular set of samples. The author of (Fisher 1936) derived the LDA approach for a two-class problem, under the assumptions that the classes have normal distributions and identical covariance matrices. The assumption of identical covariance matrices implies that the Bayes classifier is linear. Therefore, LDA provides a projection of the data points to a one-dimensional subspace where the Bayes classification error is the smallest. The KDA method (Shawe-Taylor and Cristianini 2004) is the kernel version of the LDA algorithm, which is somewhat similar to the KRR algorithm.

It appears that sometimes the covariance of the input vectors with the targets is more important than the variance of the vectors, for regression problems. The *Partial Least Squares* (PLS) approach is based on the covariance to guide feature selection, before performing least squares regression in the derived feature space. More precisely, PLS is used to find the fundamental relations between the input matrix $X$ and response matrix $Y$. The kernel version of PLS is a powerful algorithm that can also be used for classification problems.

For a particular classification problem, some kernel methods may be more suitable than others. The accuracy level depends on many aspects such as class distribution, the number of classes, data noise, size of the training data, and so on. For example, the KRR classifier can be used with success for problems with well-balanced classes, while the Kernel Partial Least Squares (KPLS) classifier is more suitable for many class problems. In some particular cases, when the number of classes is greater than two, there is a serious problem with the regression methods. More precisely, some classes can be masked by others. The KDA classifier is able to improve accuracy by avoiding the masking problem (Hastie and Tibshirani 2003). More details about SVM, KRR, KDA and KPLS can be found in (Shawe-Taylor and Cristianini 2004). The important fact is that the optimization problems of these classifiers are solved in such a way that the coordinates of the embedded points are not needed, only their pairwise inner products which in turn are given by the kernel function $k$.

The SVM and KRR classifiers are used in Chap. 4 for handwritten character recognition and texture classification, and in Chap. 8 for native language identification. The SVM classifier is used as well in Chap. 5 for object recognition, while the KRR method is used in Chap. 9 for text categorization by topic. The KDA and KPLS methods are used in Chap. 4 for texture classification, along with the SVM and KRR methods. The KDA classifier is also used in Chap. 8 for native language identification.

### 2.4.3 Kernel Functions

Since the concept of kernel method emerged, researchers have proposed several kernels in the literature over the years. The most common kernel is the linear kernel that is obtained by computing the inner product of two vectors. The map function in this case is $\phi(x) = x$. Example 1 shows how to compute the linear kernel for two vectors of four components each.

*Example 1* Let $x = (1, 2, 4, 1)$ and $z = (5, 1, 2, 3)$ be two vectors in $\mathbb{R}^4$. The linear kernel between $x$ and $z$ is

$$\begin{aligned} k(x, z) &= \langle x, z \rangle \\ &= 1 \cdot 5 + 2 \cdot 1 + 4 \cdot 2 + 1 \cdot 3 \\ &= 18. \end{aligned}$$

Let $k_1(x, z)$ be a kernel over $X \times X$, where $x, z \in X$, and $p(x)$ is a polynomial with positive coefficients. Then the following functions are also kernels. The *polynomial kernel* is defined by $k(x, y) = p(k_1(x, z))$. Another two kernels based on the exponential function are defined as $k(x, z) = \exp(k_1(x, z))$ and $k(x, z) = \exp(-\|x - z\|^2/(2\sigma^2))$. The latter kernel is known as the *Gaussian kernel*. Such functions form the hidden units of RBF networks, and the Gaussian kernel is therefore also referred to as the *RBF kernel*. Example 2 shows how to compute the RBF kernel for two given vectors.

*Example 2* Let $x = (1, 2, 4, 1)$ and $z = (5, 1, 2, 3)$ be two vectors in $\mathbb{R}^4$, and $\sigma = 1$. The RBF kernel between $x$ and $z$ is

$$
\begin{aligned}
k(x, z) &= \exp\left(-\frac{\|x - z\|^2}{2\sigma^2}\right) \\
&= \exp\left(-\frac{\sqrt{(1-5)^2 + (2-1)^2 + (4-2)^2 + (1-3)^2}}{2 \cdot 1^2}\right) \\
&= \exp\left(-\frac{\sqrt{16 + 1 + 4 + 4}}{2}\right) \\
&= \exp\left(-\frac{5}{2}\right) \\
&\approx 0.0821.
\end{aligned}
$$

The *(histogram) intersection kernel* is given by $k(x, z) = \sum_i \min\{x_i, z_i\}$. The intersection kernel between two vectors in $\mathbb{R}^4$ is computed as indicated in Example 3.

*Example 3* Let $x = (1, 2, 4, 1)$ and $z = (5, 1, 2, 3)$ be two vectors in $\mathbb{R}^4$. The intersection kernel between $x$ and $z$ is

$$
\begin{aligned}
k(x, z) &= \sum_i \min\{x_i, z_i\} \\
&= \min\{1, 5\} + \min\{2, 1\} + \min\{4, 2\} + \min\{1, 3\} \\
&= 1 + 1 + 2 + 1 \\
&= 5.
\end{aligned}
$$

The definition of the *Hellinger's kernel* (also known as the *Bhattacharyya coefficient*) is $k(x, z) = \sum_i \sqrt{x_i \cdot z_i}$. Other examples of kernels from a broad variety of such functions are the $\chi^2$ kernel, the Jensen–Shannon kernel, or the Matern kernel. Two new kernels, namely the PQ kernel and the Spatial Non-Alignment Kernel, are presented in Chap. 5 of this book.

The following proposition shows the operations that can be used to build new kernels from existing kernels.

**Proposition 2** *Let $k_1$ and $k_2$ be two kernels over $X \times X$, $X \subseteq \mathcal{H}$, $a \in \mathbb{R}^+$, $f(\cdot)$ a real-valued function on $X$, and $B$ a symmetric positive semi-definite $n \times n$ matrix. Then the following functions are kernels:*

$$(i)\ k(x,z) = k_1(x,z) + k_2(x,z);$$
$$(ii)\ k(x,z) = ak_1(x,z);$$
$$(iii)\ k(x,z) = k_1(x,y) \cdot k_2(x,z);$$
$$(iv)\ k(x,z) = f(x) \cdot f(z);$$
$$(v)\ k(x,z) = x'Bz.$$

## 2.4.4 Kernel Normalization

Another example of creating a new kernel from an existing one is provided by normalizing the existing kernel. Given a kernel $k(x, y)$ that corresponds to the feature map $\phi$, the normalized kernel $\hat{k}(x, y)$ corresponds to the feature map given by

$$x \longmapsto \phi(x) \longmapsto \frac{\phi(x)}{\|\phi(x)\|}.$$

Researchers have found that data normalization helps to improve machine learning performance for various applications. Since the range of values of raw data can have large variation, classifier objective functions will not work properly without normalization. Features are usually normalized through a process called *standardization*, which makes the values of each feature in the data have zero-mean and unit-variance. By normalization, each feature has an approximately equal contribution to the distance between two samples.

The kernel normalization can also be done directly on the kernel matrix. To obtain a normalized kernel matrix, each component is divided by the square root of the product of the two corresponding diagonal components, as follows:

$$\hat{K}_{ij} = \frac{K_{ij}}{\sqrt{K_{ii} \cdot K_{jj}}}. \tag{2.1}$$

This is equivalent to normalizing the kernel function as follows:

$$\hat{k}(x_i, x_j) = \frac{k(x_i, x_j)}{\sqrt{k(x_i, x_i) \cdot k(x_j, x_j)}}. \tag{2.2}$$

An interesting study that gives a good insight into how different kernels should be normalized is (Vedaldi and Zisserman 2010). The authors state that $\gamma$-homogeneous kernels should be $L_\gamma$-normalized. For example, the linear kernel or the Jensen–Shannon kernel should be $L_2$-normalized, while the Hellinger's kernel should be $L_1$-normalized.

**Algorithm 3**: Generic Algorithm of Kernel Methods

1 **Input**:
2 $\mathcal{S} = \{(x_i, t_i) \mid x_i \in \mathbb{R}^m, t_i \in \{+1, -1\}, i \in \{1, 2, \ldots, n\}\}$ – the set of $n$ training samples and labels;
3 $Z = \{z_i \mid z_i \in \mathbb{R}^m, i \in \{1, 2, \ldots, l\}\}$ – the set of $l$ test samples;
4 $k$ – a kernel function;
5 $\mathcal{C}$ – a binary kernel classifier.
6 **Initialization**:
7 $K \leftarrow \mathbf{0}_n$;
8 $K^{test} \leftarrow \mathbf{0}_{l,n}$;
9 **Computation**:
10 **for** $x_i \in X$ **do**
11     **for** $x_j \in X$ **do**
12         $K_{ij} \leftarrow k(x_i, x_j)$;
13 **for** $z_i \in Z$ **do**
14     **for** $x_j \in X$ **do**
15         $K_{ij}^{test} \leftarrow k(z_i, x_j)$;
16 $(\alpha, b) \leftarrow$ the dual weights of $\mathcal{C}$ trained on $K$ with the labels $T$;
17 $Y \leftarrow K^{test}\alpha + b$;
18 **Output**:
19 $Y = \{y_i \mid y_i \in \{+1, -1\}, i \in \{1, 2, \ldots, l\}\}$ – the set of predicted labels for the test samples in $Z$.

### 2.4.5 *Generic Kernel Algorithm*

A generic kernel method in the dual form is described in Algorithm 3. The following notations are used in the algorithm. The zero matrix of $l \times n$ components is denoted by $\mathbf{0}_{l,n}$, and the square zero matrix is denoted by $\mathbf{0}_n$. The column vector of dual weights is denoted by $\alpha$ and the bias value is denoted by $b$. In Step 16, the kernel classifier $\mathcal{C}$ assigns a weight to each training sample. Thus, the vector of weights $\alpha$ contains $n$ values, such that the weight $\alpha_i$ corresponds to the training sample $x_i$. When the test kernel matrix $K^{test}$ of $l \times n$ components is multiplied with the vector $\alpha$ in Step 17, the result is a vector of labels $Y$ of $l$ components. These are the predicted labels for the test samples in $Z$, such that $y_i$ corresponds to test sample $z_i$. A particular kernel method can be obtained from Algorithm 3 by specifying the kernel function $k$ and by choosing the kernel classifier $\mathcal{C}$.

### 2.4.6 *Multiple Kernel Learning*

Different kernel representations can be obtained from the same data. The idea of combining all these kernels is natural when one wants to improve the performance of a classifier. When multiple kernels are combined, the features are actually embedded in a higher dimensional space. As a consequence, the search space of linear patterns grows, which helps the classifier to select a better a discriminant function. The concept of learning using multiple kernels is known as *multiple kernel learning* (MKL).

The most natural way of combining two kernels is to sum them up. Summing up kernels or kernel matrices is equivalent to feature vector concatenation. However, the feature vectors are usually high-dimensional vectors, and the concatenation of such vectors is not a viable solution in terms of space and time. In this case, the kernel trick can be employed to obtain the kernel matrices that must be summed up. Another possibility to obtain a combination is to multiply the kernels. An interesting remark is that multiplying sparse kernel matrices (component-wise) will produce an even more sparse kernel matrix, which might not be desirable in some cases, since patterns simply disappear. These two methods of combining kernels are also given in Proposition 2.

Another option is to combine kernels by kernel alignment (Cristianini et al. 2001). Instead of simply summing kernels, kernel alignment assigns weights for each of the two kernels based on how well they are aligned with the ideal kernel $YY'$ obtained from labels. The work of (Cortes et al. 2013) presents a new algorithm for multi-class classification with multiple kernels, which is based on a natural notion of the multi-class margin of a kernel. The algorithm shows improvements over the performance of state-of-the-art algorithms in binary and multi-class classification with multiple kernels. A review of MKL algorithms is presented in (Gonen and Alpaydin 2011).

MKL based on kernel sum is used in Chap. 5 and in Chap. 9 to obtain representations that include spatial information. MKL based on kernel alignment and on kernel sum is used in Chap. 8 to combine string kernels for native language identification. As such, MKL can be viewed as another concept that yields improved performance across the studied domains, namely computer vision and text mining.

## 2.5 Cluster Analysis

A form of unsupervised learning used in data mining is clustering. Unlike supervised learning, it has the advantage that training labels are not required, thus having more general applications. Clustering has long played an important role in a wide variety of fields, such as biology, statistics, pattern recognition, information retrieval, machine learning, data mining, psychology, and other social sciences.

Clustering is the task of assigning a set of objects into groups (termed clusters) such that the objects in the same cluster are more similar to each other than to those in other clusters. Objects are clustered based only on the information found in the data that describes the objects and their relationships. Pairwise relationships are usually described through a similarity or dissimilarity function. The goal of clustering is to maximize the similarity of objects within groups, and in the same time, to minimize the similarity of objects from different groups. The greater the similarity within a group and the greater the difference between groups, the better or more distinct the clustering. The clusters should capture the natural structure hidden in the data. An important remark is that the appropriate clustering algorithm and parameter settings, such as the distance function to use, the density threshold, or the number of expected clusters, all depend on individual data sets.

There are various clustering algorithms that differ significantly in their notion of what constitutes a cluster. Popular notions of clusters include groups with low distances among the cluster members, dense areas of the data space, intervals or particular statistical distributions. Clustering methods can be roughly divided into several categories, such as hierarchical clustering methods, centroid-based methods, distribution-based methods, grid-based methods, and density-based methods. This section discusses only the first two categories of clustering methods, which are also used in some of the experiments presented in this book. However, a complete reference of the major clustering methods is given in (Han et al. 2011). It is important to mention that some of the proposed approaches do not necessarily fall in one of these categories, such as the subspace clustering method of (Kailing et al. 2004), and some of them use mixed models (McCallum et al. 2000).

In recent years considerable effort has been made to improve the performance of the existing clustering algorithms. The work of (Huang 1998) proposes an extension to the k-means algorithm for clustering large data sets with categorical values. A clustering method that aims to identify spatial structures that may be present in the data is proposed in (Ng and Han 2002).

An unsupervised data mining algorithm used to perform hierarchical clustering over particularly large data sets is presented in (Zhang et al. 1996). The advantage of this algorithm is its ability to incrementally and dynamically cluster incoming, multidimensional metric data points in an attempt to produce the best quality clustering with a given set of resources.

With the recent need to process larger and larger data sets (also known as big data), the willingness to trade semantic meaning of the generated clusters for computational performance has been increasing. This led to the development of pre-clustering methods such as canopy clustering (McCallum et al. 2000), which can process huge data sets efficiently, but the resulting clusters are only a rough pre-partitioning of the data set. These partitions can be subsequently analyzed with existing slower methods such as k-means clustering.

For high-dimensional data, many of the existing methods fail due to the curse of dimensionality, which renders particular distance functions problematic in high-dimensional spaces. This led to new clustering algorithms for high-dimensional data that focus on subspace clustering and correlation clustering (Kriegel et al. 2009). An example of subspace clustering algorithm is SUBCLU (Kailing et al. 2004) which aims at automatically identifying subspaces of the feature space in which clusters exist. This algorithm is able to detect arbitrarily shaped and positioned clusters in subspaces. Another similar algorithm is CLIQUE (Agrawal et al. 1998) which identifies dense clusters in subspaces of maximum dimensionality. Ideas from density-based clustering methods have been adopted to subspace clustering (Achtert et al. 2006, 2007) and correlation clustering (Bohm et al. 2004).

**Algorithm 4**: K-Means Clustering Algorithm

1 **Input**:
2 $X = \{x_i \mid x_i \in \mathbb{R}^m, i \in \{1, 2, \ldots, n\}\}$ – the set of $n$ training samples;
3 $k$ – the number of clusters.
4 **Initialization**:
5 $C^{(1)} \leftarrow \{c_i^{(1)} \mid c_i^{(1)} \in \mathbb{R}^m$ is a random sample of $X, i \in \{1, 2, \ldots, k\}\}$;
6 **Computation**:
7 **while** $C^{(t)} \neq C^{(t+1)}$ **do**
8 **for** $i \in \{1, 2, \ldots, k\}$ **do**
9 $S_i^{(t)} \leftarrow \{x_l \mid \|x_l - c_i^{(t)}\|^2 \leq \|x_l - c_j^{(t)}\|^2, \forall j \in \{1, 2, \ldots, k\}, x_l \in X\}$;
10 $c_i^{(t+1)} \leftarrow \frac{1}{|S_i^{(t)}|} \sum_{x_l \in S_i^{(t)}} x_l$;
11 $t \leftarrow t + 1$;
12 **Output**:
13 $C$ – the set of $k$ centroids resulted after the algorithm has converged.

### 2.5.1 *K-Means Clustering*

The k-means clustering technique is a simple method of cluster analysis which aims to partition a set of objects into $k$ clusters in which each object belongs to the cluster with the nearest mean. The k-means clustering is formally described in Algorithm 4. The iteration index is denoted by $t$. The number of samples in a cluster $S_i$ is denoted by $|S_i|$. It is important to note that the clusters $S_i, \forall i \in \{1, 2, \ldots, k\}$ form a partition of $X$. This means that $S_i \cap S_j = \emptyset, \forall i, j \in \{1, 2, \ldots, k\}$ such that $i \neq j$. In other words, each sample must be assigned to a single cluster.

Algorithm 4 begins with choosing $k$ initial centroids (Step 5), where $k$ is an a priori parameter, namely, the number of desired clusters. Each sample is then assigned to the nearest centroid (Step 9), and each group of samples assigned to a centroid represents a cluster. The centroid of each cluster is then updated based on the samples that belong to that cluster (Step 10). The assignment and update steps are repeated until no point changes clusters (Step 7). Alternatively, the algorithm can be stopped before the clusters converge, namely when a maximum number of iterations set a priori is reached. The k-means algorithm aims at minimizing an objective function, given by

$$J = \sum_{j=1}^{k} \sum_{i=1}^{n} \|x_i^{(S_j)} - c_j\|^2,$$

where $x_i^{(S_j)}$ is a vector in cluster $S_j$ and $c_j$ is the cluster centroid (or mean vector). The alternating optimization procedure that minimizes this objective function is also given in (Hastie and Tibshirani 2003). An important remark is that the underlying euclidean distance can be replaced with another distance measure in the objective function. There are many practical situations when this could be useful, for example to obtain a better clustering. Given a distance measure $\Delta$, the objective function can be more generally expressed as follows:

$$J = \sum_{j=1}^{k} \sum_{i=1}^{n} \Delta(x_i^{(S_j)}, c_j).$$

It is interesting to mention that the k-means algorithm generates a Voronoi partitioning of the data. Each cluster is a Voronoi cell determined by the cluster centroid. In Chap. 5, the k-means algorithm is used to obtain visual words from vector quantized image descriptors.

### 2.5.2 *Hierarchical Clustering*

Hierarchical clustering creates a hierarchical decomposition of a given set of data objects. Hierarchical methods can be divided into two main categories: *agglomerative* methods and *divisive* methods. Agglomerative methods start at the bottom and recursively join clusters two by two at each level, until a single cluster is obtained. On the other hand, divisive methods start at the top and recursively divide a cluster into two new clusters at each level, until objects are completely divided into separate clusters. Divisive methods are not generally available and have rarely been applied due to the difficulty of taking the right decision of dividing clusters at a high level. Thus, many hierarchical clustering techniques are variations of a single (agglomerative) algorithm: starting with individual objects as clusters, successively join the two nearest clusters until only one cluster remains. These techniques connect objects to form clusters based on their distance. An important remark is that hierarchical algorithms do not provide a single partitioning of the data set, but an extensive hierarchy of clusters that merge with each other at certain distances. This structure can be represented using a *dendrogram*.

A hierarchical clustering approach is formally presented in Algorithm 5. Apart from the choice of a distance function $\Delta$, another decision is needed for the linkage criterion $\Delta^{\text{link}}$ to be used. The most popular choices are the single linkage, the complete linkage, or the average linkage. The linkage criterion is used in Step 12 of the algorithm. In the single linkage method, the similarity between two clusters is measured by the similarity of the closest pair of data points from different clusters. To use the single linkage criterion, in Step 12 of Algorithm 5, $\Delta^{\text{link}}$ should be assigned as follows:

$$\Delta^{\text{link}}_{S_i,S_j} \leftarrow \min\{\Delta(x, z) \mid \forall x \in S_i, \forall z \in S_j\},$$

**Algorithm 5**: Hierarchical Clustering based on Linkage

1 **Input**:
2 $X = \{x_i \mid x_i \in \mathbb{R}^m, i \in \{1, 2, \dots, n\}\}$ – the set of $n$ training samples;
3 $\Delta$ – a distance measure.
4 **Initialization**:
5 $S_i \leftarrow \{x_i\}, \forall i \in \{1, 2, \dots, n\}$;
6 $\mathcal{S} \leftarrow \{S_1, S_2, \dots, S_n\}$;
7 $\mathcal{P} \leftarrow \emptyset$;
8 **Computation**:
9 **for** $k \in \{1, 2, \dots, n-1\}$ **do**
10 **for** $S_i \in \mathcal{S}$ **do**
11 **for** $S_j \in \mathcal{S} \setminus \{S_i\}$ **do**
12 compute $\Delta^{link}_{S_i,S_j}$;
13 $(S_i, S_j) \leftarrow \text{argmin}\{\Delta^{link}_{S_i,S_j}, \forall S_i, S_j \in \mathcal{S}, \text{ such that } S_i \neq S_j\}$;
14 $S_{n+k} \leftarrow S_i \cup S_j$;
15 $\mathcal{S} \leftarrow \mathcal{S} \setminus \{S_i, S_j\}$;
16 $\mathcal{S} \leftarrow \mathcal{S} \cup \{S_{n+k}\}$;
17 $\mathcal{P} \leftarrow \mathcal{P} \cup \{(S_i, S_j, \Delta^{link}_{i,j})\}$;
18 **Output**:
19 $\mathcal{P} = \{(S_i, S_j, \Delta^{link}) \mid S_i \text{ and } S_j \text{ are the clusters joined at distance } \Delta^{\text{link}}\}$ – the set of triplets resulted after the algorithm has converged to a single cluster.

where $S_i$ and $S_j$ are two clusters from $\mathcal{S}$. For every pair of clusters $S_i$ and $S_j$, $\Delta^{\text{link}}_{S_i,S_j}$ represents the minimum distance between pairs of data points (samples) taken one from $S_i$ and one from $S_j$. The complete linkage takes the similarity of the furthest pair of data points from different clusters. In Algorithm 5, Step 12 becomes

$$\Delta^{\text{link}}_{S_i,S_j} \leftarrow \max\{\Delta(x, z) \mid \forall x \in S_i, \forall z \in S_j\}.$$

The average linkage takes the average similarity between all the pairs of data points from different clusters. To use the average linkage criterion, Step 12 of Algorithm 5 should be modified as follows:

$$\Delta^{\text{link}}_{S_i,S_j} \leftarrow \text{avg}\{\Delta(x, z) \mid \forall x \in S_i, \forall z \in S_j\}.$$

An interesting remark is that the number of clusters is always $2 \cdot n - 1$, where $n$ is the number of samples to be clustered. The algorithm ends when the last cluster $S_{2 \cdot n-1}$ (that contains all the samples) is formed.

Some hierarchical clustering techniques based on Local Rank Distance are used in the experiments presented in Chap. 7. They employ the average linkage criterion in order to form clusters of strings.

## References


Achtert E, Bohm C, Kriegel HP, Kroger P, Muller-Gorman I, Zimek A (2006) Finding hierarchies of subspace clusters. In: Proceedings of PKDD, pp 446–453

Achtert E, Bohm C, Kriegel HP, Kroger P, Muller-Gorman I, Zimek A (2007) Detection and visualization of subspace cluster hierarchies. In: Proceedings of DASFAA, pp 152–163

Agrawal R, Gehrke J, Gunopulos D, Raghavan P (1998) Automatic subspace clustering of high dimensional data for data mining applications. SIGMOD Rec 27(2):94–105

Altschul S, Gish W, Miller W, Myers E, Lipman D (1990) Basic local alignment search tool. J Mol Biol 215(3):403–410

Belongie S, Malik J, Puzicha J (2002) Shape matching and object recognition using shape contexts. IEEE Trans Pattern Anal Mach Intell 24(4):509–522

Bohm C, Kailing K, Kroger P, Zimek A (2004) Computing clusters of correlation connected objects. In: Proceedings of the 2004 ACM SIGMOD, pp 455–466

Borg I, Groenen PJF (2005) Modern multidimensional scaling: theory and applications. Springer, Berlin

Bottou L, Vapnik V (1992) Local learning algorithms. Neural Comput 4:888–900

Cazzanti L, Gupta MR, Koppal AJ (2008) Generative models for similarity-based classification. Pattern Recognit 41(7):2289–2297

Cazzanti L, Gupta MR (2007) Local similarity discriminant analysis. In: Proceedings of ICML, pp 137–144

Chen Y, Garcia EK, Gupta MR, Rahimi A, Luca C (2009) Similarity-based classification: concepts and algorithms. J Mach Learn Res 10:747–776

Chimani M, Woste M, Bocker S (2011) A closer look at the closest string and closest substring problem. In: Proceedings of ALENEX, pp 13–24

Cortes C, Mohri M, Rostamizadeh A (2013) Multi-class classification with maximum margin multiple kernel. J Mach Learn Res 28(3):46–54

Cortes C, Vapnik V (1995) Support-vector networks. Mach Learn 20(3):273–297

Cover T, Hart P (1967) Nearest neighbor pattern classification. IEEE Trans Inf Theory 13(1):21–27

Cristianini N, Shawe-Taylor J, Elisseeff A, Kandola JS (2001) On kernel-target alignment. In: Proceedings of NIPS, pp 367–373

Dasarathy BV (1991) Nearest neighbor (NN) norms: pattern classification techniques. IEEE Computer Society Press, Los Alamitos

Devroye L, Györfi L, Lugosi GA (1996) Probabilistic theory of pattern recognition. Springer, New York

Dinu LP, Ionescu RT (2012a) Clustering based on rank distance with applications on DNA. In: Proceedings of ICONIP, vol 7667, pp 722–729

Dinu LP, Ionescu RT (2012b) An efficient rank based approach for closest string and closest substring. PLoS ONE 7(6):e37576

Faragó A, Linder T, Lugosi G (1993) Fast nearest-neighbor search in dissimilarity spaces. IEEE Trans Pattern Anal Mach Intell 15(9):957–962

Fisher RA (1936) The use of multiple measurements in taxonomic problems. Ann Eugenics 7(7):179–188

Fix E, Hodges J (1951) Discriminatory analysis, non-parametric discrimination: consistency properties. Technical report, USAF School of Aviation and Medicine, Randolph Field, TX, 1951. Technical Report 4

Gonen M, Alpaydin E (2011) Multiple kernel learning algorithms. J Mach Learn Res 12:2211–2268

Graepel T, Herbrich R, Bollmann-Sdorra P, Obermayer K (1998) Classification on pairwise proximity data. In: Proceedings of NIPS, pp 438–444

Graepel T, Herbrich R, Scholkopf B, Smola A, Bartlett P, Muller K, Obermayer K, Williamson R (1999) Classification on proximity data with LP-machines. In: Proceedings of ICANN, vol 1, pp 304–309

Grauman K, Darrell T (2005) The pyramid match kernel: discriminative classification with sets of image features. In: Proceedings of ICCV, vol 2, pp 1458–1465
Han J, Kamber M, Pei J (2011) Data mining: concepts and techniques, 3rd edn. Morgan Kaufmann Publishers Inc., San Francisco
Hastie T, Tibshirani R (2003) The elements of statistical learning. Springer, New York. ISBN 0387952845
Huang Z (1998) Extensions to the k-means algorithm for clustering large data sets with categorical values. Data Mining Knowl Discov 2(3):283–304
Ionescu RT, Popescu M (2013) Kernels for visual words histograms. In: Proceedings of ICIAP, vol 8156, pp 81–90
Ionescu RT, Popescu M, Cahill A (2014) Can characters reveal your native language? A language-independent approach to native language identification. In: Proceedings of EMNLP, pp 1363–1373
Kailing K, Kriegel HP, Kroger P (2004) Density-connected subspace clustering for high-dimensional data. In: Proceedings of SDM
Kriegel HP, Kroger P, Zimek A (2009) Clustering high-dimensional data: a survey on subspace clustering, pattern-based clustering, and correlation clustering. ACM Trans Knowl Discov Data 3(1):1:1–1:58
Lazebnik S, Schmid C, Ponce J (2006) Beyond bags of features: spatial pyramid matching for recognizing natural scene categories. In: Proceedings of CVPR, vol 2, pp 2169–2178
Liao L, Noble WS (2003) Combining pairwise sequence similarity and support vector machines for detecting remote protein evolutionary and structural relationships. J Comput Biol 10(6):857–868
Lipman DJ, Pearson WR (1985) Rapid and sensitive protein similarity searches. Science 227:1435–1441
Lodhi H, Saunders C, Shawe-Taylor J, Cristianini N, Watkins CJCH (2002) Text classification using string kernels. J Mach Learn Res 2:419–444
Manning CD, Raghavan P, Schütze H (2008) Introduction to information retrieval. Cambridge University Press, New York
McCallum A, Nigam K, Ungar LH (2000) Efficient clustering of high-dimensional data sets with application to reference matching. In: Proceedings of ACM SIGKDD, pp 169–178
Nadaraya EA (1964) On estimating regression. Theory Probab Appl 9:141–142
Ng RT, Jiawei H (2002) CLARANS: a method for clustering objects for spatial data mining. IEEE Trans Knowl Data Eng 14(5):1003–1016
Pekalska E, Duin RPW (2002) Dissimilarity representations allow for building good classifiers. Pattern Recognit Lett 23(8):943–956
Popescu M, Dinu LP (2007) Kernel methods and string kernels for authorship identification: the federalist papers case. In: Proceedings of RANLP
Popescu M, Grozea C (2012) Kernel methods and string kernels for authorship analysis. CLEF (Online Working Notes/Labs/Workshop)
Popescu M, Ionescu RT (2013) The story of the characters, the DNA and the native language. In: Proceedings of the Eighth Workshop on Innovative Use of NLP for Building Educational Applications, pp 270–278
Popov YV (2007) Multiple genome rearrangement by swaps and by element duplications. Theoret Comput Sci 385(1–3):115–126
Rubner Y, Tomasi C, Guibas LJ (2000) The Earth Mover's distance as a metric for image retrieval. Int J Comput Vis 40(2):99–121
Sanderson C, Guenter S (2006) Short text authorship attribution via sequence kernels, Markov chains and author unmasking: An investigation. In: Proceedings of EMNLP, pp 482–491
Shapira D, Storer JA (2003) Large edit distance with multiple block operations. In: Proceedings of SPIRE, vol 2857, pp 369–377
Shawe-Taylor J, Cristianini N (2004) Kernel methods for pattern analysis. Cambridge University Press, Cambridge

Simard P, LeCun Y, Denker JS, Victorri B (1996) Transformation invariance in pattern recognition, tangent distance and tangent propagation. Neural Networks: Tricks of the Trade
Vapnik V (2006) Estimation of dependencies based on empirical data (Information Science and Statistics), 2nd edn. Springer, New York
Vedaldi A, Zisserman A (2010) Efficient additive kernels via explicit feature maps. In: Proceedings of CVPR, pp 3539–3546
Vezzi F, Fabbro CD, Tomescu AI, Policriti A (2012) rNA: a fast and accurate short reads numerical aligner. Bioinformatics 28(1):123–124
Vladimir V, Chervonenkis A (1971) On the uniform convergence of relative frequencies of events to their probabilities. Theory Probab Appl 16(2):264–280
Zhang T, Ramakrishnan R, Livny M (1996) BIRCH: an efficient data clustering method for very large databases. SIGMOD Rec 25(2):103–114. ISSN 0163–5808
Zhang B, Srihari SN (2004) Fast k-nearest neighbor classification using cluster-based trees. IEEE Trans Pattern Anal Mach Intell 26(4):525–528, 2004

# Part I
# Knowledge Transfer from Text Mining to Computer Vision

# Chapter 3
# State-of-the-Art Approaches for Image Classification

## 3.1 Introduction

This chapter is a concise overview of the main state-of-the-art methods in computer vision, and the aim is to provide some context for the models that are used throughout this work for various computer vision tasks. *Computer vision* is a field that studies methods for acquiring, processing, analyzing, and understanding images and video. Such methods attempt to solve different tasks including object recognition, scene reconstruction, event detection, video tracking, image retrieval, image segmentation, motion estimation, image restoration, and many others. Historically, the main successful lines of research in computer vision include face recognition with eigenfaces (Turk and Pentland 1991) and with Haar features (Viola and Jones 2001, 2004), shape matching using shape context (Belongie et al. 2002), object recognition and matching with SIFT features (Lowe 1999, 2004), human detection with HOG features and SVM (Dalal and Triggs 2005), followed by object detection with Deformable Part Models (Felzenszwalb et al. 2010). More recently, convolutional neural networks (Krizhevsky et al. 2012; Simonyan and Zisserman 2014; Szegedy et al. 2015) have demonstrated significant improvements in object recognition and detection.

One of the most important tasks in computer vision is *object recognition*. This task deals with building computer systems that attempt to identify objects represented in digitized images or video, thus enabling robots to see. Machine learning methods represent the state-of-the-art approach for the object recognition problem. The mainstream approach is to treat object recognition as a classification task. Therefore, the problem is also referred to as *image classification* or *image categorization*. A preliminary step in order to obtain state-of-the-art image categorization performance is feature detection and extraction. The goal is to obtain a better and more compact representation of the image. This is usually done by detecting interest points in the image using edge detectors or corner detectors, among others. The next step is to extract image descriptors from the nearby regions of interest points. For example, the *bag of*



R.T. Ionescu and M. Popescu, *Knowledge Transfer between Computer Vision and Text Mining*, Advances in Computer Vision and Pattern Recognition,
DOI 10.1007/978-3-319-30367-3_3

*visual words* model (Csurka et al. 2004; Fei-Fei and Perona 2005; Leung and Malik 2001; Sivic et al. 2005; Zhang et al. 2007) is one of the state-of-the-art methods that employs interest point detection and feature extraction in a preliminary phase. It then builds a vocabulary of visual words by clustering local image descriptors extracted from images. The bag of visual words model has demonstrated impressive levels of performance for image categorization (Zhang et al. 2007), image retrieval (Philbin et al. 2007), and even facial expression recognition (Ionescu et al. 2013).

As discussed in Chap. 2, similarity-based learning methods are also used as state-of-the-art methods in computer vision. Researchers have proposed several distance measures for images such as the Tangent distance (Simard et al. 1996), the Earth Mover's distance (Rubner et al. 2000), or the shape matching distance (Belongie et al. 2002).

The chapter is organized as follows. State-of-the-art image distance measures are presented in Sect. 3.2. An overview of image descriptors and interest point detectors is given in Sect. 3.4. The bag of visual words model and and some of its variations are discussed in Sect. 3.5. Finally, deep learning approaches are presented in Sect. 3.6.

## 3.2 Image Distance Measures

Similarity measures of images can be categorized as follows: pixelwise comparison of intensities, morphological measures that define the distance between images by the distance between their level sets, and measures based on the gray value distributions of the image.

An image can be represented in a vector space, for example, as a color histogram. The similarity between vector representations of images is measured by usual practical distances: $L_p$-metrics, weighted editing metrics, Tanimoto distance, cosine distance, Mahalanobis distance and its extensions, Earth Mover's distance. Among probability distances, the ones that are most used are Bhattacharyya 2, Hellinger, Kullback–Leibler, Jeffrey and for histograms $\chi^2$, Kolmogorov–Smirnov, Kuiper distances (Deza and Deza 1998). Many distance measures are specific for a single type of images, such as color images, binary (black and white) images, grayscale images, and so on.

### 3.2.1 Color Image Distances

The basic assumption of colorimetry, supported experimentally by Indow (1991), is that the perceptual color space admits a metric, the true color distance. This metric is expected to be locally euclidean. Another assumption is that there is a continuous mapping for the metric space of light stimuli to this metric space. However, a uniform color space, where equal distances in the color space correspond to equal differences in the color, was not obtained. Despite of this fact, several color distances have been proposed in different color spaces such as RGB, CIE $L^*u^*v^*$, CIE $L^*a^*b^*$, HSV, or CMY. The main color image distances are the average color distance, the histogram intersection quasi-distance, and the histogram quadratic distance.

For a given $3D$ color space and a list of $n$ colors, let $(c_{i1}, c_{i2}, c_{i3})$ be a representation of the $i$-th color of the list in this space. In a color histogram $x = (x_1, \ldots, x_n)$, a component $x_i$ represents the number of pixels of color $i$. The average color of $x$ is the vector $(\bar{x_1}, \bar{x_2}, \bar{x_3})$, where $\bar{x_j} = \sum_{i=1}^{n} x_i c_{ij}$ of the pixels in the image. For example, in the RGB color space, the average color vector contains the average red, green, and blue values. The average color distance between two color histograms (Hafner et al. 1995) is the euclidean distance of their average colors.

Given two color histograms $x = (x_1, \ldots, x_n)$ and $y = (y_1, \ldots, y_n)$, where $x_i$ and $y_i$ represent number of pixels in the bin $i$, the Swain–Ballard's histogram intersection quasi-distance between them is defined by

$$1 - \frac{\sum_{i=1}^{n} \min\{x_i, y_i\}}{\sum_{i=1}^{n} x_i}.$$

For $L_1$-normalized histograms the above quasi-distance becomes the usual $L_1$-metric, namely

$$\sum_{i=1}^{n} |x_i - y_i|.$$

Given two color histograms $x = (x_1, \ldots, x_n)$ and $y = (y_1, \ldots, y_n)$ (where $n$ is usually 256 or 64) representing the color percentages of two images, their histogram quadratic distance is the Mahalanobis distance, defined by

$$\sqrt{(x - y)' A (x - y)},$$

where $A = (a_{ij})$ is a symmetric positive-definite matrix, and weight $a_{ij}$ is some (perceptually justified) similarity between colors $i$ and $j$. One of the similarities used is given by

$$a_{ij} = 1 - \frac{d_{ij}}{\max_{1 \leqslant p,q \leqslant n} d_{pq}},$$

where $d_{ij}$ is the euclidean distance between vectors representing colors $i$ and $j$ in some 3D color space.

### 3.2.2 *Grayscale Image Distances*

Let $f(x)$ and $g(x)$ denote the brightness values of two digital grayscale images $f$ and $g$ at the pixel $x \in X$, where $X$ is a raster of pixels. Any distance between point-weighted sets $(X, f)$ and $(X, g)$ (for example, the Earth Mover's distance) can be applied for measuring distances between $f$ and $g$. The most used distances, which are sometimes called errors, between images $f$ and $g$ are the root-mean-square error, the signal-to-noise ratio, and the normalized Hamming distance.

The root-mean-square error is defined by

$$\sqrt{\frac{1}{|X|}\sum_{x\in X}(f(x) - g(x))^2}.$$

Another variant is to use the $L_1$-norm $|f(x) - g(x)|$ instead of the $L_2$-norm.

The signal-to-noise ratio is defined by

$$\sqrt{\frac{\sum_{x\in X} g(x)^2}{\sum_{x\in X}(f(x) - g(x))^2}}.$$

The normalized Hamming distance, also known as the pixel misclassification error rate, is defined by

$$\frac{1}{|X|}|\{x \in X : f(x) \neq g(x)\}|.$$

### 3.2.3 *Earth Mover's Distance*

Given two distributions, the Earth Mover's distance (Rubner et al. 2000) is the least amount of work needed to transform earth or mass (which is properly spread in space) from one distribution to the other (a collection of holes in the same space). The Earth Mover's distance is a discrete form of the Monge–Kantorovich distance. Instead of histograms, the distance is based on *signatures*, which are variable-sized descriptions of distributions. A signature is a set of the main clusters of a distribution. Each cluster is represented by its mean (or mode) and by the fraction of pixels that belong to that cluster.

### 3.2.4 *Tangent Distance*

Tangent distance (Simard et al. 1996) is a distance measure that is invariant with respect to specific transformations such as small distortions and translations of the image. If an image is considered as a point in a high dimensional pixel space, then an evolving distortion of the image traces out a curve in pixel space. Taken together, all these distortions define a low-dimensional manifold in pixel space. For small distortions, this manifold can be approximated by a tangent plane. Tangent distance measures the closeness between the tangent planes of two images.

### *3.2.5 Shape Match Distance*

The shape matching distance (Belongie et al. 2002) is based on an algorithm for finding correspondences between shapes. Shapes are represented by a set of points sampled from the output of an edge detector. To describe the coarse distribution of the entire shape with respect to a given point on the shape, the method introduces the *shape context* descriptor, which describes a point by its context. Finding correspondences between two shapes is then equivalent to finding a bipartite graph match between sample points on the two shapes that have the most similar shape context.

## 3.3 Patch-based Techniques

*Image patches* (or simply patches) denote squared subimages extracted from an image. The parameters that determine a patch uniquely are the horizontal and vertical location within the image, and its size. For a given location and size, the patch can be extracted by simply determining which image pixels are located within that particular square. Patches belong to the category of local features, which means that they describe properties of a certain region of an image. In contrast to that, global features provide information about an image as a whole.

For numerous computer vision applications, the image can be analyzed at the patch level rather than at the individual pixel level or global level. Patches contain contextual information and have advantages in terms of computation and generalization. For example, patch-based methods produce better results and are much faster than pixel-based methods for texture synthesis (Efros and Freeman 2001). However, patch-based techniques are still heavy to compute with current machines (Barnes et al. 2011).

A paper that describes a patch-based approach for rapid image correlation or template matching is that of Guo et al. (2007). By representing a template image with an ensemble of patches, the method is robust with respect to variations such as local appearance variation, partial occlusion, and scale changes. Rectangle filters are applied to each image patch for fast filtering based on the integral image representation.

An approach to object recognition was proposed by Deselaers et al. (2005), where image patches are clustered using the EM algorithm for Gaussian mixture densities and images are represented as histograms of the patches over the (discrete) membership to the clusters. Patches are also regarded in the work of Paredes et al. (2001), where they are classified by a nearest neighbor-based voting scheme.

The work of Agarwal and Roth (2002) describes a method where images are represented by binary feature vectors that encode which patches from a codebook appear in the images and what spatial relationship they have. The codebook is obtained by clustering patches from training images whose locations are determined by interest point detectors.

In the work of Passino et al. (2007), an image classification system based on a Conditional Random Field model is proposed. The model is trained on simple features obtained from a small number of semantically representative image patches.

The patch transform, proposed by Cho et al. (2010), represents an image as a bag of overlapping patches sampled on a regular grid. This representation allows users to manipulate images in the patch domain, which then seeds the inverse patch transform to synthesize a modified image.

In the work of Barnes et al. (2011), a new randomized algorithm for quickly finding approximate nearest neighbor matches between image patches is introduced. This algorithm forms the basis for a variety of applications including image retargeting, completion, reshuffling, object detection, digital forgery detection, and video summarization.

## 3.4 Image Descriptors

Beside patches, another popular class of local image features are image descriptors. They describe elementary visual features of the contents of images, such as the shape, the color, the contrast, and so on.

Image descriptors are usually extracted using interest point detectors. The most widely used detector is the Harris detector (Harris and Stephens 1988). Based on the concept of automatic scale selection (Lindeberg 1998), the authors of (Mikolajczyk et al. 2001) created robust and scale-invariant feature detectors, called Harris–Laplace and Hessian–Laplace. Lowe (1999, 2004) approximated the Laplacian of Gaussian (LoG) by a difference of Gaussians (DoG) filter. Among the other scale-invariant interest point detectors proposed in the literature are the salient region detectors proposed by Kadir and Brady (2001), and the edge-based region detector proposed by Jurie and Schmid (2004).

The most famous image descriptor is probably the Scale-Invariant Feature Transform (SIFT) (Lowe 1999, 2004). The SIFT descriptor converts each extracted patch to a 128-dimensional vector containing a 3D histogram of gradient locations and orientations. Each image is then represented as a set of vectors of same dimension, where the order of different vectors is of no importance. The SIFT descriptor is invariant to image scaling, translation, rotation, illumination changes, and affine or 3D projection. It has been used in a wide variety of applications, even for object matching in videos (Sivic and Zisserman 2003).

Researchers have developed improved variants of the SIFT descriptor. The work of Ke et al. (2004) proposes the PCA-SIFT descriptor that uses the image gradient patch and applies PCA to reduce the size of the SIFT descriptor. In the paper of Mikolajczyk and Schmid (2005), another variant of the SIFT descriptor, termed

GLOH, is proposed. The GLOH descriptor proved to be even more distinctive with the same number of dimensions. However, GLOH is computationally more expensive. The SURF descriptor (Bay et al. 2008) approximates or even outperforms previously proposed schemes, yet can be computed and compared much faster.

Shape context (Belongie et al. 2002) is similar to the SIFT descriptor, but is based on edges. Shape context is a 3D histogram of edge point locations and orientations. Edges are extracted by the Canny detector (Canny 1986).

In the work of Dalal and Triggs (2005), the use of grids of Histograms of Oriented Gradients (HOG) descriptors for human detection is proposed. The technique counts occurrences of gradient orientation in localized portions of an image. Using the appearance (Lowe 1999) and shape (Dalal and Triggs 2005) descriptors together with the image spatial layout, the work of Bosch et al. (2007) proposes two representations: a pyramid histogram of visual words (PHOW) descriptor for appearance and a pyramid HOG (PHOG) descriptor for shape. This method is similar to that of edge orientation histograms, SIFT descriptors, and shape contexts, but differs in that it is computed on a dense grid of uniformly spaced cells and uses overlapping local contrast normalization for improved accuracy.

Several other image descriptors have been proposed in the literature. Examples, that are also evaluated by Mikolajczyk and Schmid (2005), are spin images, steerable filters, differential invariants, complex filters, moment invariants, and cross-correlation of sampled pixel values.

## 3.5 Bag of Visual Words

In computer vision, the *bag of visual words* (BOVW) model can be applied to image classification and related tasks, by treating image descriptors as words. This model is an example of knowledge transfer from text mining. In text mining and information retrieval, the bag of words model represents a text as an unordered collection of words. In a similar fashion, a bag of *visual words* is a sparse vector of occurrence counts of a vocabulary of local image features. This representation can also be described as a histogram of visual words. In text, the vocabulary of words can easily be constructed by taking all the words that appear in a corpus of text documents. In image, the vocabulary is usually obtained by vector quantizing image features into visual words. One of the most popular methods to build a vocabulary is to apply a k-means clustering on the image features (Leung and Malik 2001). Then, each cluster center becomes a visual word in the vocabulary. Recent papers have demonstrated the advantage of using a vocabulary tree (Nister and Stewenius 2006) or a randomized forest of k-d trees (Philbin et al. 2007) to reduce search cost in the quantization stage.

One of the early approaches of building a vocabulary of features is that of Leung and Malik (2001). The main idea is to construct a vocabulary of prototype tiny surface patches, called 3D textons. Textons obtained by k-means clustering are used for texture classification. The work of Sivic et al. (2005) also builds a vocabulary of visual words using vector quantized SIFT descriptors. The vector quantization is done via a probabilistic Latent Semantic Analysis (pLSA). Others have used similar descriptors for object classification (Csurka et al. 2004), but in a supervised setting. There are many other successful approaches to obtain visual words from image data (Deselaers et al. 2005; Perronnin and Dance 2007; Perronnin et al. 2010; Winn et al. 2005; Xie et al. 2010).

The method proposed by Winn et al. (2005) classifies regions according to the proportions of different visual words. An optimally compact visual dictionary is learned by pairwise merging of visual words from an initially large dictionary. The final visual words are described by Gaussian Mixture Models (GMM).

In the work of Xie et al. (2010), a novel texture classification method via patch-based sparse texton learning is proposed. The dictionary of textons is learned by applying sparse representation to image patches in the training data set.

Fisher Vectors (Perronnin and Dance 2007) can be derived as an approximate yet improved case of the general Fisher Kernel. In the Fisher Vectors framework, a Gaussian Mixture Model is employed to fit the distribution of descriptors. The improved Fisher Vectors approach of Perronnin et al. (2010) is based on using a nonlinear additive kernel and on applying $L_2$-normalization before using the representation in a linear classification model.

### *3.5.1 Encoding Spatial Information*

An important remark is that the classical BOVW model ignores any spatial relationships between image features, similar to its text correspondent, the bag of words model, that completely disregards grammar, word order, and syntactic groups. Despite of this fact, visual words showed a high discriminatory power and have been used for region or image level classification (Csurka et al. 2004; Fei-Fei and Perona 2005; Zhang et al. 2007). Nevertheless, researchers have demonstrated that the performance can be improved by including spatial information. Several approaches of adding spatial information to the BOVW model have been proposed (Koniusz and Mikolajczyk 2011; Krapac et al. 2011; Lazebnik et al. 2006; Sánchez et al. 2012; Uijlings et al. 2009). The spatial pyramid (Lazebnik et al. 2006) is one of the most popular frameworks of using the spatial information. In this framework, the image is gradually divided into spatial bins. The frequency of each visual word is recorded in a histogram for each bin. The final feature vector for the image is a concatenation of these histograms. To reduce the dimension of the feature representation induced by the spatial pyramid, researchers have tried to encode the spatial information at a lower level (Koniusz and Mikolajczyk 2011; Sánchez et al. 2012). The Spatial Coordinate Coding scheme (Koniusz and Mikolajczyk 2011) applies spatial location and angular information at descriptor level. Krapac et al. (2011) model the spatial location of the image regions assigned to visual words using Gaussian Mixture Models, which is related to a soft-assign version of the spatial pyramid representation. A similar approach is proposed by Sánchez et al. (2012), but the change is made at the low-level feature representation, enabling the model to be extended to other encoding methods. It is worth mentioning that in the work of Krapac et al. (2011), the spatial mean and the variance of image regions associated with visual words are used to define a GMM. In the SNAK framework described in Chap. 5, the spatial mean and the standard deviation of visual words are also used, but in a completely different way, by embedding them into a kernel function. Another way of using spatial information is to consider the location of objects in the image, which can be determined either by using manually annotated bounding boxes (Uijlings et al. 2009) or by using the objectness measure (Ionescu et al. 2014; Sánchez et al. 2012). More recently, the work of Lopez-Monroy et al. (2015) shows how to encode spatial information with the joint use of visual words and multidirectional sequences of visual words, called visual $n$-grams. Their approach is inspired by the popular idea of $n$-gram representations used in natural language processing. This is yet another successful example of knowledge transfer from natural language processing to computer vision.

## 3.6 Deep Learning

A broad variety of object recognition methods (Krizhevsky et al. 2012; Simonyan and Zisserman 2014; Szegedy et al. 2015) are based on deep learning (Bengio 2009; Montavon et al. 2012; LeCun et al. 2015). Deep learning is a way to transform one representation into another, by better disentangling the factors of variation that explain the observed data. Such algorithms are aimed at discovering multiple levels of representation, or a hierarchy of features. The main approach in this area is represented by the convolutional neural networks (Krizhevsky et al. 2012; Simonyan and Zisserman 2014; Szegedy et al. 2015). Deep networks minimize a non-convex loss function, thus obtaining impressive levels of performance when very large training data is available. For example, the convolutional neural network of Krizhevsky et al. (2012) won the ImageNet Large Scale Visual Recognition Challenge 2012. However, a lot of training data and time are usually needed to train such deep models. Indeed, the network of Krizhevsky et al. (2012), consisting of 650,000 neurons, 832 million synapses, and 60 million parameters, was trained with backpropagation on GPU for almost one week. Not surprisingly, more recently developed deep networks are based on even larger architectures (Simonyan and Zisserman 2014; Szegedy et al. 2015). For instance, the state-of-the-art network of Szegedy et al. (2015) has no more than 22 layers. This makes it about three times larger than the network of Krizhevsky et al. (2012), which is based on 8 layers. Another motivation that supports the good results of deep networks is that handcrafted models are replaced by trainable models, which intuitevely should work better. However, deep models produce acceptable performance when there are at least 5,000 samples per class available for training according to Goodfellow et al. (2015). This comes hand in hand with the recent burst of *big data*, mostly made available through the Internet. Nevertheless, there are many problems in which large training sets are not available, usually because of their nature. One such example could be face recognition, as collecting 5,000 images per person is very difficult in a multiway classification setting. In spite of their general success, deep architectures are not very useful when there is less training data available for a specific task. This is where approaches such as bag of visual words or Deformable Part Models can still come into play.

## References

Agarwal S, Roth D (2002) Learning a sparse representation for object detection. In: Proceedings of ECCV, pp 113–127

Barnes C, Goldman DB, Shechtman E, Finkelstein A (2011) The PatchMatch randomized matching algorithm for image manipulation. Commun ACM 54(11):103–110

Bay H, Ess A, Tuytelaars T, Gool LV (2008) Speeded-up robust features (SURF). Comput Vis Image Underst 110(3):346–359

Belongie S, Malik J, Puzicha J (2002) Shape matching and object recognition using shape contexts. IEEE Trans Pattern Anal Mach Intell 24(4):509–522

Bengio Y (2009) Learning deep architectures for AI. Found Trends Mach Learn 2(1):1–127

Bosch A, Zisserman A, Munoz X (2007) Image classification using random forests and ferns. In: Proceedings of ICCV, pp 1–8

Canny J (1986) A computational approach to edge detection. IEEE Trans Pattern Anal Mach Intell 8(6):679–698

Cho TS, Avidan S, Freeman WT (2010) The patch transform. IEEE Trans Pattern Anal Mach Intell 32(8):1489–1501

Csurka G, Dance CR, Fan L, Willamowski J, Bray C (2004) Visual categorization with bags of keypoints. In: Workshop on statistical learning in computer vision, ECCV, pp 1–22

Dalal N, Triggs B (2005) Histograms of oriented gradients for human detection. In: Proceedings of CVPR, vol 1, pp 886–893

Deselaers T, Keyser D, Ney H (2005) Discriminative training for object recognition using image patches. In: Proceedings of CVPR, pp. 157–162

Deza E, Deza M-M (1998) Dictionary of distances. Elsevier, The Netherlands

Efros AA, Freeman WT (2001) Image quilting for texture synthesis and transfer. In: Proceedings of SIGGRAPH, pp 341–346

Fei-Fei L, Perona P (2005) A Bayesian hierarchical model for learning natural scene categories. In: Proceedings of CVPR, vol 2, pp 524–531

Felzenszwalb PF, Girshick RB, McAllester D, Ramanan D (2010) Object detection with discriminatively trained part-based models. IEEE Trans Pattern Anal Mach Intell 32(9):1627–1645

Goodfellow I, Courville A, Bengio Y (2015) Deep learning. Book in preparation for MIT Press. http://www.deeplearningbook.org/

Guo G, Dyer CR (2007) Patch-based image correlation with rapid filtering. In: Proceedings of CVPR

Hafner JL, Sawhney HS, Equitz W, Flickner M, Niblack W (1995) Efficient color histogram indexing for quadratic form distance functions. IEEE Trans Pattern Anal Mach Intell 17(7):729–736

Harris C, Stephens M (1988 )A combined corner and edge detector. In: Proceedings of the 4th Alvey Vision Conference, pp 147–151

Indow T (1991) A critical review of Luneburg's model with regard to global structure of visual space. Psychol Rev 98(3):430–453

Ionescu RT, Popescu M (2014) Objectness to improve the bag of visual words model. In: Proceedings of ICIP, pp 3238–3242

Ionescu RT, Popescu M, Grozea C (2013) Local learning to improve bag of visual words model for facial expression recognition. In: Workshop on challenges in representation learning, ICML

Jurie F, Schmid C (2004) Scale-invariant shape features for recognition of object categories. In: Proceedings of CVPR 2:90–96
Kadir T, Brady M (2001) Saliency, scale and image description. Int J Comput Vis 45(2):83–105
Ke Y, Sukthankar R (2004) PCA-SIFT: a more distinctive representation for local image descriptors. In: Proceedings of CVPR, pp 506–513
Koniusz P, Mikolajczyk K (2011) Spatial coordinate coding to reduce histogram representations, dominant angle and colour pyramid match. In: Proceedings of ICIP, pp 661–664
Krapac J, Verbeek J, Jurie F (2011) Modeling spatial layout with Fisher vectors for image categorization. In: Proceedings of ICCV, pp 1487–1494
Krizhevsky A, Sutskever I, Hinton GE (2012) ImageNet classification with deep convolutional neural networks. In: Proceedings of NIPS, pp 1106–1114
Lazebnik S, Schmid C, Ponce J (2006) Beyond bags of features: spatial pyramid matching for recognizing natural scene categories. In: Proceedings of CVPR 2:2169–2178
LeCun Y, Bengio Y, Hinton G (2015) Deep learning. Nature 521(7553):436–444, 05
Leung T, Malik J (2001) Representing and recognizing the visual appearance of materials using three-dimensional textons. Int J Comput Vis 43(1):29–44
Lindeberg T (1998) Feature detection with automatic scale selection. Int J Comput Vis 30:79–116
Lopez-Monroy A, Pastor YG, Manuel M, Escalante HJ, Cruz-Roa A, Gonzalez FA (2015) Improving the BOVW via discriminative visual n-grams and MKL strategies. Neurocomputing, doi:10.1016/j.neucom.2015.10.053. ISSN 0925-2312
Lowe DG (1999) Object recognition from local scale-invariant features. In: Proceedings of ICCV 2:1150–1157
Lowe DG (2004) Distinctive image features from scale-invariant keypoints. Int J Comput Vis 60(2):91–110
Mikolajczyk K, Schmid C (2001) Indexing based on scale invariant interest points. In: Proceedings of ICCV 1:525–531
Mikolajczyk K, Schmid C (2005) A performance evaluation of local descriptors. IEEE Trans Pattern Anal Mach Intell 27(10):1615–1630
Montavon G, Orr GB, Müller K-R (eds) (2012) Neural Networks: Tricks of the Trade. In: Lecture notes in computer science (LNCS), vol 7700, 2nd edn. Springer
Nister D, Stewenius H (2006) Scalable recognition with a vocabulary tree. In: Proceedings of CVPR 2:2161–2168
Paredes R, Perez-Cortes J, Juan A, Vidal E (2001) Local representations and a direct voting scheme for face recognition. In: Proceedings of workshop on pattern recognition in information systems, pp 71–79
Passino G, Izquierdo E (2007) Patch-based image classification through conditional random field model. In: Proceedings of MobiMedia, pp 6:1–6:6
Perronnin F, Dance CR (2007) Fisher kernels on visual vocabularies for image categorization. In: Proceedings of CVPR
Perronnin F, Sánchez J, Mensink T (2010) Improving the Fisher kernel for large-scale image classification. In: Proceedings of ECCV, pp 143–156
Philbin J, Chum O, Isard M, Sivic J, Zisserman A (2007) Object retrieval with large vocabularies and fast spatial matching. In: Proceedings of CVPR, pp 1–8
Rubner Y, Tomasi C, Guibas LJ (2000) The Earth Mover's distance as a metric for image retrieval. Int J Comput Vis 40(2): 99–121
Sánche, J, Perronnin F, de Campos T (2012) Modeling the spatial layout of images beyond spatial pyramids. Pattern Recogn Lett 33(16):2216–2223. ISSN 0167-8655
Simard P, LeCun Y, Denker JS, Victorri B (1996) Transformation invariance in pattern recognition, tangent distance and tangent propagation. Neural Networks: Tricks of the Trade
Simonyan K, Zisserman A (2014) Very deep convolutional networks for large-scale image recognition. CoRR abs/1409.1556
Sivic J, Zisserman A (2003) Video google: a text retrieval approach to object matching in videos. In: Proceedings of ICCV 2:1470–1477

Sivic J, Russell BC, Efros AA, Zisserman A, Freeman WT (2005) Discovering objects and their localization in images. In: Proceedings of ICCV, pp 370–377
Szegedy C, Liu W, Jia Y, Sermanet P, Reed S, Anguelov D, Erhan D, Vanhoucke V, Rabinovich A (2015) Going deeper with convolutions. In: Proceedings of CVPR
Turk M, Pentland A (1991) Eigenfaces for recognition. J Cogn Neurosci 3(1):71–86
Uijlings JRR, Smeulders AWM, Scha RJH (2009) What is the spatial extent of an object? In: Proceedings of CVPR, pp 770–777
Viola PA, Jones MJ (2001) Robust real-time face detection. In: Proceedings of ICCV, pp 747
Viola P, Jones MJ (2004) Robust real-time face detection. Int J Comput Vis 57(2):137–154. ISSN 0920-5691
Winn J, Criminisi A, Minka T (2005) Object categorization by learned universal visual dictionary. In: Proceedings of ICCV 2:1800–1807
Xie J, Zhang L, You J, Zhang D (2010) Texture classification via patch-based sparse texton learning. In: Proceedings of ICIP, pp 2737–2740
Zhang J, Marszalek M, Lazebnik S, Schmid C (2007) Local features and kernels for classification of texture and object categories: a comprehensive study. Int J Comput Vis 73(2):213–238

# Chapter 4
# Local Displacement Estimation of Image Patches and Textons

## 4.1 Introduction

This chapter aims to present two novel and related dissimilarity measures for images and textures, respectively. The first one is termed Local Patch Dissimilarity (LPD) and it was published by Dinu et al. (2012). This recently developed dissimilarity measure is inspired by rank distance, which is a distance measure for strings. Rank distance (Dinu and Manea 2006) has been used with very good results in biology, computational linguistics, and computer science. As many other computer vision techniques, LPD considers patches rather than pixels, in order to capture distinctive features such as edges, corners, and other primitive shapes. Moreover, patches contain contextual information and have advantages in terms of generalization.

An algorithm that computes the Local Patch Dissimilarity between two images is presented in this chapter. Because patch-based techniques are known to be computationally heavy, several ways of optimizing the LPD algorithm are presented, such as using a hash table to store precomputed patch distances or skipping the comparison of overlapping patches. Another way to avoid the problem of the higher computational time on large sets of images is to turn to local learning methods. All these ways of optimizing the LPD algorithm were also presented in the work of Ionescu and Popescu (2013).

The theoretical properties of LPD are also discussed. LPD fits best in the definition of a semi-metric with a relaxed coincidence axiom. Several experiments are conducted on two data sets using both standard machine learning methods and local learning methods. All methods are based on LPD. The obtained results come to support the fact that LPD is a very good dissimilarity measure for images.

The second dissimilarity measure presented in this chapter is based on textons. Local Texton Dissimilarity (LTD) is a dissimilarity measure designed for texture images inspired from LPD, which was recently introduced by Ionescu et al. (2014a). Instead of patches, LTD works with textons, which are represented as a set of features extracted from image patches. Similar textons are represented through similar



R.T. Ionescu and M. Popescu, *Knowledge Transfer between Computer Vision and Text Mining*, Advances in Computer Vision and Pattern Recognition,
DOI 10.1007/978-3-319-30367-3_4

features. Thus, image patches are implicitly quantized into textons. Textons provide a lighter representation of patches, allowing for a faster computational time and broader application to practical problems. Several experiments are conducted on three texture data sets. LTD shows its first application on biomass type identification, a direct application of texture classification. The other experiments are conducted on two popular texture classification data sets, namely Brodatz and UIUCTex. The proposed method benefits from a faster computational time compared to LPD and a better accuracy when used for texture classification. The performance level of the machine learning methods based on LTD is comparable to the state-of-the-art methods.

The chapter is organized as follows. The concepts behind extending rank distance to images are presented in Sect. 4.2.1. An algorithm to compute LPD is described in Sect. 4.2.2. Section 4.2.3 presents several means of optimizing the LPD algorithm in terms of speed. Theoretical properties of LPD are discussed in Sect. 4.3. Experiments with both standard and local learning methods based on LPD are presented in Sect. 4.4. The Local Texton Dissimilarity is presented in Sect. 4.5. Related work about texton-based techniques is discussed in Sect. 4.5.1. An algorithm to compute LTD is described in Sect. 4.5.3. The algorithm is based on the texture features presented in Sect. 4.5.2. Experiments with machine learning methods based on LTD are presented in Sect. 4.6. An ending discussion is given in Sect. 4.7.

## 4.2 Local Patch Dissimilarity

### *4.2.1 Extending Rank Distance to Images*

Rank distance is a measure of similarity between strings proposed by Dinu and Manea (2006). It has applications in many different fields such as computational biology (Dinu and Ionescu 2012, 2013; Dinu and Sgarro 2006), computational linguistics (Dinu and Dinu 2005; Dinu and Popescu 2009), and computer science. In a recent study on DNA comparison (Dinu and Ionescu 2012), rank distance seems to achieve better results than other string distances such as Hamming distance (Hamming 1950) or edit (Levenshtein) distance (Levenshtein 1966). Rank distance can be computed fast and benefits from some features of the edit distance.

The distance between two strings can be measured with rank distance by scanning (from left to right) both strings. Rank distance was initially defined to work on rankings, but it can naturally be extended to strings using the following observation. If a string does not contain identical symbols, it can be transformed directly into a ranking (the rank of each symbol is its position in the string). Otherwise, characters need to be annotated with indexes in order to eliminate duplicates, before computing the rank distance. For each annotated letter, rank distance measures the offset between its position in the first string and its position in the second string. Finally, all these offsets are summed up to obtain the rank distance. In other words, the rank distance measures the spatial offset between the positions of a letter in the two given strings, and then sums up these values. Intuitively, the rank distance computes the total non-alignment score between two strings. Further details about rank distance are given in Chap. 6.

There are a few aspects that need to be discussed and explained in order to extend rank distance (that works very good on strings) to images. The first concern is that the rank distance, that works on one-dimensional input (strings), should be extended to make it work on two-dimensional input (digital images). A way of extending rank distance to images can be discovered by taking an example in order to better understand how rank distance works on strings. For two strings $s_1$ and $s_2$, the characters must be annotated with indexes in order to eliminate duplicate characters. Then, the rank distance between $s_1$ and $s_2$ can easily be computed as in Example 4. It is worth noting that this example does not include unmatched characters, which require special treatment in the case of rank distance, as described in Chap. 7.

*Example 4* If $s_1 = CCGAATTACG$ and $s_2 = AGACTCTGAC$, the annotated strings are $\bar{s_1} = C_1C_2G_1A_1A_2T_1T_2A_3C_3G_2$ and $\bar{s_2} = A_1G_1A_2C_1T_1C_2T_2G_2A_3C_3$. The rank distance between $s_1$ and $s_2$ is

$$\Delta(s_1, s_2) = |1-4| + |2-6| + |3-2| + |4-1| + |5-3| + |6-5| + |7-7| \\ + |8-9| + |9-10| + |10-8| = 18.$$

In order to compute rank distance on strings, a global order is introduced by the annotation step. However, one may ask whether this global order is really necessary or whether it can be defined in another way. For example, should strings be annotated from right to left instead of left to right? Since text is unidimensional data, this question is easy to answer because there are not so many options. One can argue that strings can be annotated from left to right and from right to left, and the two distances obtained after annotation can be summed up. Nonetheless, in order to define rank distance for images (two-dimensional data), answering such questions becomes difficult. One would have to ask which is the first pixel of the image, then which is the second one? There is a very large number of possibilities to define a global order on the pixels of an image. And one may even ask if this global order is really necessary? All these questions have to be answered before extending rank distance to images.

If longer DNA strings, that contain only characters in the alphabet $\Sigma = \{A, C, G, T\}$, are considered, the local behavior can be observed without needing to introduce a global order, because the characters in DNA strings are almost randomly distributed and the frequency of the characters has a nearly uniform distribution. By considering two very long DNA strings and looking at some random aligned substrings (of the two strings):

$$...TTACGCTGAC...$$
$$...CATCTGACGA...$$

the local phenomenon, that appears by disregarding the global order, is that a certain character (in the first string) is expected to be paired with a similar character (in the second string) such that their positions are very close, the offset between their positions depending on the size of the alphabet (when characters follow a uniform distribution). In other words, rank distance can be computed (or rather approximated), without annotating the characters, just by pairing each character in one string with similar characters in the other string, that are nearby.

An interesting remark, that can be observed by looking at how rank distance actually works, is that the global order introduced by the standard definition of rank distance (for strings) is not really necessary. For images, introducing a global order is a problem in the first place because there are too many ways of defining it. However, this problem can be avoided by replicating only the local behavior of rank distance when it is used to measure the distance of strings.

Another observation that follows the concept of treating text and image in a similar fashion is the difference between characters and pixels, which are the building blocks of text and image, respectively. To extract meaning from text, one should look at words, which are formed from a few characters. To extract meaning from image, one should look at certain features such as contour, contrast, shape, color, and so on. If rank distance measures the offset between characters in two strings, the extension of rank distance for images should measure the offset between features such as contrast, contour, edges, and other primitive shapes. It is clear that these features cannot be captured in single pixels, but rather in small, overlapping rectangles of fixed size (e.g., $4 \times 4$ pixels), called patches. It is reasonable to consider patches rather than pixels, since many researchers have developed state-of-the-art methods for analyzing and editing digital images that are patch-based (Barnes et al. 2011; Efros and Freeman 2001; Guo and Dyer 2007).

### 4.2.2 *Local Patch Dissimilarity Algorithm*

Local Patch Dissimilarity is defined through the following algorithmic process. To compute the dissimilarity between two grayscale images, the LPD algorithm sums up all the spatial offsets of similar patches between the two images. More precisely, the LPD algorithm works as follows. For every patch in one image, the algorithm searches for a similar patch in the other image. First, it looks for similar patches in the same position in both images. If those patches are similar with respect to another distance that is computed between the two patches, then the algorithm sums up to 0 since there is no offset (or gap) between the patches. If the patches are not similar, the algorithm starts looking in concentric squares around the initial patch position in the second image until it finds a patch similar to the one in the first image. In other words, this spatial search gradually explores the vicinity of the patch position from the first image in the second image. The spatial offset from the initial position is increased as the algorithm continues to search for a similar patch without success. If a similar patch is found during this process, the algorithm sums up the current offset, which represents the minimum offset where a similar patch is found. The search goes on until the algorithm finds a similar patch or until the offset reaches the borders of the second image. In the latter case the algorithm sums up the latest

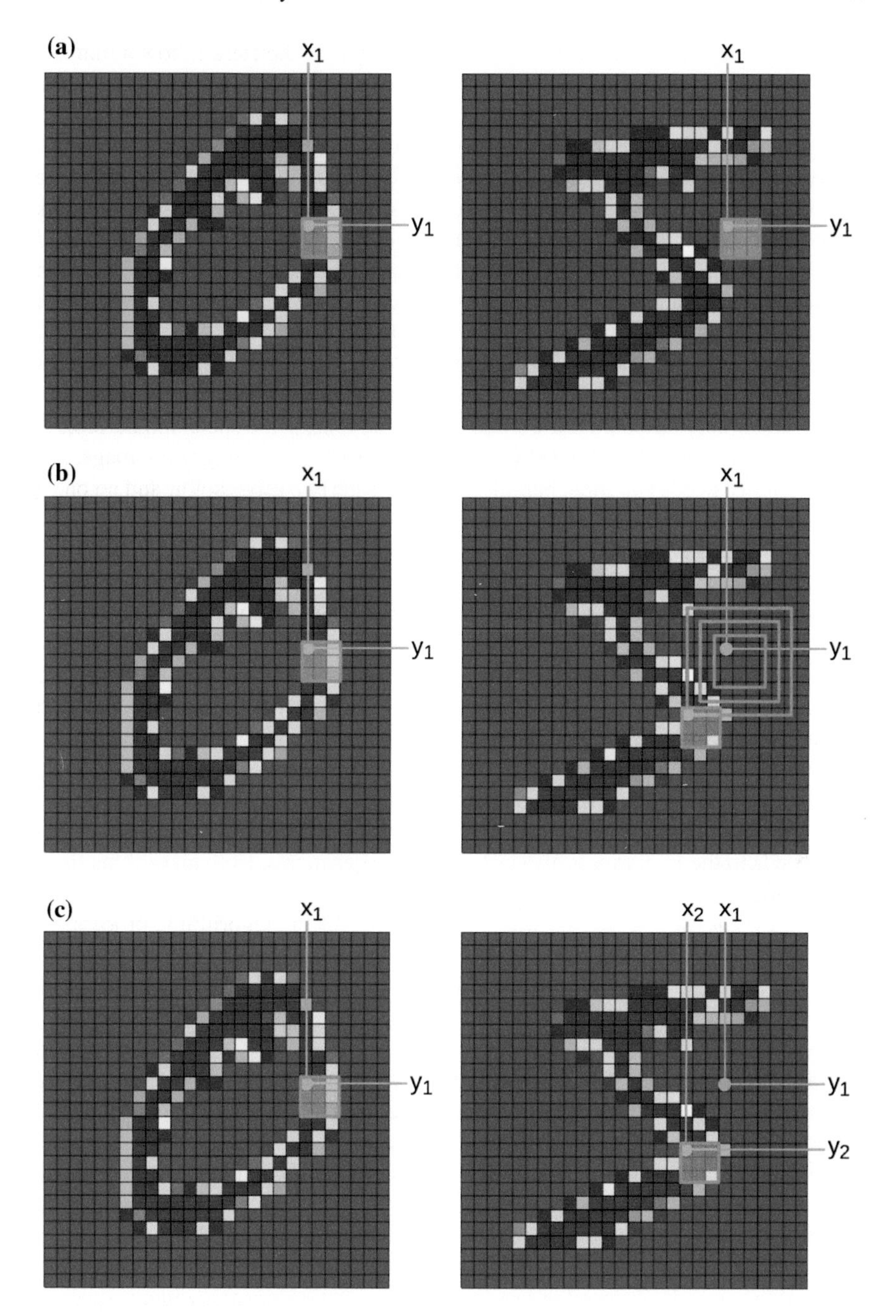


**Fig. 4.1** Two images that are compared with LPD. **a** For every position $(x_1; y_1)$ in the first image, LPD tries to find a similar patch in the second image. First, it looks at the same position $(x_1; y_1)$ in the second image. The patches are not similar. **b** LPD gradually looks around position $(x_1; y_1)$ in the second image to find a similar patch. **c** LPD sum up the spatial offset between the similar patches at $(x_1; y_1)$ from the first image and $(x_2; y_2)$ from the second image

offset, which should not be greater than the diagonal of the image. To summarize, the LPD gives an estimation of the total displacement of patches among two images. Figure 4.1 illustrates the main steps involved in the computation of LPD.

Algorithm 6 computes the LPD between grayscale images $I_1$ and $I_2$ using the underlying mean euclidean distance to compute the similarity between patches. In Algorithm 6, $P_{ij}$ represents the pixel on row $i$ and column $j$ of the patch $P$. The rank distance extension works with square patches of fixed size. Naturally, the size of the patches must be less than the size of the image. Actually, the image-patch ratio should be somehow similar to the word–character ratio. In other words, the patch should be several times smaller than the image itself.

The time complexity of the LPD algorithm is $O(h^2 \cdot w^2)$, where $h$ and $w$ represent the minimum height and width of the two compared images, respectively. The algorithm is essentially based on brute-force search and thus, some speed improvements are required. This aspect is treated in the next section.

Algorithm 6 needs two input parameters besides the two images. The square patch size parameter is the height and width measured in pixels for the patches involved in the computation of LPD. The patch similarity threshold is a number in the [0, 1] interval that determines when two patches are considered to be similar or not. These parameters need to be adjusted with respect to the image width and height, the type of information contained in the image, the noise, and so on.

It is important to mention that LPD is based on another distance between patches. Any image distance can be used to compute the similarity between patches, as long as a threshold, that determines what patches are similar and what patches are not, can be provided. Algorithm 6 determines patch similarity using the mean squared euclidean distance that corresponds to the $L_2$-norm. Another version of the LPD algorithm, that determines patch similarity using the mean euclidean distance that corresponds to the $L_1$-norm, is also tested in the experiments. Both algorithms show good results.

### *4.2.3 LPD Algorithm Optimization*

As stated by Barnes et al. (2011), patch-based algorithms are heavy to compute with current computers because these algorithms must deal with millions of patches. Some means of improving the LPD algorithm in terms of speed are discussed here. It may not occur at first look, but LPD needs to compute the similarity between many pairs of patches and for some of them, even several times. Recomputation of patch similarities can be avoided by storing the precomputed values in a hash table. Some of the tests presented in Sect. 4.4 are performed using the hash table optimization which brings an 8–10 % speed improvement.

Another optimization is to stop the search for similar patches earlier. In Algorithm 6, the search goes on until a similar patch is found or until a maximum offset is reached. It is very unlikely to find similar patches at high offset from each other. Even if two similar patches are found at a great distance from each other, it does not mean

**Algorithm 6**: Local Patch Dissimilarity Algorithm

```
1  Input:
2  I_1 – a grayscale image of h_1 × w_1 pixels;
3  I_2 – another grayscale image of h_2 × w_2 pixels;
4  p – the square patch size (the patch has p × p pixels);
5  th – the patch similarity threshold.
6  Initialization:
7  Δ ← 0;
8  h ← min{h_1, h_2} − p + 1;
9  w ← min{w_1, w_2} − p + 1;
10 Computation:
11 for x ∈ {1, 2, ..., h} do
12     for y ∈ {1, 2, ..., w} do
13         P^1 ← the patch of p × p pixels at position (x, y) in I_1;
14         d ← 0;
15         while did not find patch at offset d similar to P^1 do
16             P^2 ← a patch of p × p pixels at offset d from (x, y) in I_2;
17             s_1 ← (1/p^2) Σ_{i=1}^{p} Σ_{j=1}^{p} (P^1_ij − P^2_ij)^2;
18             if s_1 < th then
19                 Δ ← Δ + d;
20                 break;
21             if all patches at offset d were tested then
22                 d ← d + 1;
23         P^2 ← the patch of p × p pixels at position (x, y) in I_2;
24         d ← 0;
25         while did not find patch at offset d similar to P^2 do
26             P^1 ← a patch of p × p pixels at offset d from (x, y) in I_1;
27             s_2 ← (1/p^2) Σ_{i=1}^{p} Σ_{j=1}^{p} (P^2_ij − P^1_ij)^2;
28             if s_2 < th then
29                 Δ ← Δ + d;
30                 break;
31             if all patches at offset d were tested then
32                 d ← d + 1;
33 Output:
34 Δ – the dissimilarity between images I_1 and I_2.
```

Line 17: $$s_1 \leftarrow \frac{1}{p^2}\sum_{i=1}^{p}\sum_{j=1}^{p}\left(P^1_{ij} - P^2_{ij}\right)^2;$$

Line 27: $$s_2 \leftarrow \frac{1}{p^2}\sum_{i=1}^{p}\sum_{j=1}^{p}\left(P^2_{ij} - P^1_{ij}\right)^2;$$

that the compared images are actually similar to each other. In fact, this phenomenon may bring noise into the computation of LPD. To avoid this extensive search that can potentially harm the dissimilarity measure, setting a maximum offset radius much lower than the image diagonal size is a good choice. This search limitation was included in all the experiments, which resulted in a great improvement in terms of speed and accuracy. However, one must be careful not to reduce the maximum offset by too much, which can badly alter the performance and strength of the dissimilarity measure. To stop the spatial search too early would mean to disregard some similar patches that bring important information in the dissimilarity computation. In the experiments, an offset radius of 25–50 % of the image diagonal size works very well. This also brings a speed improvement of 25–30 %.

The last proposed algorithm optimization comes from the fact that LPD computes the similarity between many overlapping patches. A fast version of the LPD algorithm is to skip the comparison of overlapping patches. Basically, this means that LPD is computed on a dense grid over the image instead of the entire image. This can be achieved by increasing the offset by more than one unit, each time a similar patch is not found at the current offset. The results presented in Sect. 4.4.9 are obtained by increasing the offset with two units at every step, thus, skipping half of the comparisons between patches and speeding the LPD algorithm by a factor of two. The proposed speed improvements do not affect the time complexity, but they are very useful in practice.

## 4.3 Properties of Local Patch Dissimilarity

LPD essentially replicates only the local behavior of rank distance when it is computed on strings, disregarding any global constraints imposed by rank distance. In order to achieve this behavior, LPD is defined as a relaxed version of rank distance (Dinu and Manea 2006). Consequently, some of the distance properties are lost. Theoretical properties of LPD are studied next, as the concern of this section is to fit LPD into a standard definition. For this, the following definitions are needed.

**Definition 6** A metric on a set $X$ is a function (called the distance function or simply distance) $d : X \times X \rightarrow \mathbb{R}$. For all $x, y, z \in X$, this function is required to satisfy the following conditions:

$$(i)\ d(x, y) \geqslant 0 \text{ (non-negativity, or separation axiom);}$$
$$(ii)\ d(x, y) = 0 \text{ if and only if } x = y \text{ (coincidence axiom);}$$
$$(iii)\ d(x, y) = d(y, x) \text{ (symmetry);}$$
$$(iv)\ d(x, z) \leqslant d(x, y) + d(y, z) \text{ (triangle inequality).}$$

Note that axioms $(i)$ and $(ii)$ from Definition 6 produce positive definiteness. One can also observe that the first condition is implied by the others.

**Definition 7** A pseudometric on $X$ is a function $d : X \times X \rightarrow \mathbb{R}$ which satisfies the axioms for a metric, except that instead of the coincidence axiom only $d(x, x) = 0$ for all $x$ is required.

**Definition 8** A semi-metric on $X$ is a function $d : X \times X \rightarrow \mathbb{R}$ that satisfies the first three axioms, but not necessarily the triangle inequality.

In the LPD algorithm, one can observe that the initial value of the distance to be computed is 0. The computation can only increase this distance by adding positive offsets of similar patches. This ensures the non-negativity condition in Definition 6. It is important to note that the algorithm computes the distance by taking all the patches from the first image and by searching for similar patches in the second image. In the same manner, it takes all the patches from the second image and searches for similar patches in the first one. Thus, the same output is obtained when images are swapped. This ensures the symmetry of LPD.

Because patches are allowed to be similar under a certain threshold, distinct (non-identical) images may have an LPD equal to 0. This is a plus if images with small differences in contrast or induced by noise must be considered similar. While this helps detect similar images even with noise, it also breaks the coincidence axiom in the standard definition of a metric. However, the relaxed version in Definition 7 is still verified. The dissimilarity of two identical images computed with LPD is always 0, but allowing distinct images to also have an LPD equal to 0 breaks the triangle inequality. Although the inequality is met in most cases (some empirical results reveal that about 98.5 % of the cases do meet the triangle inequality), LPD cannot be considered a distance function from a theoretical perspective. Thus, LPD fits best in the definition of a semi-metric with a relaxed coincidence axiom as in Definition 7. Note that adjusting the dissimilarity computation, so that conditions $(ii)$ and $(iv)$ in Definition 6 are met, may be possible and could be the subject of further work.

An interesting remark is that LPD can be normalized when required. Knowing the maximum offset radius (used to stop similar patch searching), the maximum value of the dissimilarity between two images can be computed as the product between the maximum offset and the number of pairs of compared patches. Thus, LPD can be normalized to a value in the [0, 1] interval. A value closer to 0 means that images are similar, and a value closer to 1 means that images are dissimilar.

## 4.4 Experiments and Results

### *4.4.1 Data Sets Description*

Isolated handwritten character recognition has been extensively studied in the literature (Srihari 1992; Suen et al. 1992), and was one of the early successful applications of neural networks (LeCun et al. 1989). Comparative experiments on recognition of individual handwritten digits are reported by LeCun et al. (1998). While recognizing

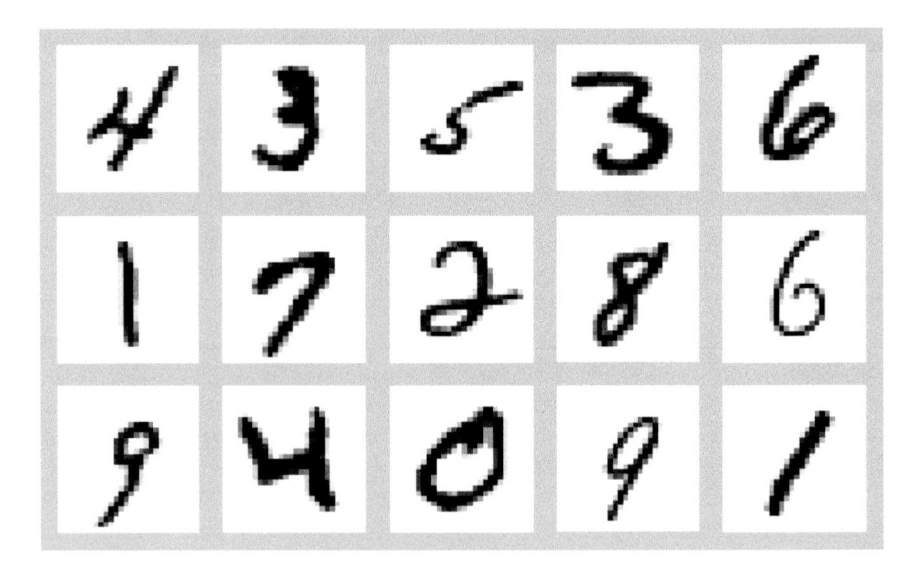

**Fig. 4.2** A random sample of 15 handwritten digits from the MNIST data set

individual digits is one of many problems involved in designing a practical recognition system, it is an excellent benchmark for comparing shape recognition methods.

The data set used for testing the dissimilarity presented in this chapter is the MNIST set, which is described in detail in the work of LeCun et al. (1998). The data set was constructed by mixing the handwritten digits collected among Census Bureau employees with the handwritten digits collected among high-school students. The regular MNIST database contains 60,000 train samples and 10,000 test samples, size normalized to $20 \times 20$ pixels, and centered by center of mass in $28 \times 28$ fields. A random sample of 15 images from this data set is presented in Fig. 4.2. The data set is available at http://yann.lecun.com/exdb/mnist/.

The second data set was collected from the Web by Lazebnik et al. (2005a) and consists of 100 images each of six different classes of birds: egrets, mandarin ducks, snowy owls, puffins, toucans, and wood ducks. Because LPD is designed for grayscale images, the images from the Birds data set are transformed to grayscale. A random sample of 12 grayscaled images from this data set is presented in Fig. 4.3. The Birds data set is available at http://www-cvr.ai.uiuc.edu/ponce_grp/data/. The Birds data set is used in the last experiment presented in this section.

### 4.4.2 Learning Methods

The LPD measure is put into a learning context in order to evaluate its performance. Therefore, LPD is employed into several learning methods and used for classification. The first classifier, that is intensively used through all the experiments, is the $k$-nearest neighbors ($k$-NN). The $k$-NN classifier was chosen because it reflects the characteristics of the dissimilarity measure, as discussed in Chap. 2.

**Fig. 4.3** A random sample of 12 images from the Birds data set. There are two images per class. Images from the same class sit next to each other in this figure

Most patch-based algorithms are heavy to compute with current computers (Barnes et al. 2011). This is also the downside of LPD, because the computational time required to apply the $k$-NN based on LPD on the entire MNIST data set is too high to even be considered in practical applications (on an Intel Core Duo 2.26 GHz processor and 4 GB of RAM memory it would take more than a few months). This time can be reduced by a factor of 15, if the computation is done on GPU, but a faster learning algorithm is still of great interest. One way to improve the time efficiency is to use a $k$-NN with filtering algorithm, which is a pure local learning technique. In the $k$-NN with filtering approach, the idea is to filter the nearest $K$ neighbors using the standard euclidean distance measure (that is much faster to compute). Then, the procedure is to select the nearest $k$ neighbors from those $K$ images using LPD. The two-step selection process is much faster to compute on a large data set (such as the MNIST data set) than a standard $k$-NN based only on LPD.

For the MNIST experiments, two state-of-the-art kernel methods are used, namely the SVM and the KRR. In the Birds experiment, the KDA classifier is also employed for the classification task. As discussed in Chap. 2, kernel methods are based on similarity. LPD can be transformed into a similarity measure. The classical way to transform a dissimilarity measure into a similarity measure is to use the RBF kernel:

$$k(I_1, I_2) = \exp\left(-\frac{\Delta_{LPD}(I_1, I_2)}{2\sigma^2}\right),$$

where $I_1$ and $I_2$ are two grayscale images. The parameter $\sigma$ is usually chosen to match the number of features so that values of $k(I_1, I_2)$ are well scaled. The number of features in an image is actually the number of pixels contained in that image.

Several classification experiments are conducted using these machine learning methods based on LPD. The experiments are organized as follows. First, the LPD parameters are tuned using a 3-NN model in a set of experiments described in Sect. 4.4.3. LPD is compared to a baseline 3-NN model based on the euclidean distance in Sect. 4.4.4. Kernel methods based on LPD are evaluated in Sect. 4.4.5. Machine learning methods based on LPD are also compared to the SVM+ model of Vapnik and Vashist (2009), in Sect. 4.4.6. The experiments mentioned so far are conducted on different subsets of the MNIST data set. The experiments described in Sects. 4.4.7 and 4.4.8 are performed on the entire MNIST data set. Finally, an experiment on the Birds data set is presented in Sect. 4.4.9.

### 4.4.3 Parameter Tuning

A set of preliminary tests are performed to adjust the parameters of LPD, such as the patch size, the patch similarity threshold, or the maximum offset radius. Patch sizes ranging from $1 \times 1$ to $10 \times 10$ pixels were considered. The patch similarity threshold was adjusted with respect to the patch size, but this parameter was also tuned.

Experiments are conducted on the first 100 and 300 images extracted from the MNIST training set. As the number of images in each set is relatively small, the tests were performed using the 10-fold cross-validation procedure. By repeating the 10-fold cross-validation procedure, variations in accuracy larger than 1 % were observed. Thus, the reported accuracy rates represent an average of 10 runs of the 10-fold cross-validation procedure, in order to obtain a better approximation of the accuracy rate.

Figures 4.4, 4.5, 4.6, 4.7, and 4.8 show the results obtained by a 3-NN model based on LPD with several patch sizes and similarity thresholds. These results are obtained on the MNIST subset of 100 images. For each patch size, a graph of the accuracy rates obtained by varying the patch similarity threshold is presented. The accuracy rates follow a Gaussian distribution, with the peak determined by a certain similarity threshold. To obtain the best accuracy rates, the patch similarity threshold must be increased as the patch size grows. It is interesting to mention that the accuracy rates obtained with pixel-sized patches presented in Fig. 4.4a are relatively good. However, the empirical results obtained with patches greater than $2 \times 2$ pixels confirm that it is better to compare patches instead of pixels.

In the following experiment, conducted on the 300 images subset, patches of $1 \times 1$ pixels and $10 \times 10$ pixels are disregarded since they give lower accuracy rates. For each remaining patch size, the patch similarity threshold that gives the highest accuracy is selected. The accuracy rates averaged over 10 runs of the 10-fold cross-validation procedure using 300 images are presented in Table 4.1. A graph with these

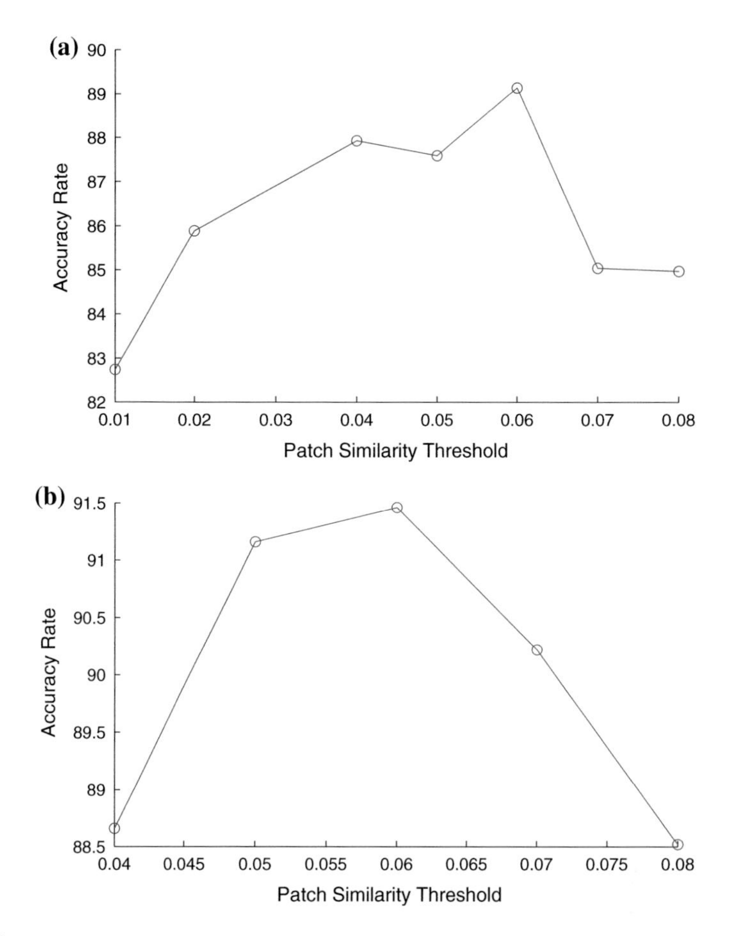


**Fig. 4.4** Average accuracy rates of the 3-NN based on LPD model with patches of 1 × 1 pixels at the *top* and 2 × 2 pixels at the *bottom*. Experiment performed on the MNIST subset of 100 images. **a** Accuracy rates with patches of 1 × 1 pixels. **b** Accuracy rates with patches of 2 × 2 pixels

accuracy rates obtained by varying the patch size is presented in Fig. 4.9. In this graph, a Gaussian distribution of the accuracy rates can be observed once more. The patch size with the highest accuracy rate is selected for the next experiments. Thus, LPD based on patches of 4 × 4 pixels will be used.

It is interesting to mention that the computational time of LPD is proportional to the patch similarity threshold. More precisely, a higher similarity threshold will give a higher probability of finding similar patches sooner, thus reducing the number of steps of the spatial search. However, the patch similarity threshold should be adjusted with respect to the accuracy of the LPD. Consequently, the rest of the experiments are conducted with a patch similarity threshold of 0.12 and 0.125, which are selected to obtain the highest accuracy with patches of 4 × 4 pixels. The similarity threshold

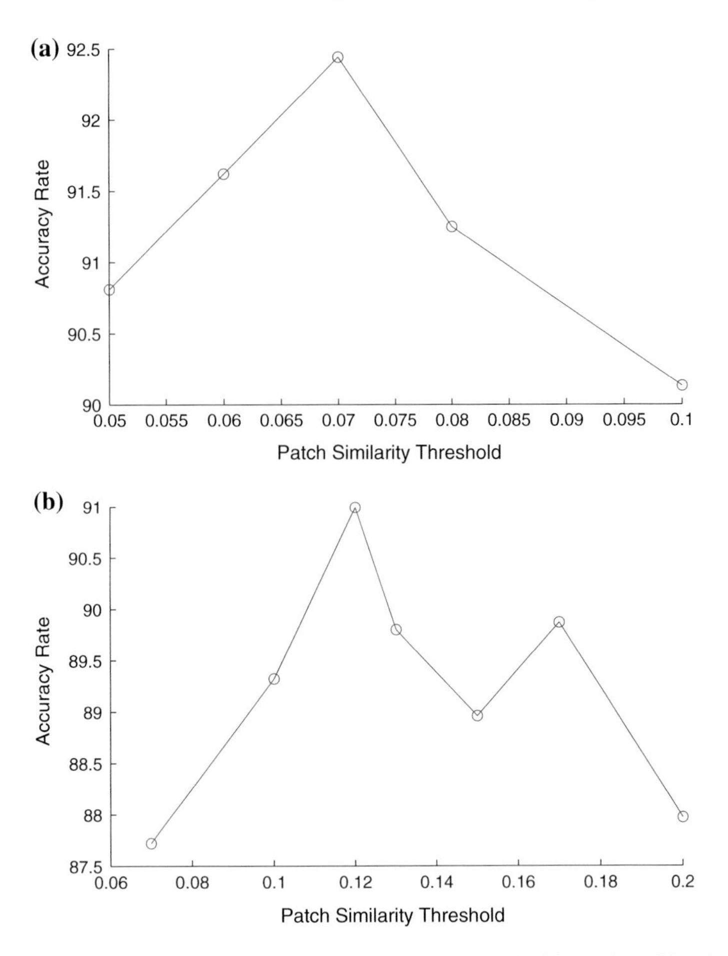


**Fig. 4.5** Average accuracy rates of the 3-NN based on LPD model with patches of $3 \times 3$ pixels at the *top* and $4 \times 4$ pixels at the *bottom*. Experiment performed on the MNIST subset of 100 images. **a** Accuracy rates with patches of $3 \times 3$ pixels. **b** Accuracy rates with patches of $4 \times 4$ pixels

of 0.125 is used for the heavy computational experiments to slightly speed up the computation without affecting the accuracy level.

Another parameter that needs to be tuned is the maximum offset radius. Several experiments were performed with various maximum offsets values such as 5, 10, 15, and 20 pixels. These experiments were performed by keeping the rest of the parameters unchanged. Patches of $4 \times 4$ pixels and a similarity threshold of 0.12 were used. The results with various maximum offsets are presented in Table 4.2. The maximum offset can be adjusted to optimize the trade-off between accuracy and time. Table 4.2 shows the average time needed to compute the dissimilarity between two images. The reported average times were measured on a computer with Intel Core i7 2.3 GHz processor and 8 GB of RAM memory using a single core. The best time

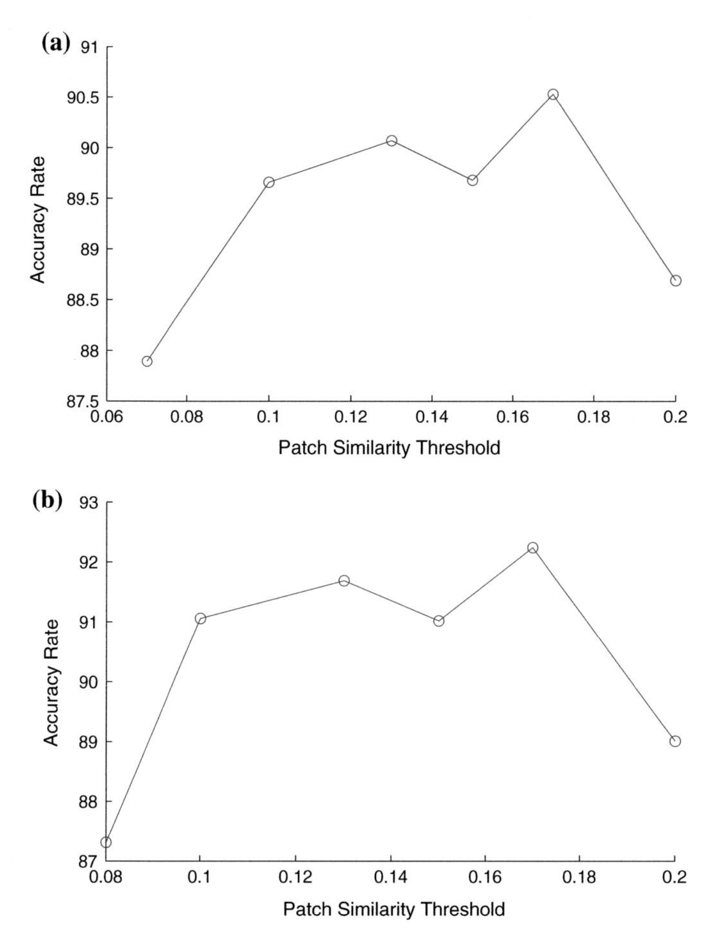


**Fig. 4.6** Average accuracy rates of the 3-NN based on LPD model with patches of 5 × 5 pixels at the *top* and 6 × 6 pixels at the *bottom*. Experiment performed on the MNIST subset of 100 images. **a** Accuracy rates with patches of 5 × 5 pixels. **b** Accuracy rates with patches of 6 × 6 pixels

is obtained using a maximum offset of 5 pixels, while the best accuracy (92.50 %) is obtained using a maximum offset of 15 pixels. Finally, the maximum offset chosen for the rest of the experiments is 15 pixels, which is close to half the height or width of the MNIST images.

### *4.4.4 Baseline Experiment*

The 3-NN classifier based on the LPD measure is compared with a baseline $k$-NN classifier. The 3-NN based on the euclidean distance measure ($L_2$-norm) between

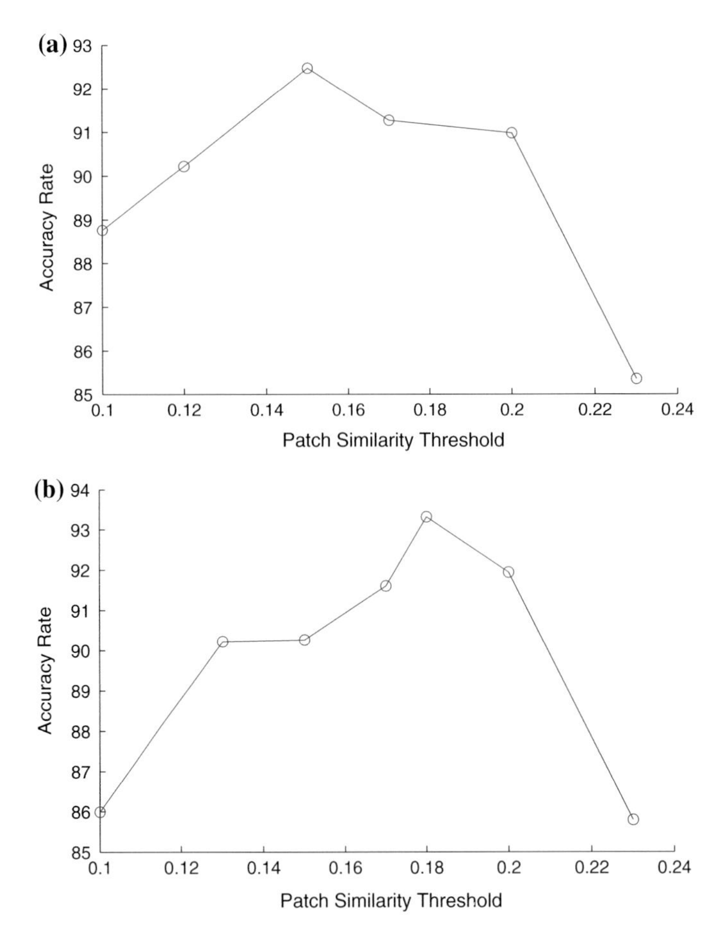


**Fig. 4.7** Average accuracy rates of the 3-NN based on LPD model with patches of $7 \times 7$ pixels at the *top* and $8 \times 8$ pixels at the *bottom*. Experiment performed on the MNIST subset of 100 images. **a** Accuracy rates with patches of $7 \times 7$ pixels. **b** Accuracy rates with patches of $8 \times 8$ pixels

input images is the chosen baseline classifier. In the work of LeCun et al. (1998) an error rate of 5.00 % on the regular test set with $k = 3$ is reported for this classifier. Other studies (Wilder 1998) report an error rate of 3.09 % on the same experiment. The experiment was recreated in this work, and an error rate of 3.09 % was obtained.

The goal of this experiment is to prove that LPD can successfully be used as a distance measure for images and that it has good results. The two classifiers, that are distinct only by the metric used, are compared using the first 100, 300, and 1000 images extracted from the MNIST data set. These subsets contain randomly distributed digits from 0 to 9 produced by different writers. Since the number of samples in each subset is relatively small, the tests were performed using the 10-fold

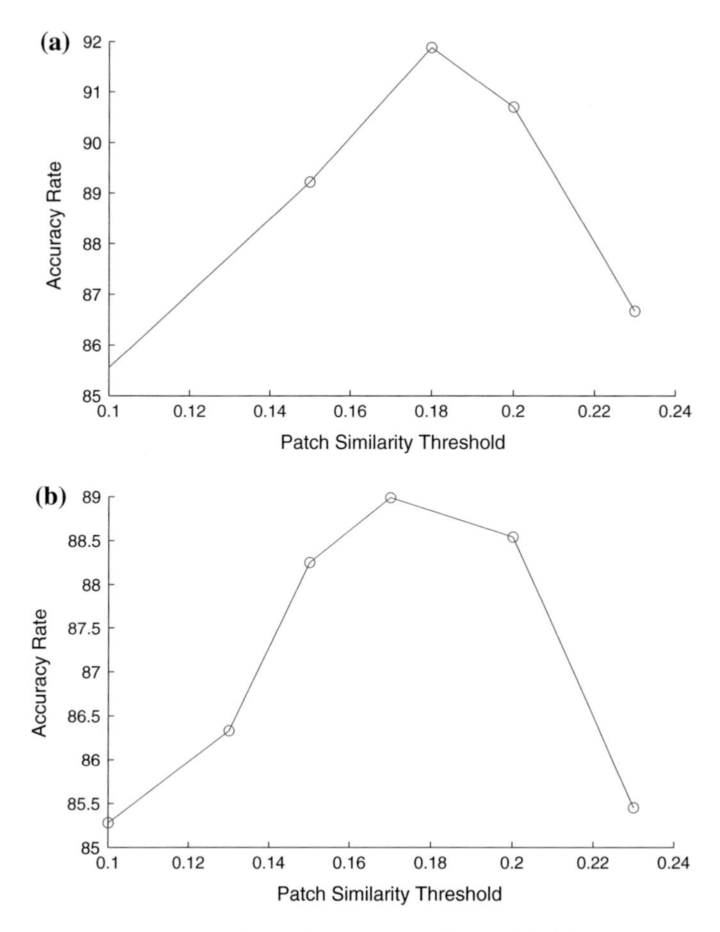


**Fig. 4.8** Average accuracy rates of the 3-NN based on LPD model with patches of $9 \times 9$ pixels at the *top* and $10 \times 10$ pixels at the *bottom*. Experiment performed on the MNIST subset of 100 images. **a** Accuracy rates with patches of $9 \times 9$ pixels. **b** Accuracy rates with patches of $10 \times 10$ pixels

cross-validation procedure which is repeated 10 times, in order to obtain the final accuracy rates.

Table 4.3 compares the accuracy of the baseline 3-NN classifier with the accuracy of the 3-NN classifier based on LPD. Results are reported on all the MNIST subsets of 100, 300, and 1000 samples, respectively. For all these subsets, the accuracy of LPD was obtained with patches of $4 \times 4$ pixels and a similarity threshold of 0.12.

The first test case requires only 100 images, but in order to obtain a more accurate result, the first 200 examples from the MNIST training set were extracted and divided into two subsets of 100 images each. The classification methods were applied on both subsets, and the reported accuracy rates were averaged on the two subsets of 100 samples. The baseline 3-NN classifier has an average accuracy of 73.33 %. The 3-NN

**Table 4.1** Results of the experiment performed on the MNIST subset of 300 images, using the 3-NN based on LPD model with patches ranging from 2 × 2 pixels to 9 × 9 pixels

| Patch size | Patch similarity threshold | Accuracy (%) |
|---|---|---|
| 2 × 2 | 0.06 | 89.25 |
| 3 × 3 | 0.07 | 90.68 |
| 4 × 4 | 0.12 | **92.50** |
| 5 × 5 | 0.17 | 91.53 |
| 6 × 6 | 0.17 | 91.45 |
| 7 × 7 | 0.15 | 91.44 |
| 8 × 8 | 0.18 | 90.38 |
| 9 × 9 | 0.18 | 89.89 |

Reported accuracy rates are averages of 10 runs

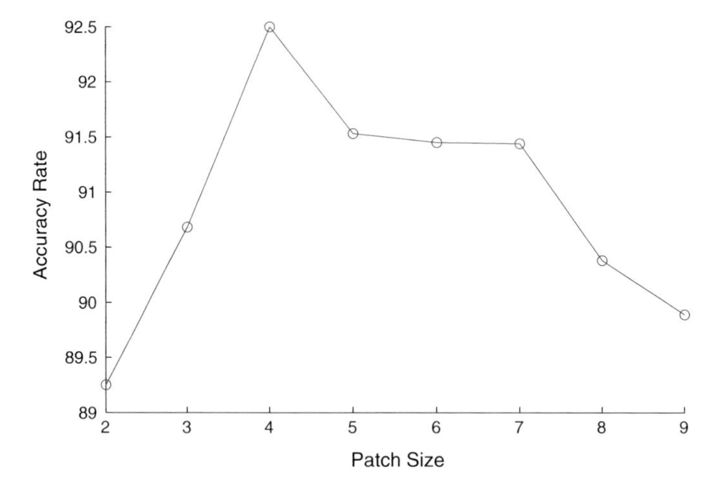


**Fig. 4.9** Average accuracy rates of the 3-NN based on LPD model with patches ranging from 2 × 2 pixels to 9 × 9 pixels. Experiment performed on the MNIST subset of 300 images

classifier based on LPD has an average accuracy of 87.53 %, which represents an improvement of 14.20 % over the baseline. The next test case uses 300 images. This time, the baseline classifier has an accuracy of 83.19 %. Using LPD, an accuracy of 92.11 % is obtained, which is 9.31 % over the baseline. The third test case uses 1000 images. The accuracy obtained with the baseline classifier in the third test case is 87.12 %. Again, the 3-NN classifier based on LPD performs better than the baseline with an accuracy of 95.53 %. This represents an improvement of 8.41 %.

One can observe that both classifiers have an increased accuracy when the number of samples is larger. The baseline method has a 9.86 % improvement from 100 images to 300 images, while the classifier based on LPD has an improvement of 4.97 %. A similar improvement in accuracy, from 300 to 1000 images, can be observed. The baseline is 3.93 % better and the classifier based on LPD is 3.03 % better.

**Table 4.2** Results of the experiment performed on the MNIST subset of 300 images, using various maximum offsets, patches of $4 \times 4$ pixels, and a similarity threshold of 0.12

| Maximum offset | Accuracy (%) | Average time per pair (seconds) |
|---|---|---|
| 5 | 90.21 | 0.11 |
| 10 | 92.11 | 0.16 |
| 15 | **92.50** | 0.19 |
| 20 | 91.98 | 0.23 |

Reported accuracy rates are averages of 10 runs. The time needed to compute the pairwise dissimilarity is measured in seconds

**Table 4.3** Baseline 3-NN versus 3-NN based on LPD

| Number of samples | Baseline 3-NN | 3-NN + LPD |
|---|---|---|
| 100 | 73.33 % $\pm$ 9.68 % | **87.53** % $\pm$ **7.38** % |
| 300 | 83.19 % $\pm$ 5.47 % | **92.50** % $\pm$ **4.23** % |
| 1000 | 87.12 % $\pm$ 2.59 % | **95.53** % $\pm$ **1.78** % |

Both the accuracy rate and standard deviation are reported for MNIST subsets of 100, 300, and 1000 images

In all test cases the 3-NN classifier based on LPD outperforms the baseline 3-NN classifier. It is worth pointing out that the accuracy of the classifier based on LPD, that is trained and tested on 1000 images (95.53 %), gets very close to the accuracy of the baseline classifier (96.91 %) trained on the full MNIST data set with 60,000 images and tested on other 10,000 images.

Note that according to the standard deviations presented in Table 4.3, it can be stated with confidence that the results of the 3-NN based on LPD are better than the baseline classifier. A student test was performed on the results obtained on 1000 images by the 3-NN based on LPD classifier on one hand, and by the baseline classifier on the other hand. The null hypothesis, that obtained results are independent random samples from normal distributions with equal means and equal variances, was rejected with a 99 % confidence interval. This means that the accuracy rates of LPD are significantly better than the baseline accuracy rates.

The pairwise similarity matrix could reveal subtle characteristics of the similarity or dissimilarity function. To analyze the LPD measure in more depth, the pairwise similarity matrix produced by the LPD measure is compared with the pairwise euclidean distance matrix. Figure 4.10 shows the similarity matrix obtained with LPD, and Fig. 4.11 shows the euclidean distance matrix. The two matrices are obtained on the same MNIST subset of 1000 images. For the visual analysis of the similarity matrices, the samples are sorted by class labels in ascending order, starting with digit 0 and finishing with digit 9. Darker regions indicate higher similarity, and lighter regions indicate lower similarity. In both the matrices, darker squares that indicate higher within-class similarities on the principal diagonal can be observed. The samples that represent the digit 1 are the most distinct from the other samples, in both matrices. However, the similarity matrix based on LPD has other distinctive classes such as

**Fig. 4.10** Similarity matrix based on LPD with patches of $4 \times 4$ pixels and a similarity threshold of 0.12, obtained by computing pairwise dissimilarities between the samples of the MNIST subset of 1000 images

digit 4, digit 6, digit 7, and digit 9. On the other hand, in the euclidean distance matrix the classes represented by digits 4 and 6 are also distinctive, but there is a greater amount of confusion between the digit 7 and digit 9 classes. The two measures also seem to mix up the digit 3 and digit 5 classes. Overall, the similarity matrices are fairly similar to each other, but the matrix produced by LPD seems to be slightly better. This is confirmed by the experiments presented in Table 4.3.

### 4.4.5 Kernel Experiment

The goal of this experiment is to combine LPD with state-of-the-art learning techniques (such as kernel methods) instead of the simple $k$-NN model, in order to improve the accuracy level. Therefore, the experiments on the MNIST subsets of 300 and 1000 images are repeated using Support Vector Machines and Kernel Ridge Regression.

In Table 4.4 the accuracy rates of the 3-NN, SVM, and KRR classifiers based on LPD are compared with the accuracy rates of the standard SVM and KRR methods,

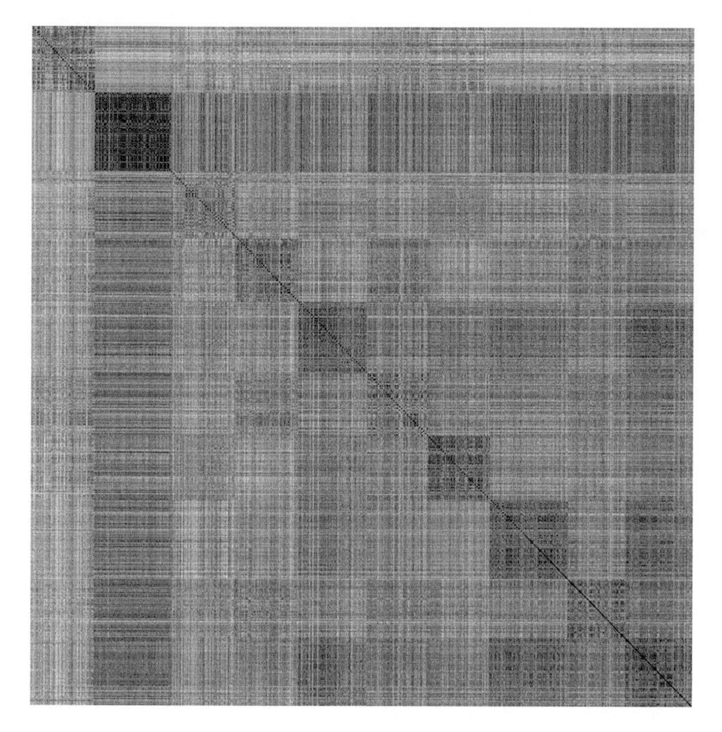

**Fig. 4.11** Euclidean distance matrix based on $L_2$-norm, obtained by computing pairwise distances between the samples of the MNIST subset of 1000 images

**Table 4.4** Accuracy rates of several classifiers based on LPD versus the accuracy rates of the standard SVM and KRR

| Method | 300 CV (%) | 1000 CV (%) | 300/700 (%) |
|---|---|---|---|
| SVM | 86.33 | 91.30 | 81.71 |
| KRR | 86.19 | 91.58 | 79.86 |
| 3-NN + LPD | 92.11 | 95.53 | 91.78 |
| SVM + LPD | **94.33** | **96.90** | **94.28** |
| KRR + LPD | 93.04 | 96.16 | 92.57 |

Tests are performed on 300 and 1000 images using cross-validation (CV), respectively. Another test is performed using a 300/700 split

which are based on the linear kernel computed from raw pixel data. The LPD kernel parameter $\sigma$ is chosen to be equal to 1000, since the square root of 1000 is close to the size of the images, which are $28 \times 28$ pixels. Results are reported on 300 and 1000 examples from the MNIST data set using 10-fold cross-validation. The last test case (identified as 300/700) divides the 1000 images into two sets: 300 images for training and 700 images for testing. The reported results on this last test case are on the 700 images used for testing.

As expected, the results slightly improve when a better learning method (such as SVM or KRR) is used rather than a simple 3-NN model. But what shows that LPD is very powerful, is that a simple 3-NN classifier based on LPD surpasses the standard SVM and KRR classifiers. The best result is obtained using the SVM combined with the LPD measure (96.90 %).

The 300/700 test case shows that LPD is a robust dissimilarity measure because the results obtained on the test set of 700 images are very close to the results reported using 10-fold cross-validation. For example, the results of the SVM based on LPD drop from 94.33 % on 300 images (using cross-validation) to only 94.28 % on the other 700 images. On the other hand, the performance of the standard SVM and KRR classifiers reported on the 300/700 test case is much worse than the performance reported using 10-fold cross-validation.

### *4.4.6 Difficult Experiment*

In the work of Vapnik and Vashist (2009) the learning using privileged information paradigm is introduced. The problem of classifying images of digits 5 and 8 in the MNIST database is considered as an application. To make it more difficult, the authors have resized the digits from $28 \times 28$ pixel to $10 \times 10$ pixel images. The authors have used 100 samples of $10 \times 10$ pixel images as the training set, 4002 samples as the validation set (for tuning the parameters of SVM and SVM+), and the rest of 1866 samples as the test set. The same split of the data set is used in this experiment.

The results of the standard SVM (based on euclidean distance) and the dSVM+ reported by Vapnik and Vashist (2009) are compared with three classifiers based on LPD. More precisely, in Table 4.5, the accuracy rates of the SVM and dSVM+ classifiers are compared with the accuracy rates of the $k$-NN, SVM, and KRR classifiers based on LPD. The number of neighbors of the $k$-NN model is 5, and it was chosen for best performance on the validation set. The parameters of the LPD are also tuned on the validation set. The best accuracy is obtained with patches of $3 \times 3$ pixels and a similarity threshold of 0.11. As the images are smaller than in the previous experiments, it is natural to obtain better performance with smaller patches of $3 \times 3$ pixels instead of $4 \times 4$ pixels. The value of the LPD kernel parameter $\sigma$ is 100, because the images are $10 \times 10$ pixels in this case.

In this experiment, it looks like the simple 5-NN classifier based on LPD is very close to the standard SVM in terms of accuracy (the difference is only 1.18 %), showing again that LPD is a very powerful dissimilarity.

**Table 4.5** Comparison of several classifiers (some based on LPD)

| SVM | dSVM+ | 5-NN + LPD | SVM + LPD | KRR + LPD |
|---|---|---|---|---|
| 92.50 % | 94.60 % | 91.32 % | **95.10** % | 95.00 % |

Results for the difficult experiment on 1866 test images

The dSVM+ gains more than 2 % in accuracy over the standard SVM using privileged information in the training process. The dSVM+ classifier is surpassed by the SVM and KRR classifiers that are based on the LPD measure. Note that these classifiers (SVM + LPD and KRR + LPD) use no privileged information. This shows that using a better dissimilarity measure (such as LPD) is sometimes more important than using privileged information to improve accuracy.

### *4.4.7 Filter-based Nearest Neighbor Experiment*

The results presented in Sect. 4.4.4 look promising, but LPD should be tested on the entire MNIST data set for a strong conclusion regarding its performance level. The problem of the $k$-NN classifier based on LPD is that it is not feasible for very large data sets, since it takes too much time to compute all the pairwise dissimilarities. To avoid this problem, local learning methods that speed up the learning algorithm can be used. Therefore, LPD can be plugged into different local learning methods and used for image classification.

In this experiment, the local learning classifier employed for tests on the entire MNIST data set is the filter-based $k$-NN model. This classifier was chosen because it reflects the characteristics of the LPD measure. For the filter-based $k$-NN approach, the idea is to filter the nearest $K$ neighbors using the standard euclidean distance measure (that is much faster to compute). Next, it selects the nearest $k$ neighbors from those $K$ images using LPD. The two-step selection process is much faster to compute on a large data set (such as the MNIST data set) than a standard $k$-NN based entirely on LPD. Results show that this method can improve accuracy and is among the top 4 $k$-NN models that reported results on the MNIST data set.

The $k$-NN based on LPD with filtering is compared with two other $k$-NN classifiers. One is the $k$-NN based on the euclidean distance measure ($L_2$-norm) between input images. This is the baseline classifier. In the paper of LeCun et al. (1998), an error rate of 5.00 % was reported on the regular test set with $k = 3$ for this classifier, while other studies (Wilder 1998) report an error rate of 3.09 % on the same experiment. The results obtained in this work also show an error rate of 3.09 % on this baseline experiment. The second classifier is the $k$-NN based on Tangent distance (Simard et al. 1996). Tangent distance is insensitive to small distortions and translations of the input image. The error rate of this classifier reported by LeCun et al. (1998) is 1.1 %, but it requires to additionally process the images by subsampling to $16 \times 16$ pixels before the classification stage, in order to obtain the reported error rate.

The accuracy of the $k$-NN based on LPD depends very much on the filtering, taking into account that the nearest $K$ images are selected using the euclidean distance in the filtering phase. If $K$ is close to $k$, a very fast classifier is obtained, but its accuracy will be near the baseline $k$-NN. As $K$ increases, the accuracy improves, but the method also becomes slower (since it has to compute LPD between more images than it had before). If $K$ is equal to the number of training examples, there is no filtering at all and the highest accuracy is obtained. However, the time to compute $k$-NN based on

**Table 4.6** Error and time of the 3-NN classifier based on LPD with filtering, for various K values

| K | Error (%) | Average time per test image (seconds) | Overall time (hours) |
|---|---|---|---|
| 3 | 2.73 | 2.2 | 6 |
| 10 | 1.78 | 2.6 | 7 |
| 30 | 1.45 | 3.7 | 10 |
| 50 | 1.38 | 4.8 | 13 |
| 100 | 1.26 | 7.6 | 21 |
| 200 | 1.15 | 13.2 | 36 |
| 500 | 1.09 | 30.5 | 84 |
| 1000 | 1.05 | 58.8 | 162 |

LPD with no filtering is too high to be considered in practice. The best solution is to choose an optimal $K$ in order to obtain a trade-off between accuracy and time. Table 4.6 shows the error rate and the execution time of the 3-NN classifier based on LPD with filtering for several $K$ values. The time was measured on a computer with Intel Core Duo 2.26 GHz processor and 4 GB of RAM memory using a single core. Empirical results show that an optimal $K$ would be somewhere between 100 and 500. When $K$ is 100, approximately 8 seconds are needed to assign a label to a test image. This is reasonable for a real-time application, since computing LPD on GPU and adding parallel processing into assigning a label can improve the time even further.

Looking at how the error gets lower as $K$ increases, it can be observed that the error tends to stabilize at some point. In other words, the error rate will not drop anymore after a certain $K$ value. That error rate is probably identical (or at least very close) to the actual error rate of a 3-NN classifier based on LPD without filtering. The limitation of LPD determines this error, but what exactly is this error rate? Fig. 4.12 gives an overview of this phenomenon and a hint about the point where the error stabilizes for both 3-NN and 6-NN classifiers. In order to make a prediction about the stabilization point of the error, the stability of the $k$-NN with filtering needs to be studied as $K$ varies. It is clear from Fig. 4.12 that the error drops with no variation when $K$ increases. Nonetheless, a greater $K$ value may cause the misclassification of some test images which are correctly classified when a lower $K$ value is used, even if, overall, more images are correctly classified. For $K = 100, 200, \ldots, 1000$, the set of misclassified images for a certain value of $K$ always includes the set of misclassified images for a greater value of $K$. Thus, the method has very stable behavior as $K$ varies. In these circumstances, the error rates of the 3-NN and 6-NN classifiers based on LPD (without filtering) can be obtained by testing the classifiers only on the previously misclassified images. By doing so, an error rate of 1.03 % is obtained for the 3-NN classifier and an error rate of 0.98 % is obtained for the 6-NN classifier. These error rates are based only on a statistical proof. But as stated before, the real proof (that of testing the 3-NN and 6-NN classifiers with no filtering on all test images) is not practical from the time perspective. The obtained error rates

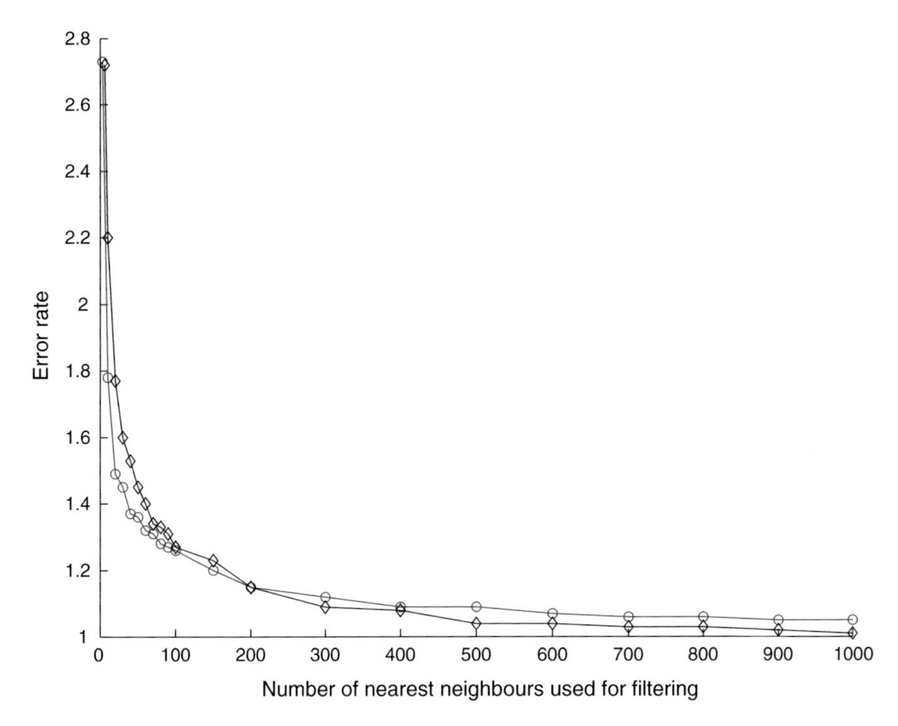


**Fig. 4.12** Error rate drops as K increases for 3-NN (○) and 6-NN (◇) classifiers based on LPD with filtering

only give a good indication of the actual ones, and they are not compared with the error rates of the other $k$-NN classifiers based on euclidean and Tangent distance, respectively.

The confusion matrix of the 3-NN based on LPD with filtering using $K = 50$ is presented in Table 4.7. The confusion matrix shows that most of the digits are fairly well distinguished by LPD, but some confusions stand out from the rest. The greatest confusions are between digits 4 and 9, between digits 3 and 5, and between digits 2 and 7, respectively. From the total amount of 138 misclassified images, there are 11 samples of digit 7 classified as digit 1, 10 samples of digit 2 classified as digit 7, and 10 samples of digit 4 classified as digit 9. The best per class accuracy of the 3-NN model is on digit 1 (99.65 %) and digit 0 (99.49 %). On the other hand, the worst accuracy is on digit 9 (97.82 %) and digit 8 (97.95 %). Thus, LPD is better on recognizing digits 1 and 0 and slightly worse at recognizing digits 9 and 8, despite the fact that there are no significant differences between the per digit classification accuracy rates. Overall, the confusion matrix shows that LPD is able to equally classify all the digits with only a few mistakes. Actually, there are only 138 misclassified images from the total of 10,000 samples.

Table 4.8 compares error rates of the three $k$-NN classification methods (distinct only by the metric used) on the MNIST test set of 10,000 samples. For the $k$-NN

**Table 4.7** Confusion matrix of the 3-NN based on LPD with filtering using $K = 50$

| Digits | 0 | 1 | 2 | 3 | 4 | 5 | 6 | 7 | 8 | 9 |
|---|---|---|---|---|---|---|---|---|---|---|
| 0 | 975 | 1 | 1 | 0 | 0 | 1 | 1 | 1 | 0 | 0 |
| 1 | 0 | 1131 | 2 | 0 | 0 | 0 | 1 | 1 | 0 | 0 |
| 2 | 6 | 2 | 1013 | 0 | 0 | 0 | 1 | 10 | 0 | 0 |
| 3 | 0 | 0 | 1 | 994 | 1 | 6 | 0 | 5 | 1 | 2 |
| 4 | 0 | 4 | 0 | 0 | 963 | 0 | 4 | 0 | 1 | 10 |
| 5 | 1 | 0 | 0 | 4 | 1 | 883 | 1 | 0 | 1 | 1 |
| 6 | 2 | 3 | 0 | 0 | 2 | 1 | 950 | 0 | 0 | 0 |
| 7 | 0 | 11 | 3 | 1 | 0 | 0 | 0 | 1009 | 0 | 4 |
| 8 | 2 | 0 | 0 | 3 | 3 | 2 | 2 | 3 | 954 | 5 |
| 9 | 0 | 3 | 1 | 2 | 6 | 3 | 1 | 4 | 2 | 987 |

**Table 4.8** Error rates on the entire MNIST data set for baseline 3-NN, $k$-NN based on Tangent distance, and $k$-NN based on LPD with filtering

| Method | Error (%) |
|---|---|
| baseline 3-NN | 3.09 |
| $k$-NN + Tangent distance | 1.1 |
| 3-NN + LPD + filter | 1.05 |
| 6-NN + LPD + filter | **1.01** |

classifier based on LPD with filtering, $k = 3$ and $k = 6$ are used. The error rates of the $k$-NN models based on LPD are obtained using the nearest $K = 1000$ images filtered by the euclidean distance. The LPD is based on patches of $4 \times 4$ pixels and a similarity threshold of 0.125. The LPD parameter tuning procedure is described in Sect. 4.4.3. The $k$-NN based on LPD with filtering model has a better accuracy than the other $k$-NN models. In fact, it is among the top 4 $k$-NN models that reported results on the MNIST data set. The 6-NN classifier based on LPD with filtering has an error rate of only 1.01 %. Note that unlike the $k$-NN with filtering based on LPD, the other three methods from top 4 need additional preprocessing steps.

### 4.4.8 Local Learning Experiment

Because LPD is computationally heavy, it is not feasible to compute a kernel matrix with high dimensions. Even with a fast similarity measure, computing and storing a large kernel matrix could pose a serious problem for the design and implementation of a kernel classifier. Therefore, the use of kernel classifiers, such as SVM and KRR based on LPD, on the entire MNIST data set should be avoided as much as possible. Instead, kernel methods can be integrated into a local learning algorithm.

The proposed approach for this experiment is very similar to the filter-based $k$-NN model. The filtering step remains unchanged. Thus, the first step is to filter the nearest $K$ neighbors using the standard euclidean distance. The second step is to train a kernel classifier using only the filtered $K$ neighbors. A new classifier is trained for each test image. The classifier will be used to predict only the label of the test image that was built for.

Before training the classifier, it is necessary to build a kernel matrix using LPD. It is not feasible for a real-time application to build a $K \times K$ kernel matrix for each test image, when $K$ is larger than 100, especially if no GPU or parallel processing is done. But, a large number of $K$ neighbors are needed in order to improve the accuracy of the kernel method. A feasible solution is to train a kernel classifier only when a majority of images with the same label greater than 60 % among the filtered $K$ neighbors is not available, and use a $k$-NN model when such a majority exists. There are two reasons for this approach. First, if such a majority exists, the kernel method will be biased toward choosing the majority class. Second, it is likely that a simple $k$-NN would also choose this majority class that is probably the right one.

This local learning algorithm was tested on the MNIST data set. Using $K = 200$, it gives an error rate of 1.07 %. From the 10,000 test samples, 983 of them had a majority class of less than 60 % of the total 200 neighbors. For each of these 983 samples, a KRR classifier based on LPD was trained. The other test labels were predicted using the 3-NN based on LPD. Interestingly, the 3-NN with filtering approach has an error rate of 1.15 % for $K = 200$ and the local learning algorithm is able to improve it to 1.07 %. Another local learning algorithm, that uses SVM instead of KRR, was tested without being able to improve the accuracy.

In conclusion, using kernel methods in a local learning context does not bring a significant improvement in accuracy to the $k$-NN with filtering approach. However, the local learning algorithm proposed here can successfully be used for handwritten digit recognition. It also benefits from a much faster computational time compared to standard kernel methods based on LPD.

### 4.4.9 Birds Experiment

In this experiment, LPD is used to classify a more general type of images available in the Birds data set. Several $k$-NN models based on different distance measures are compared. The first $k$-NN model uses the Bhattacharyya coefficient to compare spatiograms of HSV values extracted from images. This improved measure of comparing spatiograms is proposed by Conaire et al. (2007). The second $k$-NN model is based on the mean euclidean distance measure ($L_1$-norm) between input images. Another two $k$-NN classifiers based on LPD are tested. One of them uses a slightly modified version of the LPD algorithm, in that it determines similar patches using the mean euclidean distance corresponding to the $L_1$-norm, which seems to work better than the $L_2$-norm on this experiment. The other one uses a fast version of the LPD algorithm that skips half of the comparisons between patches.

For both the euclidean distance measure and the LPD measure, images from the Birds data set need to be brought to the same size in order to be compared. Thus, images are resampled to $100 \times 100$ pixels. To observe the difference between original and resized images, the $k$-NN model based on the Bhattacharyya coefficient is tested on both types of images. The other $k$-NN models are tested only on the resampled images that are also converted to grayscale. While the method proposed by Conaire et al. (2007) works on color images, the euclidean distance and LPD are naturally computed on grayscale images. Table 4.9 compares the error rates of all these $k$-NN models. The empirical results show that skipping overlapping patches does not affect the accuracy of LPD.

Next, three kernel methods based on the fast version of LPD (that skips half of the comparisons between patches) are compared with the state-of-the-art texton learning methods from the work of Lazebnik et al. (2005a). The proposed kernel methods based on LPD are the SVM, the KRR, and the KDA classifiers. Table 4.10 compares error rates of kernel methods based on LPD and texton learning methods. The kernel methods based on LPD have an accuracy similar to some of the state-of-the-art methods. However, the proposed kernel methods are more time efficient. In the work of Lazebnik et al. (2005a), a time of about 7 days for a single experiment is reported, while the kernel methods based on LPD need about one day for the same experiment on an Intel Core i7 2.3 GHz processor and 8 GB of RAM memory. Note that error rates of all models based on LPD, used in this experiment, are obtained with patches of $4 \times 4$ pixels and a similarity threshold of 0.15. The parameters were

**Table 4.9** Error rates of different $k$-NN models on Birds data set

| Method | Preprocessing | Error (%) |
|---|---|---|
| 5-NN + Bhattacharyya | None | 57.33 |
| 5-NN + Bhattacharyya | Resize | 54.67 |
| 3-NN + euclidean | Resize, grayscale | 51.00 |
| 3-NN + LPD | Resize, grayscale | **30.33** |
| 3-NN + LPD + skip | Resize, grayscale | **30.33** |

**Table 4.10** Error on Birds data set for texton learning methods of Lazebnik et al. (2005a) and kernel methods based on LPD

| Method | Error (%) |
|---|---|
| Naïve Bayes | 21.33 |
| Exp. parts | **7.67** |
| Exp. relations | 24.67 |
| Exp. parts + relations | 8.33 |
| SVM + LPD + skip | 27.33 |
| KRR + LPD + skip | 26.00 |
| KDA + LPD + skip | 24.33 |

obtained by cross-validation on the training set. In conclusion, the fast version of LPD can successfully be used as a kernel for image classification. Remarkably, the LPD approach is essentially based on raw pixel data. No higher level features are involved in the computations of LPD. Judging from this perspective, the results of LPD presented in this section are notable. Nonetheless, the next section shows that introducing high level features in the LPD pipeline can produce more robust results.

## 4.5 Local Texton Dissimilarity

This section presents a novel texture dissimilarity measure based on textons, termed Local Texton Dissimilarity (LTD). It represents a development adapted to textures of the more general LPD measure. Instead of patches, LTD works with textons, which are represented as a set of features extracted from image patches. It is reasonable to work with textons rather than patches, since many state-of-the-art texture classification methods are based on textons. Some of these state-of-the-art methods are presented in Sect. 4.5.1.

The algorithm proposed by Dinu et al. (2012) and presented in Sect. 4.2.1 compares image patches using the mean euclidean distance. LTD differs in the way it compares these patches. First, texture-specific features are extracted from each patch. A patch can then be represented by a feature vector. Similar patches are represented through similar features. Thus, image patches are implicitly quantized into textons. Textons are compared using the Bhattacharyya coefficient between their feature vectors. Section 4.5.2 describes the features extracted from image patches, while the LTD algorithm (Ionescu et al. 2014a) is presented in Sect. 4.5.3.

### *4.5.1 Texton-based Methods*

The work of Lazebnik et al. (2005b) presents a texture representation that is invariant to geometric transformations, based on descriptors defined on affine invariant regions. A probabilistic part-based approach for texture and object recognition is presented in the work of Lazebnik et al. (2005a). Textures are represented using a part dictionary obtained by quantizing the appearance of salient image regions.

In the paper of Leung and Malik (2001), texture images are classified by using 3D textons, which are cluster centers of filter response vectors corresponding to different lighting and viewing directions of images. Varma and Zisserman (2005) model textures by the joint distribution of filter responses. This distribution is represented by the frequency histogram of textons. For most texton-based techniques, the textons are usually learned by k-means clustering. In the work of Xie et al. (2010), a novel texture classification method via patch-based sparse texton learning is proposed. The dictionary of textons is learned by applying sparse representation to image patches in the training data set.

### 4.5.2 Texture Features

Before computing LTD between texture images, a set of several image features is extracted from each patch to obtain the texton representation. There are 9 features extracted from patches, which are described next. An interesting remark is that the more features are added to the texton representation, the better the accuracy of the LTD method gets. Intuitively, a broad diversity of features can provide a better description of the image, which in turn helps to correctly classify the respective image. However, a lighter representation, such as the one based on 9 features, results in a faster and more efficient algorithm. One may choose to add or remove features in order to obtain the desired trade-off between accuracy and speed. The texton representation based on the 9 features that are about to be presented next provides reasonable accuracy levels in several experiments presented in Sect. 4.6.

The first two statistical features extracted are the mean and the standard deviation. These two basic features can be computed indirectly, in terms of the image histogram. The shape of an image histogram provides many clues to characterize the image, but the features obtained from an image histogram are not always adequate to discriminate textures, since they are unable to indicate local intensity differences. Thus, more complex features should be used.

One of the most powerful statistical methods for textured image analysis is based on features extracted from the gray-level co-occurrence matrix (GLCM), proposed by Haralick et al. (1973). The GLCM is a second-order statistical measure of image variation and it gives the joint probability of occurrence of gray levels of any two pixels, separated spatially by a fixed vector distance. Smooth texture gives a co-occurrence matrix with high values along diagonals for small distances. The range of gray-level values within a given image determines the dimensions of a co-occurrence matrix. Thus, 4 bits gray-level images give $16 \times 16$ co-occurrence matrices. Relevant statistical features for texture classification can be computed from a GLCM. The features proposed by Haralick et al. (1973), which show a good discriminatory power, are the contrast, the energy, the entropy, the homogeneity, the variance, and the correlation. Among these features that show a good discriminatory power, LTD uses only four of them, namely the contrast, the energy, the homogeneity, and the correlation.

Another feature that is relevant for texture analysis is the fractal dimension. It provides a statistical index of complexity comparing how detail in a fractal pattern changes with the scale at which it is measured. The fractal dimension is usually approximated. The most popular method of approximation is box counting (Falconer 2003). The idea behind the box counting dimension is to consider grids at different scale factors over the fractal image, and count how many boxes are filled over each grid. The box counting dimension is computed by estimating how this number changes as the grid gets finer, by applying a box counting algorithm. An efficient box counting algorithm for estimating the fractal dimension was proposed by Popescu et al. (2013). The idea of the algorithm is to skip the computation for coarse grids, and count how many boxes are filled only for finer grids. LTD includes this efficient variant of box counting in the texton representation.

Daugman (1985) found that cells in the visual cortex of mammalian brains can be modeled by Gabor functions. Thus, image analysis by the Gabor functions is similar to perception in the human visual system. A set of Gabor filters with different frequencies and orientations may be helpful for extracting useful features from an image.

The local isotropic phase symmetry measure (LIPSyM) of Kuse et al. (2011) takes the discrete time Fourier transform of the input image, and filters this frequency information through a bank of Gabor filters. Kuse et al. (2011) also note that local responses of each Gabor filter can be represented in terms of energy and amplitude. Thus, Gabor features, such as the mean squared energy and the mean amplitude, can be computed through the phase symmetry measure for a bank of Gabor filters with various scales and rotations. These features are relevant because Gabor filters have been found to be particularly appropriate for texture representation and discrimination.

Finally, textons are represented by the mean and the standard deviation of the patch, the contrast, the energy, the homogeneity, and the correlation extracted from the GLCM, the (efficient) box counting dimension, and the mean squared energy and the mean amplitude extracted by using Gabor filters. These texton features can be extracted from all images before comparing them with LTD. Thus, the LTD computation can be divided into two main steps, one for texton feature extraction, and one for dissimilarity computation. After the feature extraction step, features should be normalized. In practice, the described features work best on squared image patches having a power of two size.

It is worth mentioning that many texture classification methods are based on extracting texture-specific features from images, including features extracted from the GLCM (Haralick et al. 1973), or the fractal dimension (Falconer 2003). Some of the recent methods that are based on such features are (Backes et al. 2012; de Almeida et al. 2010). A hybrid approach for texture-based image classification using GLCM and self-organizing maps is proposed by de Almeida et al. (2010). Backes et al. (2012) present a novel approach based on the fractal dimension for color texture analysis.

### 4.5.3 *Local Texton Dissimilarity Algorithm*

To quantify the LTD between two grayscale texture images, the sum of all the offsets of similar textons between the two images must be computed. The LTD algorithm is briefly described next. For every texton in one image, the algorithm searches for a similar texton in the other image. First, it looks for similar textons in the same position in both texture images. If those textons are similar, it sums up 0 since there is no spatial offset between textons. If the textons are not similar, the algorithm starts exploring the vicinity of the initial texton position in the second image to find a texton

**Algorithm 7**: Local Texton Dissimilarity Algorithm

1 **Input**:
2 $I_1$ – a grayscale texture image of $h_1 \times w_1$ pixels;
3 $I_2$ – another grayscale texture image of $h_2 \times w_2$ pixels;
4 $n$ – the number of features that represent a texton;
5 gridStep – the skip step that generates a dense grid over the image;
6 offsetStep – the skip step used for comparing patches at different offsets;
7 $w$ – a vector of feature weights;
8 $th$ – the texton similarity threshold.
9 **Initialization**:
10 $\Delta \leftarrow 0$;
11 $h \leftarrow \min\{h_1, h_2\} - p + 1$;
12 $w \leftarrow \min\{w_1, w_2\} - p + 1$;
13 **Computation**:
14 **for** $x = 1$*:gridStep:*$h$ **do**
15   **for** $y = 1$*:gridStep:*$w$ **do**
16     $T^1 \leftarrow$ the texton of $n$ features at position $(x, y)$ in $I_1$;
17     $d \leftarrow 0$;
18     **while** *did not find texton at offset d similar to* $T^1$ **do**
19       $T^2 \leftarrow$ a texton of $n$ features at offset $d$ from $(x, y)$ in $I_2$;
20       $s_1 \leftarrow \frac{1}{n} \sum_{i=1}^{n} \left( w_i \cdot \sqrt{T_i^1} - w_i \cdot \sqrt{T_i^2} \right)^2$;
21       **if** $s_1 < th$ **then**
22         $\Delta \leftarrow \Delta + d$;
23         break;
24       **if** *all patches at offset d were tested* **then**
25         $d \leftarrow d+$ offsetStep;
26     $T^2 \leftarrow$ the texton of $n$ features at position $(x, y)$ in $I_2$;
27     $d \leftarrow 0$;
28     **while** *did not find texton at offset d similar to* $T^2$ **do**
29       $T^1 \leftarrow$ a texton of $n$ features at offset $d$ from $(x, y)$ in $I_1$;
30       $s_2 \leftarrow \frac{1}{n} \sum_{i=1}^{n} \left( w_i \cdot \sqrt{T_i^2} - w_i \cdot \sqrt{T_i^1} \right)^2$;
31       **if** $s_2 < th$ **then**
32         $\Delta \leftarrow \Delta + d$;
33         break;
34       **if** *all patches at offset d were tested* **then**
35         $d \leftarrow d+$ offsetStep;
36 **Output**:
37 $\Delta$ – the dissimilarity between texture images $I_1$ and $I_2$.

similar to the one in the first image. If a similar texton is found during this process, it sums up the offset between the two textons. The spatial search goes on until a similar texton is found or until a maximum offset is reached. The maximum texton offset must be set a priori. The computation of LTD is similar the algorithm presented in Sect. 4.2.2. In practice, the computation described so far is too heavy for a large set of images. To speed up the algorithm, textons are extracted and compared using a dense grid over the image. This is similar to the idea of skipping overlapping patches to optimize LPD.

Algorithm 7 computes the LTD between grayscale texture images $I_1$ and $I_2$, using the underlying Bhattacharyya coefficient to compute the similarity between texton feature vectors. Algorithm 7 needs a handful of other input parameters besides the two images. The number of features $n$ gives the size of the feature vector. In this work, the 9 features described in Sect. 4.5.2 were used. In Algorithm 7, $T_i$ represents the $i$-th feature of the texton $T$, and $w_i$ represents the weight associated to the $i$-th feature. By assigning different weights, the more discriminant features can become more important than the others, while the less informative features can be demoted.

The results of the LTD algorithm can further be improved by adding more features or probably by using completely different features. The parameter that generates a dense grid over the image, and the skip step used for comparing patches at different offsets can be used in order to speed up the LTD algorithm without losing too much accuracy. These parameters induce a sparse representation of the images. Using a sparse representation is indeed necessary, since patch-based algorithms are heavy to compute with current computers because they usually manipulate millions of patches, according to Barnes et al. (2011). The texton similarity threshold is a value in the [0, 1] interval, that determines when two textons are considered to be similar. All these parameters need to be adjusted with regard to the data set size and to the image dimensions, in order to obtain a good trade-off between accuracy and speed.

## 4.6 Texture Experiments and Results

In the experiments, LTD is evaluated with different kernel methods to show that good performance levels are due to the use of LTD. Two data sets of texture images are used to assess the performance of several kernel methods based on LTD, namely the Brodatz data set and the UIUCTex data set. All the experiments presented in this work aim at showing that LTD has general applications for texture classification, and that LTD is indeed a robust dissimilarity measure.

A potential application of LTD discussed in this work is biomass type identification. A method to determine the biomass type and quality has practical motivations for the biomass industry. Such methods are of great importance when one in the biomass industry needs to produce another energy product, such as biofuel or bioenergy, for example. Is the quality of biomass appropriate to efficiently obtain a certain bioproduct? What is the right biomass conversion method for a certain type of biomass? Answering such questions can help reduce the operating costs of biomass power plants. The common methods to determine the biomass quality are expensive or time-consuming. For instance, the biomass quality can be determined using expensive devices, such as mass spectrometers (Abdelnur et al. 2013; Agblevor et al. 1994) that can measure the concentration of carbon, or complex chemical tests (Obernberger and Thek 2004) that may need several days to complete. For this reason, biomass quality control is very often bypassed. In this book, biomass type identification is treated as a texture classification problem. As such, LTD is used in combination with several machine learning methods for biomass texture classification. The approach based on LTD can produce results much faster than the standard tools for biomass quality control. Another classification experiment is conducted on a data set of biomass texture images. Notably, the biomass texture classification task has been extensively studied by Ionescu et al. (2015b).

It is important to mention that, in a general sense, *biomass* refers to the biological material from living, or recently living organisms. In this work, the term *biomass* refers to a renewable energy source, that can be directly converted into another type of energy product. The Biomass Texture data set used in this book is a collection of close-up photos of different samples of three types of biomass: municipal solid waste, corn, and wheat. The goal is to build a classifier that is able to distinguish between these three types of biomass. This is a totally different approach and understanding of the biomass classification problem, compared to other researches. Usually, biomass classification refers to land cover type or forest biomass classification. Land cover classification (Dash et al. 2007) and forest biomass estimation (Wulder et al. 2008) are active research topics in the area of remote sensing.

### 4.6.1 Data Sets Description

The first data set used for testing the dissimilarity based on textons is the Brodatz data set (Brodatz 1966). This data set is probably the best-known benchmark used for texture classification, but also one of the most difficult, since it contains 111 classes with only 9 samples per class. Samples of $213 \times 213$ pixels are cut using a 3 by 3 grid from larger images of $640 \times 640$ pixels. Figure 4.13 presents three sample images per class of three classes randomly selected from the Brodatz data set. The Brodatz data set can be downloaded from http://www.ux.uis.no/~tranden/brodatz.html.

The second experiment is conducted on the UIUCTex data set of Lazebnik et al. (2005b). It contains 1000 texture images of $640 \times 480$ pixels representing different types of textures such as bark, wood, floor, water, and more. There are 25 classes of 40 texture images per class. Textures are viewed under significant scale, viewpoint, and illumination changes. Images also include nonrigid deformations. This data set is available for download at http://www-cvr.ai.uiuc.edu/ponce_grp. Figure 4.14 presents four sample images per class of four classes representing bark, brick, pebbles, and plaid.

**Fig. 4.13** Sample images from three classes of the Brodatz data set

The third experiment is conducted on a data set of biomass texture images available at http://biomass.herokuapp.com. It contains 270 images of $512 \times 512$ pixels representing close-up photos of three types of biomass resulted after the processing of wheat, municipal waste, and corn, respectively. Photos where taken at different zoom levels, under various lighting conditions. Figure 4.15 shows a few random samples of biomass images from the Biomass Texture data set. There are 90 images per class. The goal is to build a classifier that is able to identify the three types of biomass: wheat, municipal waste, and corn, respectively.

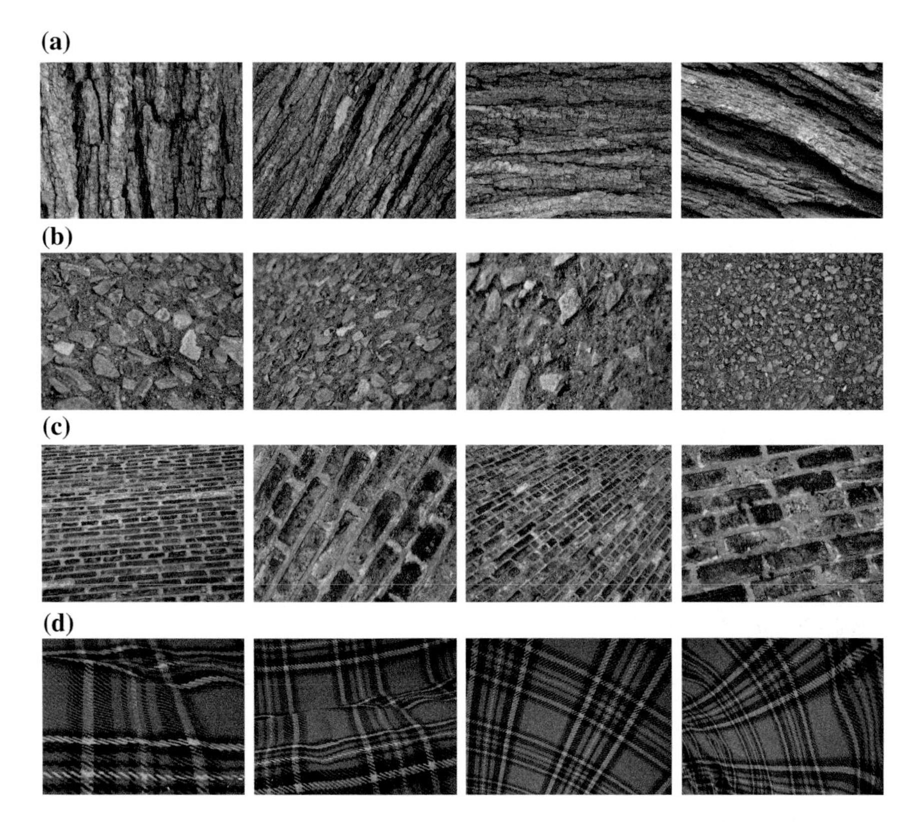


**Fig. 4.14** Sample images from four classes of the UIUCTex data set. Each image is showing a textured surface viewed under different poses. **a** Bark. **b** Pebbles. **c** Brick. **d** Plaid

## *4.6.2 Learning Methods*

In order to employ LTD in a texture classification task, it should be plugged into a similarity-based learning method. In the experiments, several similarity-based classifiers are used in combination with LTD. The first classifier proposed for the evaluation is the $k$-NN. It is chosen because it directly reflects the discriminatory power of the dissimilarity measure. Several state-of-the-art kernel methods are also used, namely the KRR, the SVM, the KDA, and the KPLS. LTD can be transformed into a similarity measure by using the RBF kernel, in a similar fashion to LPD:

$$k(I_1, I_2) = \exp\left(-\frac{\Delta_{LTD}(I_1, I_2)}{2\sigma^2}\right),$$

where $I_1$ and $I_2$ are two grayscale texture images. The parameter $\sigma$ is usually chosen to match the number of features so that values of $k(I_1, I_2)$ are well scaled.

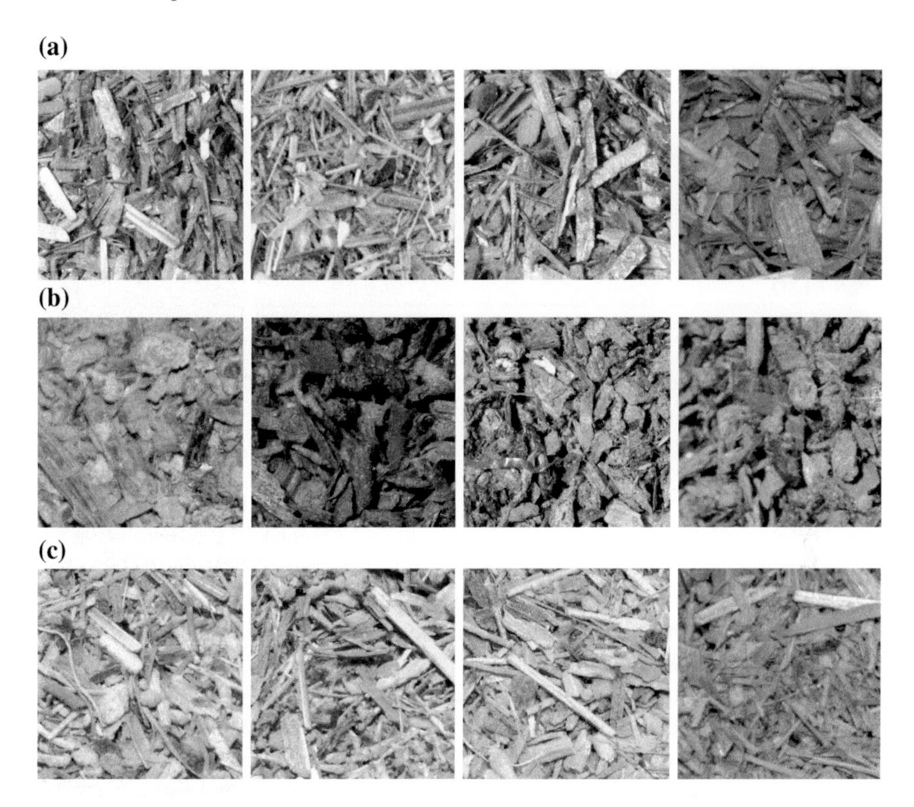


**Fig. 4.15** Sample images from the biomass texture data set. **a** Wheat. **b** Waste. **c** Corn

### *4.6.3 Brodatz Experiment*

The baseline method proposed for this experiment is a 1-NN model that is based on the Bhattacharyya coefficient computed on the 9 texture features described in Sect. 4.5.2. The features are extracted from entire images. The second proposed model is a 1-NN classifier based on LTD. The baseline is useful to assess the performance gained by the use of LTD. The other proposed classifiers are the KRR, the KPLS, the SVM, and the KDA, all based on LTD. The KDA method is particularly suitable for problems with many classes, such as Brodatz.

In the work of Lazebnik et al. (2005b), the accuracy rate reported on the Brodatz data set using 3 training samples per class is 88.15 %. Table 4.11 compares accuracy rates of the proposed classifiers with the accuracy rate of the approach described by Lazebnik et al. (2005b), using the same setup with 3 random samples per class for training. The accuracy rates presented in Table 4.11 are actually averages of accuracy rates obtained over 20 runs for each method. The 1-NN based on LTD model has a far better accuracy than the baseline, proving that LTD helps the learning method to achieve better results. All the kernel methods based on LTD are above the state-of-

**Table 4.11** Accuracy rates on the Brodatz data set using 3 random samples per class for training

| Method | Accuracy |
|---|---|
| Baseline 1-NN | 77.68 % $\pm$ 1.4 |
| Best of Lazebnik et al. (2005b) | 88.15 % |
| 1-NN + LTD | 85.41 % $\pm$ 1.2 |
| KRR + LTD | 89.43 % $\pm$ 1.1 |
| SVM + LTD | 89.48 % $\pm$ 0.8 |
| KPLS + LTD | 89.57 % $\pm$ 0.9 |
| KDA + LTD | **90.87** % $\pm$ 0.8 |

The learning methods based on LTD are compared with a state-of-the-art method. The reported accuracy rates are averaged on 20 trials using 3 random samples per class for training and the other 6 for testing

the-art classifier. The best classifier among them is KDA, which has an accuracy of 90.87 %. It is 5.46 % better than the 1-NN based on LTD, and 2.72 % better that the approach presented in the paper of Lazebnik et al. (2005b). Therefore, LTD can be regarded as a good dissimilarity measure for texture classification. Combined with suitable learning methods, LTD gives results comparable to other methods proposed in literature (Lazebnik et al. 2005b).

Despite the fact that better texture classification methods exist (Ionescu et al. 2015a; Zhang et al. 2007), the classifiers based on LTD can also be improved by adding more features to the texton representation. Another alternative is to combine the kernel based on LTD with other kernels through MKL for improved accuracy. Table 4.12 shows better accuracy rates when LTD is combined with other state-of-the-art kernels. The first MKL approach is based on combining LTD with the bag of visual words framework based on the PQ kernel described in Chap. 5. The second MKL approach is based on combining LTD with the framework of Ionescu et al. (2015a) termed TRIPAF, and with the bag of visual words based on the PQ kernel. These MKL approaches are compared with two state-of-the-art methods (Nguyen et al. 2011; Zhang et al. 2007). Both MKL approaches employ the KDA classifier for training. The best MKL approach, which includes the kernel based on LTD, yields

**Table 4.12** Accuracy rates of several MKL approaches that include LTD compared with state-of-the-art methods on the Brodatz data set

| Model | Accuracy |
|---|---|
| Best model of (Zhang et al. 2007) | 95.90 % $\pm$ 0.6 |
| Best model of (Nguyen et al. 2011) | 96.14 % $\pm$ 0.4 |
| KDA based on BOVW + LTD | 93.76 % $\pm$ 0.7 |
| KDA based on BOVW + TRIPAF + LTD | **97.25** % $\pm$ 0.5 |

The BOVW model is based on the PQ kernel. The reported accuracy rates are averaged on 20 trials using 3 random samples per class for training and the other 6 for testing. The best accuracy rate is highlighted in bold

the highest accuracy on the Brodatz data set, namely 97.25 %. This is almost 1 % better than the state-of-the-art method of Nguyen et al. (2011). Remarkably, none of the individual components of the best MKL approach reach impressive results when used alone (Ionescu et al. 2015a, 2014b), but they complement each other perfectly in the MKL context.

In this experiment, LTD was computed on patches of $32 \times 32$ pixels, using a similarity threshold of 0.02 and a maximum offset of 80 pixels. Patches were extracted on a dense grid with a gap of 32 pixels. Feature weighting can improve accuracy by almost 1 %. Thus, adjusting feature weights is not very important, but it helps the classifier. However, the feature weights were manually adjusted to increase the importance of Gabor features and fractal dimension by a factor of two, and to decrease the importance of the mean and the standard deviation by a factor of two. The weights were tuned on the baseline 1-NN model, which also uses feature weighting in the reported results. The parameter $\sigma$ of the LTD kernel was chosen to be $10^{-3}$. All the parameters were chosen by cross-validation on a subset of the Brodatz data set. An interesting remark is that these parameters do not change by too much on the other data sets.

Using these parameters, it takes less than 1 second to compute LTD between two images on a computer with Intel Core Duo 2.26 GHz processor and 4 GB of RAM memory using a single core. Reported accuracy rates can probably be improved by using a more dense grid and a greater maximum offset, but the LTD computation will also take more time. However, with the current parameters, LTD is much faster than LPD, which takes about 5 minutes to compare two images from the Brodatz data set with similar parameters, without skipping overlapping patches.

The pairwise similarity matrix of LTD between Brodatz samples, illustrated in Fig. 4.16, is analyzed next. It reveals that the LTD measure is able to discriminate most of the classes from the Brodatz data set. This fact is indicated in Fig. 4.16 by the tiny darker squares along the diagonal of the similarity matrix. These squares represent a higher within-class similarity. The fact that LTD discriminates most of the classes by itself is also suggested by the good performance of the 1-NN model based on LTD, compared to the 7.73 % lower performance of the baseline 1-NN.

### 4.6.4 UIUCTex Experiment

In this experiment, the same classifiers evaluated on the Brodatz data set are also evaluated on the UIUCTex data set. More precisely, the evaluated classifiers are the baseline 1-NN model based on the Bhattacharyya coefficient, the 1-NN classifier based on LTD, and the kernel classifiers based on LTD, namely the KRR, the KPLS, the SVM, and the KDA. These classifiers are compared with the state-of-the-art classifier of Lazebnik et al. (2005b). The best accuracy level of the state-of-the-art classifier on the UIUCTex data set, reported in the paper of Lazebnik et al. (2005b) using 20 training samples per class, is 97.41 %.

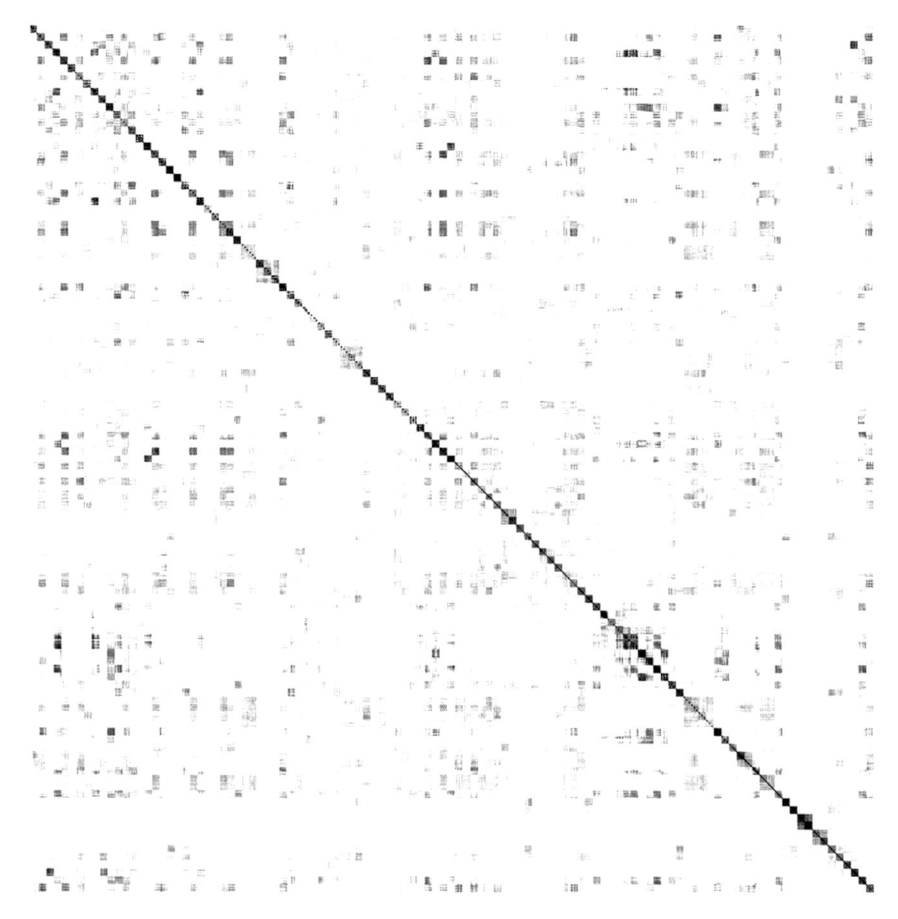

**Fig. 4.16** Similarity matrix based on LTD with patches of $32 \times 32$ pixels and a similarity threshold of 0.02, obtained by computing pairwise dissimilarities between the texture samples of the Brodatz data set

Table 4.13 compares accuracy rates of the classifiers based on LTD with the accuracy rate of the state-of-the-art classifier of Lazebnik et al. (2005b), using the same setup with 20 random samples per class for training. The accuracy rates are averaged over 20 runs for each method. The accuracy of the 1-NN model based on LTD is 9.32 % better than accuracy of the baseline 1-NN, proving again that LTD is able to achieve much better results. However, the accuracy of the 1-NN based on LTD is far behind the state-of-the-art classifier. Even the kernel methods reach accuracy rates that are roughly 4 % lower than the state-of-the-art classifier. The best classifier based on LTD is the KDA, with an accuracy of 94.13 %, which is 3.28 % lower than the state-of-the-art method. The accuracy of these kernel methods depend on LTD, which depends in turn on the features extracted from images to obtain textons. Better features will result in a dissimilarity measure capable of making finer distinctions,

**Table 4.13** Accuracy rates on the UIUCTex data set using 20 random samples per class for training

| Method | Accuracy |
|---|---|
| Baseline 1-NN | 79.34 % ± 1.3 |
| Best of Lazebnik et al. (2005b) | **97.41** % |
| 1-NN + LTD | 88.66 % |
| KRR + LTD | 93.51 % ± 0.9 |
| SVM + LTD | 93.62 % ± 1.2 |
| KPLS + LTD | 93.79 % ± 0.9 |
| KDA + LTD | 94.13 % ± 1.0 |

The learning methods based on LTD are compared with a state-of-the-art method. The reported accuracy rates are averaged on 20 trials using 20 random samples per class for training and the other 20 for testing

and, consequently, in a better kernel classifier. Nonetheless, even with the 9 features proposed in Sect. 4.5.2, LTD seems to give results that are fairly close to the state-of-the-art method.

All in all, better accuracy rates can be obtained when LTD is combined with other state-of-the-art kernels. The same MKL approaches evaluted in the Brodatz experiment are also evaluated here. The first MKL approach is based on combining LTD with the bag of visual words framework based on the PQ kernel, while the second MKL approach is based on combining LTD with TRIPAF (Ionescu et al. 2015a), and with the bag of visual words based on the PQ kernel. As in the previous experiment on Brodatz, the two MKL approaches are compared with two state-of-the-art methods for texture classification (Nguyen et al. 2011; Zhang et al. 2007). Both MKL approaches use KDA in the learning phase. Yet again, the empirical results presented in Table 4.14 prove that the MKL approaches obtain better performance than the approach based on LTD alone. The best MKL approach yields an accuracy rate of 98.44 %, which is slightly above the method of Nguyen et al. (2011) and almost equally below the state-of-the-art method of Zhang et al. (2007). This result confirms that LTD can reach state-of-the-art performance when used in a multiple kernel learning context.

**Table 4.14** Accuracy rates of several MKL approaches that include LTD compared with state-of-the-art methods on the UIUCTex data set

| Model | Accuracy |
|---|---|
| Best model of (Zhang et al. 2007) | **98.70** % ± 0.9 |
| Best model of (Nguyen et al. 2011) | 97.84 % ± 0.3 |
| KDA based on BOVW + LTD | 97.92 % ± 0.6 |
| KDA based on BOVW + TRIPAF + LTD | 98.44 % ± 0.4 |

The BOVW model is based on the PQ kernel. The reported accuracy rates are averaged on 20 trials using 20 random samples per class for training and the other 20 for testing. The best accuracy rate is highlighted in bold

In this experiment, LTD was computed on patches of 64 × 64 pixels, using a similarity threshold of 0.02 and a maximum offset of 240 pixels. Patches were extracted on a dense grid with a gap of 64 pixels. The same feature weights as in the Brodatz experiment were used. The parameter $\sigma$ of the LTD kernel was chosen to be $10^{-3}$. All the parameters were chosen by cross-validation on a subset of the UIUCTex data set.

The pairwise similarity matrix of LTD between UIUCTex samples, shown in Fig. 4.17, seems to create confusions between many classes. Darker squares along the diagonal are not so perceivable as in the Brodatz case. It can be easily observed that LTD is less capable of making fine distinctions between images from different classes. It may be that the 9 feature used for representing textons are less suitable for analyzing texture images with significant scale, viewpoint and illumination changes,

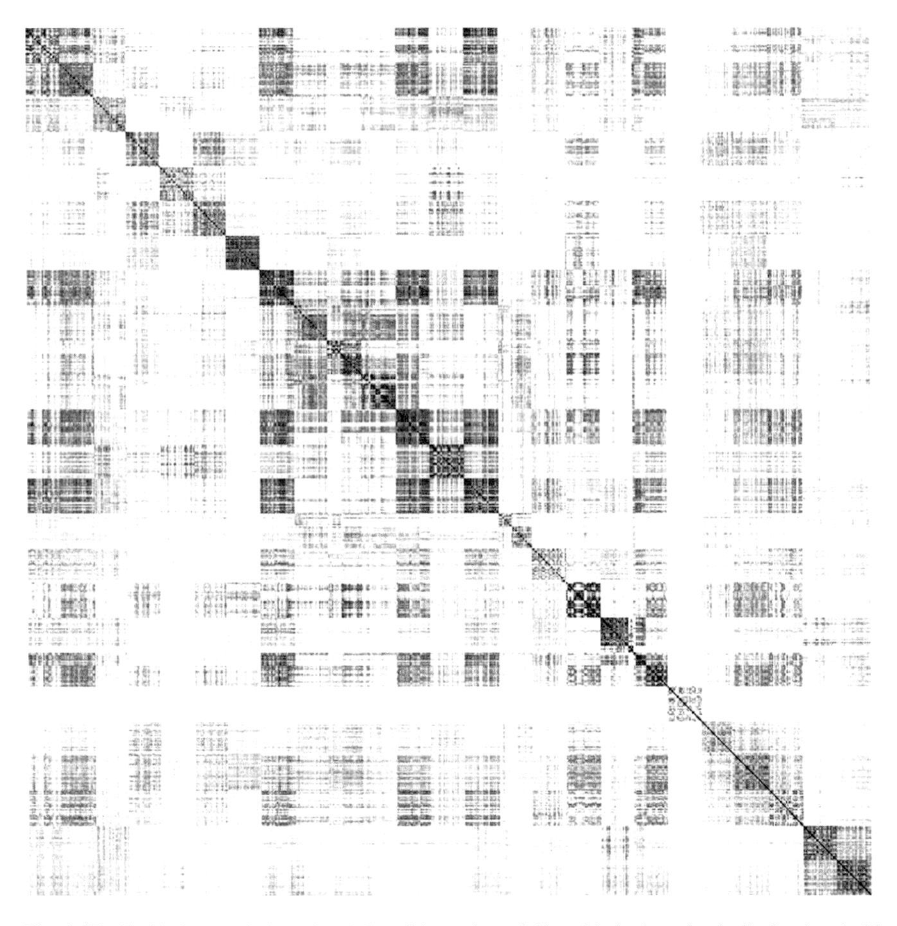

**Fig. 4.17** Similarity matrix based on LTD with patches of 64 × 64 pixels and a similarity threshold of 0.02, obtained by computing pairwise dissimilarities between the texture samples of the UIUCTex data set

such as those from the UIUCTex data set. Naturally, adding other features suitable for this type of images could improve the accuracy.

### 4.6.5 Biomass Experiment

The classifiers evaluated in this experiment are the baseline 1-NN model based on the Bhattacharyya coefficient, the 1-NN classifier based on LTD, and the kernel classifiers based on LTD, namely the KRR, the KPLS, the SVM, and the KDA. These classifiers must identify the three classes of biomass from the Biomass Texture data set.

Table 4.15 presents the accuracy rates of the proposed classifiers using three different setup procedures. The first setup is to use 20 random samples per class for training and the rest of 70 samples for testing. The second setup is to use 30 random samples per class for training and 60 samples for testing. The last setup is to use 40 random samples per class for training and the remaining 50 samples for testing. The accuracy rates are averaged over 50 runs for each method. As expected, the accuracy of each method improves when more training samples are used. For example, the accuracy of the baseline method grows by 6.83 % from 20 training samples to 40 training samples. However, the classifiers based on LTD are more stable, since the accuracy of each classifier grows only by roughly 3–4 % from 20 training samples to 40 samples. The learning methods based on LTD show a significant improvement in accuracy over the baseline. The best classifier based on LTD is KPLS. In all the test cases, the KPLS based on LTD has an accuracy of at least 10 % better than the accuracy of the baseline 1-NN. Overall, the kernel classifiers achieve roughly similar accuracy levels. The empirical results show again that LTD is a powerful dissimilarity measure for texture classification.

In this experiment, LTD was computed on patches of $64 \times 64$ pixels, using a similarity threshold of 0.02 and a maximum offset of 256 pixels. Patches were extracted on a dense grid with a gap of 64 pixels. Again, feature weights were adjusted to increase the importance of Gabor features and fractal dimension by a factor of two,

**Table 4.15** Accuracy rates on the Biomass Texture data set using 20, 30 and 40 random samples per class for training and 70, 60 and 50 for testing, respectively

| Method | 20/70 Accuracy (%) | 30/60 Accuracy (%) | 40/50 Accuracy (%) |
|---|---|---|---|
| Baseline 1-NN | 80.35 | 84.72 | 87.18 |
| 1-NN + LTD | 88.09 | 90.20 | 91.28 |
| KRR + LTD | 93.72 | 96.40 | 97.64 |
| SVM + LTD | 93.98 | 96.58 | 97.72 |
| KPLS + LTD | **94.48** | **96.90** | **97.97** |
| KDA + LTD | 94.08 | 96.40 | 97.67 |

and to decrease the importance of the mean and the standard deviation by a factor of two. The parameter $\sigma$ of the LTD kernel was chosen to be $10^{-3}$. All the parameters were chosen by cross-validation on a subset of the Biomass Texture data set.

## 4.7 Discussion

In this chapter, a novel dissimilarity measure for images, which is based on patches was presented. Several methods of improving its speed were also discussed, such as using a hash table to store already computed patch distances, and skipping the comparison step of some overlapping patches.

Empirical results showed that LPD can be used for real-time image classification, especially when local learning methods are preferred instead of standard machine learning algorithms. Local learning methods based on LPD were proposed and tested on the popular MNIST data set. The experiments show that local learning algorithms perform very well in terms of accuracy and time. The error rate achieved by the $k$-NN based on LPD with filtering approach is 1.01 % on the MNIST data set, which makes it one of the top 4 $k$-NN models that report results on this data set.

There are other ways of avoiding the problem of high computational time that are not studied in this book. For example, another local learning method to solve this problem is to rescale images to a smaller size and compute LPD on those images. For the $k$-NN with filtering approach, the nearest $K$ neighbors can be filtered using the smaller images, and then, the nearest $k$ neighbors are selected from those remaining $K$, using the original images.

Furthermore, a step towards improving the computational time and the accuracy of the dissimilarity was already taken through the development of LTD. Instead of comparing patches, LTD compares textons, which are represented as a set of texture-specific features extracted from images. The texture experiments presented in this chapter showed that LTD is a robust dissimilarity measure, achieving state-of-the-art accuracy levels in several texture classification tasks when combined with other methods through MKL. In future work, the LTD measure can further be improved by adding more features to the texton feature set, or by changing the features completely to make them suitable for the approached task.

## References

Abdelnur PV, Vaz BG, Rocha JD, de Almeida MBB, Teixeira MAG, Pereira RCL (2013) Characterization of bio-oils from different pyrolysis process steps and biomass using high-resolution mass spectrometry. Energy Fuels 27(11):6646–6654

Agblevor FA, Davis MF, Evans RJ (1994) Molecular beam mass spectrometric characterization of biomass pyrolysis products for fuels and chemicals. Preprints of Papers: Am Chem Soc Div Fuel Chem 39(3):840–845. ISSN 0569–3772

Backes AR, Casanova D, Bruno OM (2012) Color texture analysis based on fractal descriptors. Pattern Recognit 45(5):1984–1992
Barnes C, Goldman DB, Shechtman E, Finkelstein A (2011) The patchmatch randomized matching algorithm for image manipulation. Commun ACM 54(11):103–110
Brodatz P (1966) Textures: a photographic album for artists and designers., Dover pictorial archives-Dover Publications, New York
Conaire CO, O'Connor NE, Smeaton AF (2007) An improved spatiogram similarity measure for robust object localisation. In: Proceedings of ICASSP 1:1069–1072
Dash J, Mathur A, Foody GM, Curran PJ, Chipman JW, Lillesand TM (2007) Land cover classification using multi-temporal MERIS vegetation indices. Int J Remote Sens 28(6):1137–1159
Daugman JG (1985) Uncertainty relation for resolution in space, spatial frequency, and orientation optimized by two-dimensional visual cortical filters. J Opt Soc Am A 2(7):1160–1169
de Almeida CWD, de Souza RMCR, Candeias ALB (2010) Texture classification based on co-occurrence matrix and self-organizing map. In: Proceedings of SMC, pp. 2487–2491, Oct 2010
Dinu A, Dinu LP (2005) On the syllabic similarities of romance languages. In: Proceedings of CICLing 3406:785–788
Dinu LP, Ionescu RT (2012) An efficient rank based approach for closest string and closest substring. PLoS ONE 7(6):e37576
Dinu LP, Ionescu RT (2013) Clustering based on median and closest string via rank distance with applications on DNA. Neural Comput Appl 24(1):77–84
Dinu LP, Manea F (2006) An efficient approach for the rank aggregation problem. Theor Comput Sci 359(1–3):455–461
Dinu LP, Popescu M (2009) Language independent kernel method for classifying texts with disputed paternity. In: Proceedings of ASMDA
Dinu LP, Sgarro A (2006) A Low-complexity distance for DNA strings. Fundamenta Informaticae 73(3):361–372
Dinu LP, Ionescu RT, Popescu M (2012) Local Patch Dissimilarity for images. In: Proceedings of ICONIP 7663:117–126
Efros AA, Freeman WT (2001) Image quilting for texture synthesis and transfer. In: Proceedings of SIGGRAPH, pp. 341–346
Falconer K (2003) Fractal geometry: mathematical foundations and applications, 2 edn. Wiley. ISBN 0470848626
Guo G, Dyer CR (2007) Patch-based image correlation with rapid filtering. In: Proceedings of CVPR
Hamming RW (1950) Error detecting and error correcting codes. Bell Syst Tech J 26(2):147–160
Haralick RM, Shanmugam K, Dinstein I (1973) Textural features for image classification. IEEE Trans Syst, Man Cybern 3(6):610–621
Ionescu RT, Popescu M (2013) Speeding up Local Patch Dissimilarity. In: Proceedings of ICIAP 8156:1–10
Ionescu RT, Popescu AL, Popescu D, Popescu M (2014a) Local Texton Dissimilarity with applications on biomass classification. In: Proceedings of VISAPP
Ionescu RT, Popescu AL, Popescu M (2014b) Texture classification with the PQ kernel. In: Proceedings of WSCG
Ionescu RT, Popescu AL, Popescu D (2015a) Texture classification with patch autocorrelation features. In: Proceedings of ICONIP 9489:1–11
Ionescu RT, Popescu AL, Popescu M, Popescu D (2015b) BiomassID: a biomass type identification system for mobile devices. Comput Electron Agric 113:244–253
Kuse M, Wang Y-F, Kalasannavar V, Khan M, Rajpoot N (2011) Local isotropic phase symmetry measure for detection of beta cells and lymphocytes. J Pathol Inf 2(2):2
Lazebnik S, Schmid C, Ponce J (2005a) A maximum entropy framework for part-based texture and object recognition. In: Proceedings of ICCV 1:832–838
Lazebnik S, Schmid C, Ponce J (2005b) A sparse texture representation using local affine regions. IEEE Trans Pattern Anal Mach Intell 27(8):1265–1278

LeCun Y, Jackel LD, Boser B, Denker JS, Graf HP, Guyon I, Henderson D, Howard RE, Hubbard W (1989) Handwritten digit recognition: applications of neural net chips and automatic learning. IEEE Commun 41–46
LeCun Y, Bottou L, Bengio Y, Haffner P (1998) Gradient-based learning applied to document recognition. In: Proceedings of the IEEE 86(11):2278–2324
Leung T, Malik J (2001) Representing and recognizing the visual appearance of materials using three-dimensional textons. Int J Comput Vis 43(1):29–44
Levenshtein VI (1966) Binary codes capable of correcting deletions, insertions and reverseals. Cybern Control Theory 10(8):707–710
Nguyen H-G, Fablet R, Boucher J-M (2011) Visual textures as realizations of multivariate log-Gaussian Cox processes. In: Proceedings of CVPR, pp. 2945–2952
Obernberger I, Thek G (2004) Physical characterisation and chemical composition of densified biomass fuels with regard to their combustion behaviour. Biomass Bioenergy 27(6):653–669
Popescu AL, Popescu D, Ionescu RT, Angelescu N, Cojocaru R (2013) Efficient fractal method for texture classification. In: Proceedings of ICSCS, Aug 2013
Simard P, LeCun Y, Denker JS, Victorri B (1996) Transformation invariance in pattern recognition, tangent distance and tangent propagation. Neural Networks: Tricks of the Trade
Srihari SN (1992) High-performance reading machines. In: Proceedings of the IEEE (Special issue on Optical Character Recognition) 80(7):1120–1132
Suen CY, Nadal C, Legault R, Mai TA, Lam L (1992) Computer recognition of unconstrained handwritten numerals. In: Proceedings of the IEEE (Special issue on Optical Character Recognition) 80(7):1162–1180
Vapnik V, Vashist A (2009) A new learning paradigm: learning using privileged information. Neural Netw 22(56):544–557. ISSN 0893–6080
Varma M, Zisserman A (2005) A statistical approach to texture classification from single images. Int J Comput Vis 62(1–2):61–81
Wilder KJ (1998) Decision tree algorithms for handwritten digit recognition. Electronic Doctoral Dissertations for UMass Amherst, Jan 1998. http://scholarworks.umass.edu/dissertations/AAI9823791
Wulder MA, White JC, Fournier RA, Luther JE, Magnussen S (2008) Spatially explicit large area biomass estimation: three approaches using forest inventory and remotely sensed imagery in a GIS. Sensors 8(1):529–560
Xie J, Zhang L, You J, Zhang D (2010) Texture classification via patch-based sparse texton learning. In: Proceedings of ICIP, pp. 2737–2740
Zhang J, Marszalek M, Lazebnik S, Schmid C (2007) Local features and kernels for classification of texture and object categories: a comprehensive study. Int J Comput Vis 73(2):213–238

# Chapter 5
# Object Recognition with the Bag of Visual Words Model

## 5.1 Introduction

The classical problem in computer vision is that of determining whether or not the image data contains some specific object, feature, or activity. Particular formulations of this problem are image classification, object recognition, object detection. Computer vision researchers have recently developed sophisticated methods for such image-related tasks. Among the state-of-the-art models are discriminative classifiers using the *bag of visual words* (BOVW) representation (Sivic et al. 2005; Zhang et al. 2007) and spatial pyramid matching (Lazebnik et al. 2006), generative models (Fei-Fei et al. 2007), or part-based models (Felzenszwalb et al. 2010; Lazebnik et al. 2005). The BOVW model, which represents an image as a histogram of local features, has demonstrated impressive levels of performance for image categorization (Zhang et al. 2007), image retrieval (Philbin et al. 2007), or related tasks (Ionescu et al. 2013).

This chapter is focused on improving the BOVW model in several ways. Usually, kernel methods are used to compare image histograms. Popular choices, besides the linear kernel, are the intersection, Hellinger's, $\chi^2$ and Jensen–Shannon (JS) kernels. Nevertheless, there is no reason to limit the choice of kernels to these options when other kernels are available. The final goal, that is to improve the results for image classification and related tasks, can be achieved by trying different kernels that could possibly work better. In this chapter, two recently introduced kernels for visual words are presented. Both of them come from ideas that emerged in the area of text analysis.

The first kernel studied in this chapter, namely the PQ kernel, was introduced by Ionescu and Popescu (2013). It is inspired from a class of similarity measures for ordinal variables, more precisely Goodman and Kruskal's gamma and Kendall's tau. These ordinal measures were previously used in text analysis for authorship identification by Dinu and Popescu (2009). The underlying idea of PQ is to treat the visual word histograms as ordinal data, in which data is ordered but cannot be assumed to have equal distance between values. In this case, a histogram will be



R.T. Ionescu and M. Popescu, *Knowledge Transfer between Computer Vision and Text Mining*, Advances in Computer Vision and Pattern Recognition,
DOI 10.1007/978-3-319-30367-3_5

regarded as a ranking of visual words according to their frequencies in that histogram. Usage of the ranking of visual words instead of the actual values of the frequencies may seem as a loss of information, but the process of ranking can actually make PQ more robust, acting as a filter and eliminating the noise contained in the values of the frequencies. This work proves that PQ is a kernel and it also shows how to build its feature map. Moreover, this chapter describes an algorithm to compute the PQ kernel in $O(n \log n)$ time, based on merge sort. The algorithm was initially presented in the work of Ionescu and Popescu (2015a).

The second kernel presented in this chapter is the Spatial Non-Alignment Kernel (SNAK), which was recently introduced by Ionescu and Popescu (2015b). The standard BOVW model ignores the spatial information contained in the image, but researchers have demonstrated that the object recognition performance can be improved by including spatial information (Lazebnik et al. 2006; Sánchez et al. 2012; Uijlings et al. 2009). Perhaps the most popular approach is the spatial pyramid representation (Lazebnik et al. 2006), which divides the image into spatial bins. The SNAK framework is another general approach that encodes the spatial information in a much better and efficient way. More precisely, SNAK embeds the spatial information into a kernel function as briefly described next. For each visual word, the average position and the standard deviation is computed based on all the occurrences of the visual word in the image. These are computed with respect to the center of the object, which is determined with the help of the objectness measure (Alexe et al. 2010, 2012). The pairwise similarity of two images is then computed by taking into account the difference between the average positions and the difference between the standard deviations of each visual word in the two images. In other words, the SNAK kernel includes the spatial distribution of the visual words in the similarity of two images. Various kernel functions can be plugged into the SNAK framework. Interestingly, the core operations of the SNAK framework are inspired by rank distance (Dinu and Manea 2006) and its extensions (Dinu et al. 2012; Ionescu 2013), which essentially measure the local displacement among two strings or two images, respectively. Likewise, SNAK measures the local displacement of visual words.

Object class recognition experiments are conducted in order to assess the performance of different kernels, including PQ and SNAK, on two benchmark data sets of images, more precisely, the Pascal VOC data set and the Birds data set. The idea behind the evaluation is to use the same framework and variate only the feature maps induced by different kernels. The empirical evaluation shows that the PQ kernel is able to improve the mean average precision on both data sets. Furthermore, the results indicate that SNAK significantly improves the object recognition performance of every evaluated kernel. Compared to the spatial pyramid, SNAK improves performance while consuming less space and time. Therefore, SNAK can be considered a good candidate to replace the widely used spatial pyramid representation.

This chapter also describes an improved variant of the BOVW model that yields state-of-the-art performance for classifying human facial expression from low-resolution images. The proposed model was also presented by Ionescu et al. (2013) as an approach to the Facial Expression Recognition (FER) Challenge of the ICML 2013 Workshop in Challenges in Representation Learning (WREPL). The BOVW model is a rather general approach for image categorization, because it does not use any particular characteristics of the image. More precisely, this approach treats images representing faces, objects, or textures in the same manner. The method developed for the FER Challenge stems from this generic approach. However, the model has to be specifically adapted to the data set provided by the WREPL organizers. First, histograms of visual words are replaced with normalized presence vectors to eliminate the noise introduced by word frequencies. For facial expression recognition, the presence of a visual word is more important than its frequency. Second, local multiple kernel learning was used to predict class labels of test images, in order to reduce both the image variation and the labeling noise in the resulting training sets.

Preliminary experiments were performed to validate the modified BOVW approach. The empirical results show that presence vectors improve accuracy by roughly 1 %, while local learning improves performance by almost 2–3 %. Several kernel methods are also evaluated in the experiments. The SVM classifier performs better than the KDA and the KRR. Experiments show that spatial information also helps to improve recognition performance by almost 2–3 %. Presence vectors that record different spatial information are combined to improve accuracy even further. The improved BOVW method was fairly successful, it ranked fourth in the FER Challenge, with an accuracy of 67.484 % on the final (private) test set, as the work of Goodfellow et al. (2015) also reports. Further details about the Facial Expression Recognition Challenge are provided by Goodfellow et al. (2015).

The chapter is organized as follows. Section 5.2 presents the classical BOVW framework used for image retrieval, image categorization, and related tasks. The PQ kernel for histograms of visual words is discussed in Sect. 5.3. The SNAK framework is described in Sect. 5.4. Object recognition experiments conducted on two benchmark data sets are presented in Sect. 5.5. Section 5.6 presents the BOVW learning model adapted to facial expression recognition. The local learning approach is presented in Sect. 5.7. Experiments conducted on the Facial Expression Recognition Challenge data set are presented in Sect. 5.8. Finally, a discussion about the developments presented in this chapter is given in Sect. 5.9.

## 5.2 Bag of Visual Words Model

In computer vision, the BOVW model can be applied to image classification and related tasks, by treating image descriptors as words. As discussed in Chap. 3, a bag of visual words is a vector of occurrence counts of a vocabulary of local image features. This representation can also be described as a histogram of visual words. The vocabulary is usually obtained by vector quantizing image features into visual words.

The BOVW model can be divided in two major steps. The first step is for feature detection and representation. The second step is to train a kernel method in order to predict the class label of a new image. The entire process, that involves both training and testing stages, is illustrated in Fig. 5.1.

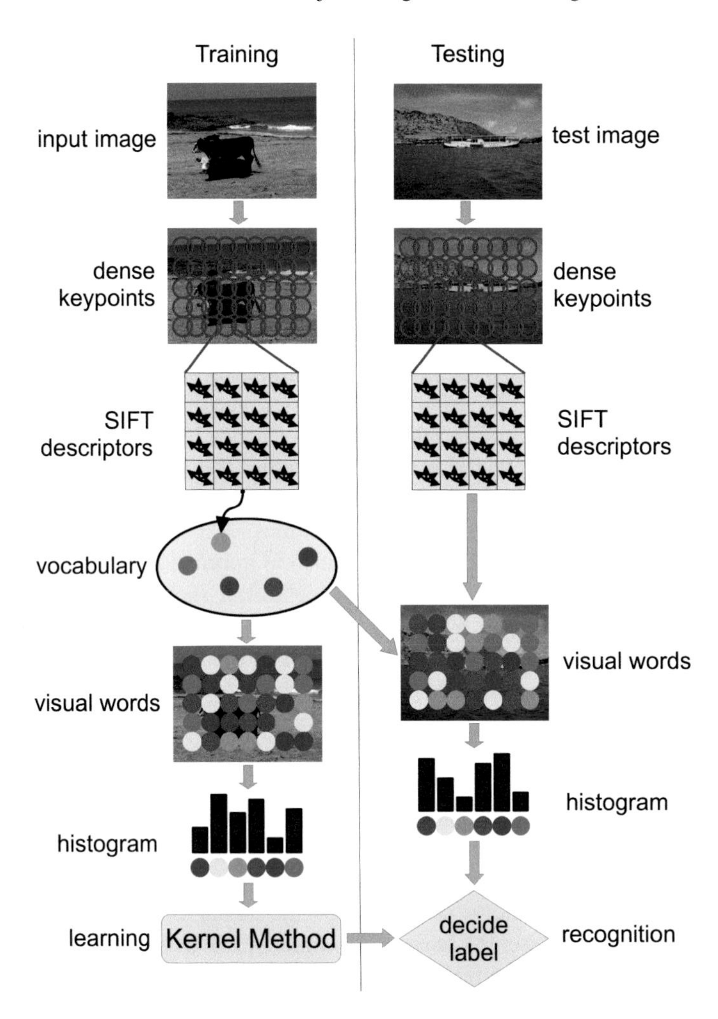


**Fig. 5.1** The BOVW learning model for object class recognition. The feature vector consists of SIFT features computed on a regular grid across the image (dense SIFT) and vector quantized into visual words. The frequency of each visual word is then recorded in a histogram. The histograms enter the training stage. Learning is done by a kernel method

The feature detection and representation step works as described next. Features are detected using a regular grid across the input image. At each interest point, a SIFT feature (Lowe 1999, 2004) is computed. This approach is known as dense SIFT (Bosch et al. 2007; Dalal and Triggs 2005). Next, SIFT descriptors are vector quantized into visual words and a vocabulary (or codebook) of visual words is obtained. The vector quantization process is done by k-means clustering (Leung and Malik 2001), and visual words are stored in a randomized forest of k-d trees (Philbin et al. 2007) to reduce search cost. The frequency of each visual word in an image is then recorded in a histogram. The histograms of visual words enter the training step. Typically, a kernel method is employed for training the model. Several kernels can be used, such as the linear kernel, the intersection kernel, the Hellinger's kernel, the $\chi^2$ kernel or the Jensen–Shannon kernel. In this chapter, two novel kernels that provide better results in practice are being used, namely the PQ kernel described in Sect. 5.3 and the SNAK kernel presented in Sect. 5.4.

After the model is trained, it can be used for new images. Given a test image, features are extracted and quantized into visual words from the vocabulary that was already obtained in the training stage. The histogram of visual words that represents the test image is compared with the histograms learned in the training stage. The system can return either a label (or a score) for the test image or a ranked list of images similar to the test image, depending on the application. For object class recognition a label (or a score) is enough, while for image retrieval a ranked list of images is more appropriate. No matter the application, the training stage of the BOVW model can be done offline. For this reason, the time that is necessary for vector quantization and learning is not of great importance. What matters most in the context of image retrieval (and also in the context of object class recognition) is to return the result for a new (test) image as quickly as possible.

The performance level of the described model depends on the number of training images, but also on the number of visual words. The number of visual words is a parameter that must be set a priori. Usually, the accuracy gets better as the number of visual words is greater, but there is always a saturation point that depends on the data set.

An interesting remark is that the described model ignores spatial relationships between image features. A good way to improve performance is to include spatial information in the BOVW model and one way to achieve this is the spatial pyramid representation (Lazebnik et al. 2006). The spatial pyramid works by dividing the image into spatial bins. The frequency of each visual word is then recorded in a histogram for each bin. The final feature vector for the image is a concatenation of these histograms. Spatial pyramids are used in the BOVW model for facial expression recognition presented in Sect. 5.6. Nevertheless, a more efficient way of encoding spatial information is the SNAK framework described in Sect. 5.4.

## 5.3 PQ Kernel for Visual Word Histograms

All common kernels used in computer vision treat histograms of visual words either as finite probability distributions, for instance, the Jensen–Shannon kernel, either as quantitative random variables whose values are the frequencies of different visual words in the respective images, for instance, the Hellinger's kernel (Bhattacharyya coefficient) and the $\chi^2$ kernel. Even the linear kernel can be seen as the Pearson's correlation coefficient if the two histograms are standardized.

However, the histograms of visual words can also be treated as ordinal data, in which data is ordered but cannot be assumed to have equal distance between values.

In this case, the values of histograms will be ranks of visual words according to their frequencies in the image, rather than the actual values of these frequencies.

An entire set of correlation statistics for ordinal data is based on the number of concordant and discordant pairs among two variables. The number of concordant pairs among two variables (or histograms) $X, Y \in \mathbb{R}^n$ is:

$$P = |\{(i,j) \ : \ 1 \leq i < j \leq n, \ (x_i - x_j)(y_i - y_j) > 0\}|.$$

In the same manner, the number of discordant pairs is:

$$Q = |\{(i,j) \ : \ 1 \leq i < j \leq n, \ (x_i - x_j)(y_i - y_j) < 0\}|.$$

*Goodman and Kruskal's gamma* (Upton and Cook 2004) is defined as:

$$\gamma = \frac{P - Q}{P + Q}.$$

Kendall developed several slightly different types of ordinal correlation as alternatives to gamma. *Kendall's tau-a* (Upton and Cook 2004) is based on the number of concordant versus discordant pairs, divided by a measure based on the total number of pairs ($n$ is the sample size):

$$\tau_a = \frac{P - Q}{\frac{n(n-1)}{2}}.$$

*Kendall's tau-b* (Upton and Cook 2004) is a similar measure of association based on concordant and discordant pairs, adjusted for the number of ties in ranks. It is calculated as the difference between $P$ and $Q$ divided by the geometric mean of the number of pairs not tied in $X$ and the number of pairs not tied in $Y$, denoted by $X_0$ and $Y_0$, respectively:

$$\tau_b = \frac{P - Q}{\sqrt{(P + Q + X_0)(P + Q + Y_0)}}.$$

The three correlation statistics described so far are very related. If $n$ is fixed and $X$ and $Y$ have no ties, then $P$, $X_0$ and $Y_0$ are completely determined by $n$ and $Q$. Actually, all are based on the difference between $P$ and $Q$, normalized differently. Following this observation, the PQ kernel between two histograms $X$ and $Y$ is defined as:

$$k_{PQ}(X, Y) = 2(P - Q).$$

The name of this kernel is derived from the original notations of Kendall (1948) for the number of concordant pairs $P$ and the number of discordant pairs $Q$. The following theorem proves that PQ is indeed a kernel, by showing how to build its feature map.

**Theorem 2** *The function denoted by $k_{PQ}$ is a kernel function.*

*Proof* To prove that $k_{PQ}$ is a kernel, the explicit feature map induced by $k_{PQ}$ is provided next.

Let $X, Y \in \mathbb{R}^n$ be two histograms of visual words. Let $\Psi$ be defined as follows:

$$\Psi : \mathbb{R}^n \to \boldsymbol{M}_{n,n} \quad \Psi(X) = (\Psi(X)_{i,j})_{1 \le i \le n, 1 \le j \le n},$$

with

$$\Psi(X)_{i,j} = \begin{cases} 1, & \text{if } x_i > x_j \\ 0, & \text{if } x_i = x_j \\ -1, & \text{if } x_i < x_j \end{cases}, \forall 1 \le i, j \le n.$$

Note that $\Psi$ associates to each histogram a matrix that describes the order of its elements.

If matrices are treated as vectors, then the following equality is true:

$$k_{PQ}(X, Y) = 2(P - Q) = \langle \Psi(X), \Psi(Y) \rangle ,$$

where $\langle \cdot, \cdot \rangle$ denotes the scalar product. This proves that $k_{PQ}$ is a kernel and provides the explicit feature map induced by $k_{PQ}$. □

Another approach inspired from rank correlation measures is the WTA hash proposed by Yagnik et al. (2011). For $K = 2$, the WTA hash is closely related to the PQ kernel. However, there are two important differences. The first one is that WTA hash works with a random selection of pairs from the feature set. The second one is that, unlike the PQ kernel, the WTA hash ignores equal pairs. In terms of feature representation, the PQ kernel represents a histogram with a feature vector containing $\{-1, 0, 1\}$ (0 for equal pairs), while the WTA hash with $K = 2$ uses only $\{1, 0\}$. The empirical results of Ionescu and Popescu (2013) show that these differences have direct consequences to the performance level, creating an even greater gap between the two methods. More precisely, the PQ kernel obtains better results than the WTA hash.

According to Vedaldi and Zisserman (2010), the feature vectors of $\gamma$-homogeneous kernels should be $L_\gamma$-normalized. Being linear in the feature space, PQ is a 2-homogeneous kernel and its feature vectors should be $L_2$-normalized. Therefore, in the experiments, the PQ kernel is $L_2$-normalized:

$$\hat{k}_{PQ}(X, Y) = \frac{k_{PQ}(X, Y)}{\sqrt{k_{PQ}(X, X) \cdot k_{PQ}(Y, Y)}}.$$

Treating visual words frequencies as ordinal variables means that in the calculation of the distance (or similarity) measure, the ranks of visual words according to their frequencies in the image will be used, rather than the actual values of these frequencies. Usage of the ranking of visual words in the calculation of the distance (or similarity) measure, instead of the actual values of the frequencies, may seem as a loss of information, but the process of ranking can actually make the measure more robust, acting as a filter and eliminating the noise contained in the values of the frequencies. For example, the fact that a specific visual word has rank 2 (it is the second most frequent feature) in one image, and rank 4 (it is the fourth most frequent feature) in another image can be more relevant than the fact that the respective feature appears 34 times in the first image, and only 29 times in the second.

It is important to note that for vocabularies with more than 1000 words, the kernel trick should be employed to directly obtain the PQ kernel matrix instead of computing its feature maps, since there is a quadratic dependence between the size of the feature map and the number of visual words. This will greatly reduce the space cost of the PQ kernel. The problem of the time complexity remains to be solved. Usually, rank correlation statistics, such as Kendall's tau (Upton and Cook 2004), have been avoided because they are thought to be computationally expensive. Statements like "the Cayley and Hamming distances are computed in linear time rather than quadratic time like Kendall's tau" from Ziegler et al. (2012) can be encountered even in recent publications. Actually, a rank correlation measure based on counting concordant and discordant pairs can be computed in $O(n \log n)$ time based on merge sort (Knight 1966). Another algorithm to compute Kendall's tau in $O(n \log n)$ is based on AVL Trees (Christensen 2005). The paper of Christensen (2005) is also symptomatic for the common conception that computing Kendall's tau needs $O(n^2)$ time. Indeed, the work of Christensen (2005) seems to ignore the previous work of (Knight 1966), by stating that: "Traditional algorithms for the calculation of Kendall's tau between two data sets of $n$ samples have a calculation time of $O(n^2)$".

An algorithm to compute the PQ kernel in $O(n \log n)$ time, where $n$ is the number of visual words, is described next. While the PQ feature map proposed by Ionescu and Popescu (2013) has a quadratic dependence on the number of visual words, the fast algorithm based on merge sort leverages the use of the PQ kernel (in the dual form) for large visual word vocabularies. The algorithm presented here follows the work of Knight (1966), which proposed a similar algorithm for computing the Kendall's tau measure. The algorithm is based on the key insight of Kendall which, in his book (Kendall 1948), proves that the number of discordant pairs $Q$ between two rankings is equal to the number of interchanges (or swaps), denoted by $s$, required to transform one ranking into the other. Another important observation is that, given a pair of indexes $(i, j)$, interchanging the corresponding values in the same time in the two histograms $X$ and $Y$, respectively, does not change the number of discordant pairs. Thus, sorting the two histograms $X$ and $Y$ in the same time, using as sorting criteria first the values of $X$ and second (for ties in $X$) the values of $Y$, will end up with a new pair of variables $X'$ and $Y'$ that will have the same number of discordant pairs as $X$ and $Y$. Now, $X'$ is completely sorted while $Y'$ is not. To represent the same ranking as $X'$, $Y'$ must also be sorted. This implies that the number of swaps needed to sort $Y'$ will be the number of discordant pairs between $X'$ and $Y'$, which, in turn, is the number of discordant pairs between $X$ and $Y$. Even if bubble sort is the obvious solution, it is not required to be used for computing the number of swaps that are needed to sort $Y'$. Instead, a slightly modified merge sort algorithm can sort $Y'$ while computing $s$. Algorithm 8 computes $k_{PQ}$ between two histograms $X, Y \in \mathbb{R}^n$

based on these observations. Steps 1 and 5 compute the number of discordant pairs $Q$ between $X$ and $Y$, given that $Q = s$.

The number of concordant pairs $P$ is completely determined by the number of discordant pairs, the total numbers of pairs denoted by $t = n(n-1)/2$, the number of equal pairs in $X$ denoted by $e_X$, the number of equal pairs in $Y$ denoted by $e_Y$, and the number of pairs that are equal in both $X$ and $Y$ denoted by $e_{X \wedge Y}$. More precisely, $P$ can be expressed as follows:

$$P = t + e_{X \wedge Y} - e_X - e_Y - s.$$

**Algorithm 8**: PQ Kernel Algorithm

1 **Input**:
2 $X, Y \in \mathbb{R}^n$ – two histograms of visual words.
3 **Computation**:
4 Sort $X$ and $Y$ in the same time using merge sort, according to the values in $X$ in ascending order. If two values in $X$ are equal, sort them according to $Y$ in ascending order.
5 Compute the total number of pairs as $t = n(n-1)/2$.
6 Compute the number of equal pairs in $X$ as $e_X$.
7 Compute the number of pairs that are equal in both $X$ and $Y$ as $e_{X \wedge Y}$.
8 Sort $Y$ using merge sort in ascending order and sum the differences of swap positions into $s$.
9 Compute the number of equal pairs in $Y$ as $e_Y$.
10 Finally, compute $k_{PQ}(X, Y) = 2(t + e_{X \wedge Y} - e_X - e_Y - 2s)$.
11 **Output**:
12 $k_{PQ}$ – the PQ kernel between the histograms $X$ and $Y$.

Thus, the difference between $P$ and $Q$ can also be expressed in terms of $t$, $e_X$, $e_Y$, and $e_{X \wedge Y}$, as in step 7 of Algorithm 8.

The analysis of the computational complexity of Algorithm 8 is straightforward. The time for steps 2 and 7 is constant. Steps 1 and 5 are based on merge sort which is known to work in $O(n \log n)$ time. Because $X$ and $Y$ are already sorted, steps 3, 4, and 6 can be computed in linear time with respect to the number of visual words $n$. Consequently, the time complexity of Algorithm 8 is $O(n \log n)$. It is worth mentioning that open source MATLAB and C/C++ implementations of this algorithm are provided at http://pq-kernel.herokuapp.com.

## 5.4 Spatial Non-Alignment Kernel

A simple yet powerful framework for including spatial information into the BOVW model is presented next. This framework is termed *Spatial Non-Alignment Kernel* (SNAK) and it is based on measuring the spatial non-alignment of visual words in two images using a kernel function. To some extent, it is similar the Local Patch

Dissimilarity, only that it replaces patches with visual words obtained by vector quantization. Thus, SNAK can be interpreted as a rather more elaborate extension of rank distance (Dinu and Manea 2006) to images, so its roots are in string and text processing.

In the SNAK framework, additional information for each visual word needs to be stored first in the feature representation of an image. More precisely, the average position and the standard deviation of the spatial distribution of all the descriptors that belong to a visual word are computed. These statistics are computed independently for each of the two image coordinates. The SNAK feature vector includes the average coordinates and the standard deviation of a visual word together with the frequency of the visual word, resulting in a feature space that is five times greater than the original feature space corresponding to the histogram representation. The size of the feature space is identical to a spatial pyramid based on two levels, but it is roughly four times smaller than a spatial pyramid based on three levels.

Let $U$ represent the SNAK feature vector of an image. For each visual word at an index $i$, $U$ will contain 5-tuples as defined below:

$$u(i) = \left(h^u(i), m^u_x(i), m^u_y(i), s^u_x(i), s^u_y(i)\right).$$

The first component of $u(i)$ represents the visual word's frequency. The following two components ($m_x(i)$ and $m_y(i)$) represent the mean (or average) position of the $i$-th visual word on each of the two coordinates $x$ and $y$, respectively. The last two components ($s_x(i)$ and $s_y(i)$) represent the standard deviation of the $i$-th visual word with respect to the two coordinates $x$ and $y$. If the visual word $i$ does not appear in the image ($h^u(i) = 0$), the last four components are undefined. In fact, these four values are not being used at all, if $h^u(i) = 0$.

Using the above notations, the SNAK kernel between two feature vectors $U$ and $V$ can be defined as follows:

$$k_{\text{SNAK}}(U, V) = \sum_{i=1}^{n} \exp\left(-c_1 \cdot \Delta_{\text{mean}}(u(i), v(i))\right) \cdot \exp\left(-c_2 \cdot \Delta_{std}(u(i), v(i))\right), \tag{5.1}$$

where $n$ is the number of visual words, $c_1$ and $c_2$ are two parameters with positive values, $u(i)$ is the 5-tuple corresponding to the $i$-th visual word from $U$, $v(i)$ is the 5-tuple corresponding to the $i$-th visual word from $V$, and $\Delta_{\text{mean}}$ and $\Delta_{std}$ are defined as follows:

$$\Delta_{\text{mean}}(u, v) = \begin{cases} \left(m^u_x - m^v_x\right)^2 + \left(m^u_y - m^v_y\right)^2, & \text{if } h^u, h^v > 0 \\ \infty, & \text{otherwise} \end{cases}$$

$$\Delta_{std}(u, v) = \begin{cases} \left(s^u_x - s^v_x\right)^2 + \left(s^u_y - s^v_y\right)^2, & \text{if } h^u, h^v > 0 \\ \infty, & \text{otherwise} \end{cases}$$

where $m_x$, $m_y$, $s_x$, and $s_y$ are components of the 5-tuples $u$ and $v$. If a visual word does not appear in at least one of the two compared images, its contribution to $k_{\text{SNAK}}$ is zero, since $\Delta_{\text{mean}}$ and $\Delta_{\text{std}}$ are infinite.

It can be easily demonstrated that SNAK is a kernel function. Indeed, the proof that $k_{\text{SNAK}}$ is a kernel comes out immediately from the following observation. For a given visual word $i$ and two 5-tuples $u$ and $v$, the equations below represent two RBF kernels:

$$\exp\left(-c_1 \cdot \Delta_{\text{mean}}(u(i), v(i))\right)$$
$$\exp\left(-c_2 \cdot \Delta_{\text{std}}(u(i), v(i))\right),$$

and their product is also a kernel. By summing up the RBF kernels corresponding to all the 5-tuples inside the SNAK feature vectors $U$ and $V$, the $k_{\text{SNAK}}$ function is obtained. From the additive property of kernel functions given in Proposition 2, it results that $k_{\text{SNAK}}$ is also a kernel function.

An interesting remark is that $k_{\text{SNAK}}$ can be seen as a sum of separate kernel functions, each corresponding to a visual word that appears in both images. This is a fairly simple approach that can be easily generalized and combined with many other kernel functions. The following equation shows how to combine SNAK with another kernel $k^*$ that takes into account the frequency of visual words:

$$\begin{aligned} k^*_{\text{SNAK}}(U, V) = \sum_{i=1}^{n} & k^*(h^u(i), h^v(i)) \\ & \cdot \exp\left(-c_1 \cdot \Delta_{\text{mean}}(u(i), v(i))\right) \cdot \exp\left(-c_2 \cdot \Delta_{std}(u(i), v(i))\right). \end{aligned} \tag{5.2}$$

Equation (5.2) can be used to combine SNAK with other kernels at the visual word level individually. Certainly, using the above equation, SNAK can be combined with kernels such as the linear kernel, the Hellinger's kernel, or the intersection kernel. The following equation is a particularization of Eq. (5.2) for the intersection kernel:

$$\begin{aligned} k^{\cap}_{\text{SNAK}}(U, V) = \sum_{i=1}^{n} & \min\left\{h^u(i), h^v(i)\right\} \\ & \cdot \exp\left(-c_1 \cdot \Delta_{\text{mean}}(u(i), v(i))\right) \cdot \exp\left(-c_2 \cdot \Delta_{std}(u(i), v(i))\right). \end{aligned}$$

Moreover, being a kernel function, SNAK can be combined with any other kernel using various approaches specific to kernel methods, such as multiple kernel learning.

### 5.4.1 Translation and Size Invariance

Intuitively, the SNAK kernel measures the distance between the average positions of the same visual word in two images. SNAK can be used to encode spatial information for various classification tasks, but some improvements based on task-specific information are possible. One such example is object class recognition. If the objects appear in roughly the same locations, the SNAK approach would work fine. However, this restriction may be often violated in practice. Any object can appear in any part of the image, and a visual word describing some part of the object can therefore appear in a different location in each image. Due to this fact, SNAK is not invariant to translations of the object. If the object's location in each image is known a priori, the average position of the visual word can be computed with respect to the object's location, by translating the origin of the coordinate system over the center of the object. The exact location of the object is not known in practice, but it can be approximated using the *objectness* measure (Alexe et al. 2010, 2012). This measure quantifies how likely it is for an image window to contain an object. By sampling a reasonable number of windows and by accumulating their probabilities, a pixelwise objectness map of the image can be produced. The objectness map provides a meaningful distribution of the (interesting) image regions that indicate locations of objects. Furthermore, the center of mass of the objectness map provides a good indication of where the center of the object might be. The SNAK framework employs the objectness measure to determine the object's center in order to use it as the origin of the coordinate system of the image. The range of the coordinate system is normalized by dividing the $x$-axis coordinates by the width of the image and the $y$-axis coordinates by the height of the image. For each image, the coordinate system has a range from $-1$ to 1 on each axis. Normalizing the coordinates ensures that the average position or the standard deviation of a visual word do not depend on the image size, and it is a necessary step to reduce the effect of size variation in a set of images. The SNAK framework is illustrated in Fig. 5.2. Although the ideal conditions for SNAK would be to have a single object per image, this is rarely the case in practice, yet it still achieves impressive performance. Nonetheless, its performance would probably get even better if a class-specific object localization or detection framework would be used instead of the objectness measure, but SNAK would also become less generally applicable, in the sense that it would need class-specific information to work.

## 5.5 Object Recognition Experiments

Object recognition experiments presented in this section compare the SNAK and the PQ kernels with state of the art kernels and spatial representations on two benchmark data sets. First, a brief description of the data sets is given. Details about the implementation of the learning model and the evaluation procedure are given next. Finally, the results of the two kernels are separately discussed.

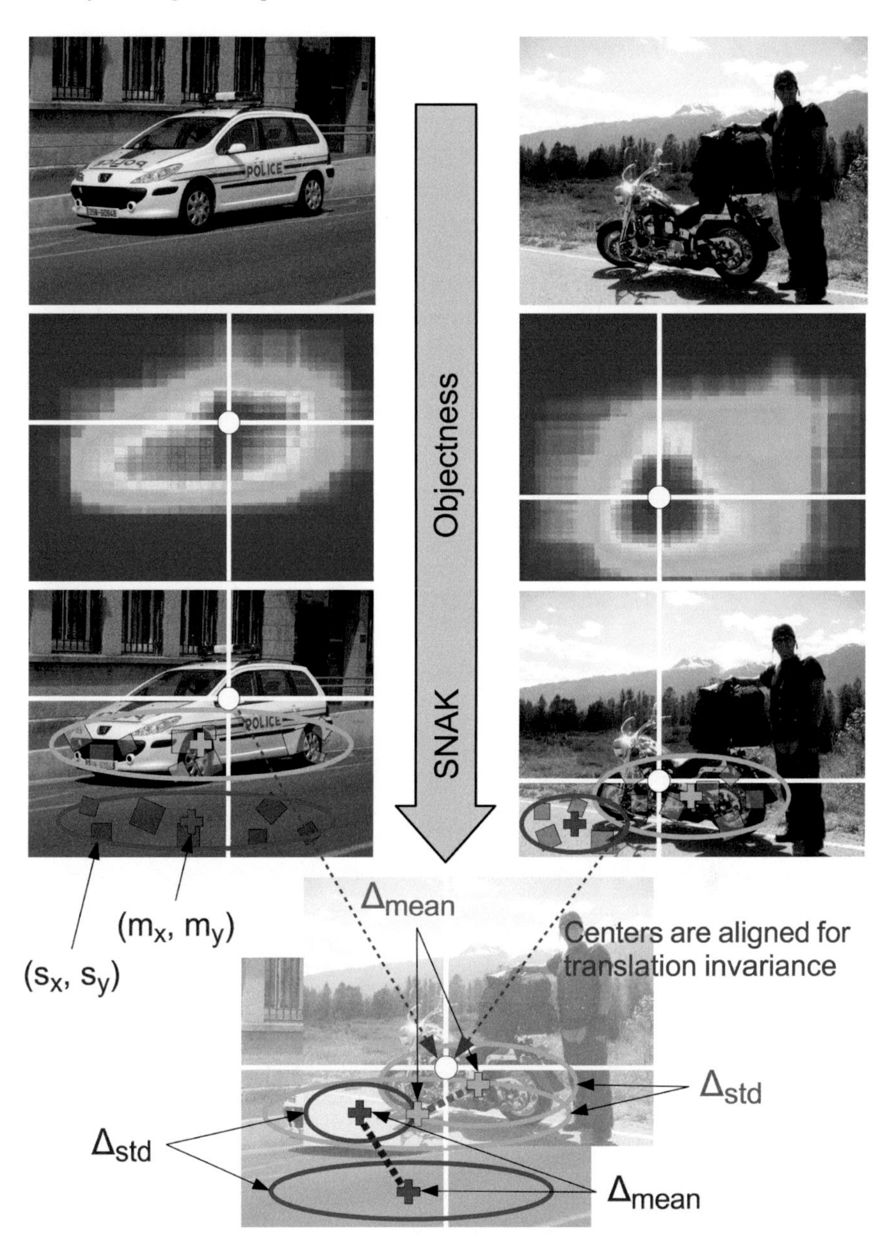


**Fig. 5.2** The spatial similarity of two images computed with the SNAK framework. First, the center of mass is computed according to the objectness map. The average position and the standard deviation of the spatial distribution of each visual word are computed next. The images are aligned according to their centers, and the SNAK kernel is computed by summing the distances between the average positions and the standard deviations of each visual word in the two images

### 5.5.1 Data Sets Description

The Pascal Visual Object Classes (VOC) challenge (Everingham et al. 2010) is a benchmark in visual object category recognition and detection, providing the vision and machine learning communities with a standard data set of images and annotation, and standard evaluation procedures. In the experiments of this work, the Pascal VOC 2007 data set is used. The reason for this choice is that this is the latest data set for which testing labels are available for download, and the experiments can be done offline. There are roughly 10 thousand images in this data set that are divided into 20 classes. As illustrated in Fig. 5.3, some images may contain objects from several classes. Thus, the class labels are not mutually exclusive. For each class, the data set provides a training set, a validation set, and a test set. The training and validation

**Fig. 5.3** A random sample of 12 images from the Pascal VOC data set. Some of the images contain objects of more than one class. For example, the image at the *top left* shows a dog sitting on a couch, and the image at the *top right* shows a person and a horse. Dog, couch, person, and horse are among the 20 classes of this data set

**Fig. 5.4** A random sample of 12 images from the Birds data set. There are two images per class. Images from the same class sit next to each other in this figure

sets have roughly 2500 images each, while the test set has about 5000 images. This data set is available at http://host.robots.ox.ac.uk/pascal/VOC/voc2007/index.html.

The second data set was collected from the Web by Lazebnik et al. (2005) and consists of 100 images each of 6 different classes of birds: egrets, mandarin ducks, snowy owls, puffins, toucans, and wood ducks. The training set consists of 300 images and the test set consists of another 300 images. For each class, the data set contains 50 positive train images and 50 positive test images. The purpose of using this data set is to assess the behavior of the proposed kernels in the context of fine-grained object recognition. Figure 5.4 shows two images from each class of the Birds data set. The data set is available at http://www-cvr.ai.uiuc.edu/ponce_grp/data/.

### 5.5.2 *Implementation and Evaluation Procedure*

Details about the particularities of the learning framework are given next. In the feature detection and representation step, a variant of dense SIFT descriptors extracted at multiple scales is used (Bosch et al. 2007). The implementation of the BOVW model is mostly based on the VLFeat library (Vedaldi and Fulkerson 2008).

The PQ kernel and the SNAK approach are independently evaluated. First of all, a set of experiments are conducted to compare various state-of-the-art kernels with the PQ kernel. The kernels proposed for this evaluation are the $L_2$-normalized linear kernel, the $L_1$-normalized Hellinger's kernel, the $L_1$-normalized intersection kernel, the $L_1$-normalized Jensen–Shannon kernel, and the $L_2$-normalized PQ kernel. In order to determine which kernel can better leverage the use of spatial information, a two-level spatial pyramid is combined with each of these kernels.

In a second set of experiments, the SNAK framework is compared with the spatial pyramid. Three kernels are proposed for evaluation, namely the $L_2$-normalized linear kernel, the $L_1$-normalized Hellinger's kernel, and the $L_1$-normalized intersection kernel. An important remark is that the intersection kernel was particularly chosen because it yields very good results in combination with the spatial pyramid according to Lazebnik et al. (2006), and it might work equally well in the SNAK framework. The three kernels proposed for evaluation are based on four different representations, three of which include spatial information. The goal of the experiments is to compare the bag of words representation with a spatial pyramid based on two levels, a spatial pyramid based on three levels, and the SNAK feature vectors. The spatial pyramid based on two levels combines the full image with $2 \times 2$ bins, and the spatial pyramid based on three levels combines the full image with $2 \times 2$ and $4 \times 4$ bins. In the SNAK framework, the linear kernel, the Hellinger's kernel, and the intersection kernel are used in turn as $k^*$ in Eq. (5.2). It is worth noting that SNAK can also be indirectly compared with the approach described in (Krapac et al. 2011), since the results reported by Krapac et al. (2011) are very similar to the spatial pyramid based on three levels.

The norms of all the evaluated kernels are chosen according to Vedaldi and Zisserman (2010), that state that $\gamma$-homogeneous kernels should be $L_\gamma$-normalized. Furthermore, it is important to mention that all these kernels are used in the dual form, that implies using the *kernel trick* to directly build kernel matrices of pairwise similarities between samples. Since the kernel trick is employed in the evaluation, it comes natural to obtain the spatial pyramid for each kernel by summing up kernel matrices obtained for each level of the pyramid. The spatial pyramid representation is usually obtained as a concatenation of visual word histograms, but in the dual representation, concatenating feature vectors is equivalent to summing up kernel matrices.

In all the experiments, the training is always done using Support Vector Machines (SVM). On the Birds data set, the SVM classifier based on the one-versus-all scheme is used for the multi-class task. The SNAK approach employs the objecteness measure to align images. The objectness measure is trained on 50 images that are neither from the Pascal VOC data set nor from the Birds data set. The objectness map is obtained by sampling 1000 windows using the NMS sampling procedure (Alexe et al. 2012). The source code used to generate the objectness heat maps is available online at http://groups.inf.ed.ac.uk/calvin/objectness/.

The experiments are conducted using 500, 1000, and 3000 visual words, respectively. The evaluation procedure on the Pascal VOC data set follows the Pascal VOC benchmark. As such, the qualitative performance of the learning model is measured by using the classifier score to rank all the test images. Next, the retrieval performance can be measured by computing a precision-recall curve. In order to represent the retrieval performance by a single number (rather than a curve), the mean average precision (mAP) is often computed. The average precision as defined by TREC is used in the Pascal VOC experiments. This is the average of the precision observed each time a new positive sample is recalled. For the experiments performed on the Birds data set, the classification accuracy is used to evaluate the various kernels and spatial representations.

### 5.5.3 PQ Kernel Results on Pascal VOC Experiment

The first experiment with the PQ kernel is on the Pascal VOC 2007 data set. For each of the 20 classes, the data set provides a training set, a validation set, and a test set. The validation set is used to validate the regularization parameter $C$ of the SVM algorithm. After validating the regularization parameter, the classifier is trained one more time on both the training and the validation sets, that have roughly 5000 images together. Table 5.1 presents the mean AP of various BOVW models obtained on the test set, by combining different vocabulary dimensions, representations, and kernels. For each model, the reported mAP represents the average score on all the 20 classes of the Pascal VOC data set.

The accuracy of the state-of-the-art kernels is well above the accuracy of the baseline linear kernel. In terms of AP, the state of the art kernels are roughly 10–15 % better than the baseline method. Among the state-of-the-art kernels, the PQ kernel gives consistently better results. Considering the standard histogram representation, the mAP of the PQ kernel is almost always 2 % above the mAP of the JS kernel, which seems to be the second-best kernel after PQ. The results of the Hellinger's and the intersection kernels are roughly 1 % lower than those of the JS kernel. Regarding the spatial pyramid representation, the results of the various kernels are significantly better compared to the results based on the standard representation. The results of each kernel improve by roughly 4 % when the spatial pyramid is used, but the score

**Table 5.1** Mean AP on Pascal VOC 2007 data set for SVM based on different kernels

| Representation | Vocabulary (words) | Lin. $L_2$ (%) | Hel. $L_1$ (%) | Int. $L_1$ (%) | JS $L_1$ (%) | PQ $L_2$ (%) |
|---|---|---|---|---|---|---|
| Histogram | 500 | 28.59 | 39.06 | 39.11 | 41.13 | **43.95** |
| Histogram | 1000 | 28.71 | 42.28 | 42.99 | 44.27 | **46.50** |
| Histogram | 3000 | 28.96 | 45.23 | 46.97 | 47.36 | **49.58** |
| Spatial pyramid | 500 | 31.17 | 44.21 | 45.17 | 46.32 | **48.54** |
| Spatial pyramid | 1000 | 31.38 | 46.94 | 48.27 | 48.45 | **50.59** |
| Spatial pyramid | 3000 | 31.85 | 49.21 | 50.78 | 50.70 | **52.35** |

Results of the kernels based on the standard histogram representation are given in the top half, and those of the kernels based on the spatial pyramid representation are given in the bottom half of the table. The best AP for each vocabulary dimension is highlighted in bold

difference between each pair of kernels remains fairly similar to the score difference that can be observed on the standard histogram representation. The mAP of the PQ kernel is always above the mAP of the JS kernel, but the difference between the scores of the two kernels is slightly attenuated. The intersection kernel seems to be able to take a greater advantage of the spatial pyramid representation, as it obtains similar results to the JS kernel.

Almost all the evaluated kernels obtain increasingly better results from 500 visual words to 1000 words, and from 1000 to 3000 words. The only exception seems to be the linear kernel, which shows a very small improvement. Overall, the best results are obtained by the SVM classifier based on the PQ kernel with a vocabulary of 3000 words. For the standard representation, the best mAP of the PQ kernel is 49.58 %, which represents an improvement of 2.22 % over the second-best score, which is obtained by the SVM based on the JS kernel. For the spatial pyramid representation, the best mAP of the PQ kernel is 52.35 %, which represents an improvement of 1.57 % over the second-best score, which is obtained by the SVM based on the intersection kernel. The empirical results on the Pascal VOC data set show that the PQ kernel is consistently better at object class recognition than the other evaluated kernels.

Even if the overall results point towards a single conclusion, it is still worthy to analyze some detailed results that can offer more information about these kernels. To server this purpose, the mean AP per class obtained by the SVM classifier using a spatial pyramid based on 3000 words and various kernels is presented in Table 5.2. When the best AP per class is considered, the PQ kernel wins most of the classes. Indeed, the PQ kernel gives the best AP on 14 out of 20 classes. The JS kernel wins 2 classes, namely the *Airplane* and the *Potted Plant* classes. Despite the fact that the intersection kernel has a lower overall accuracy than the JS kernel, it is able to win 4 of the 20 classes. An interesting remark is that the intersection kernel gives the best AP on the *Bottle* class, which seems to be the most difficult class in the Pascal VOC data set. It is worth noting that the linear kernel and the Hellinger's kernel are not able to take any class. Another interesting fact is that on 4 of the classes the PQ kernel is able to give results that are at least 3 % higher (in terms of AP) than any of the results obtained by the other kernels. These are the *Bicycle* class, the *Cow* class, the *Sheep* class, and the *TV Monitor* class, respectively. On the other hand, the PQ kernel gives significantly lower results for only one class, namely the *Dining Table* class. However, the best AP of the PQ kernel on this class is no more than 3 % lower than that of the intersection kernel, which wins the *Dining Table* class. The overall results presented in Table 5.2 point towards the conclusion that the PQ kernel is a better choice than the other state of the art kernels, since it gives the best AP on most of the classes and it also obtains the best AP averaged on all the 20 classes of the Pascal VOC data set.

Table 5.3 provides the time required by each kernel to produce the two kernel matrices for training and testing, respectively. The kernel matrix for training contains pairwise similarities of the training samples, while the kernel matrix for testing contains pairwise similarities between the test samples and the training samples. Thus, each kernel matrix has roughly 5000 rows and 5000 columns, since both the

**Table 5.2** Mean AP on the 20 classes of the Pascal VOC 2007 data set for the SVM classifier based on 3000 visual words using the spatial pyramid representation and different kernels

| Class | Lin. $L_2$ (%) | Hel. $L_1$ (%) | Int. $L_1$ (%) | JS $L_1$ (%) | PQ $L_2$ (%) |
|---|---|---|---|---|---|
| Airplane | 37.88 | 68.20 | 68.10 | **70.25** | 69.29 |
| Bicycle | 26.06 | 52.30 | 50.63 | 52.34 | **55.63** |
| Bird | 19.92 | 33.43 | **39.70** | 37.18 | 36.69 |
| Boat | 36.80 | 62.29 | 59.76 | 61.82 | **62.41** |
| Bottle | 11.74 | 16.70 | **19.03** | 16.26 | 18.74 |
| Bus | 31.97 | 53.41 | 54.37 | 54.75 | **56.46** |
| Car | 56.47 | 70.47 | 74.03 | 72.45 | **75.91** |
| Cat | 28.13 | 52.67 | 51.71 | 53.10 | **53.87** |
| Chair | 31.27 | 50.62 | 51.26 | 51.18 | **52.25** |
| Cow | 14.90 | 29.96 | 31.35 | 31.34 | **37.09** |
| Dining table | 29.68 | 41.09 | **47.21** | 44.06 | 44.31 |
| Dog | 20.70 | 33.07 | 34.59 | 34.39 | **36.63** |
| Horse | 62.81 | 73.38 | **74.40** | 74.25 | 73.54 |
| Motorbike | 28.94 | 59.67 | 60.73 | 61.31 | **63.42** |
| Person | 66.74 | 77.98 | 80.66 | 79.51 | **81.59** |
| Potted plant | 14.07 | 17.27 | 17.63 | **17.89** | 17.64 |
| Sheep | 23.69 | 33.41 | 36.37 | 38.27 | **41.28** |
| Sofa | 20.58 | 46.12 | 46.94 | 47.40 | **49.76** |
| Train | 49.78 | 68.28 | 71.93 | 71.18 | **72.35** |
| TV monitor | 24.93 | 43.91 | 45.14 | 45.15 | **48.10** |
| Overall | 31.85 | 49.21 | 50.78 | 50.70 | **52.35** |

The best AP for each class is highlighted in bold

**Table 5.3** Running time required by each kernel to compute the two kernel matrices for training and testing, respectively

| Representation | Vocabulary (words) | Lin. $L_2$ | Hel. $L_1$ | Int. $L_1$ | JS $L_1$ | PQ $L_2$ |
|---|---|---|---|---|---|---|
| Histogram | 500 | 8.9 | 9.1 | 330.2 | 1140.3 | 2797.4 |
| Histogram | 1000 | 9.7 | 9.7 | 441.9 | 2098.5 | 5703.2 |
| Histogram | 3000 | 12.0 | 12.2 | 869.7 | 6329.6 | 15882.6 |
| Spatial pyramid | 500 | 19.5 | 19.9 | 980.4 | 4817.6 | 11456.8 |
| Spatial pyramid | 1000 | 22.9 | 23.1 | 1510.4 | 8534.9 | 22335.4 |
| Spatial pyramid | 3000 | 35.8 | 36.3 | 3496.1 | 25812.5 | 67462.0 |

The time is reported separately for each kernel and for each vocabulary dimension. The time is measured in seconds

training and the test sets consist of roughly 5000 images each. The time was measured on a computer with Intel Core i7 2.3 GHz processor and 8 GB of RAM memory using a single Core. Despite of using an algorithm that requires $O(n \log n)$ time to compute the PQ kernel, it is easy to observe that the PQ kernel is still the most computationally expensive kernel among the evaluated kernels. However, the running time of the PQ kernel is now comparable to that of the JS kernel. More precisely, the PQ kernel is less than three times slower than the state of the art JS kernel, even if the JS kernel between two histograms requires $O(n)$ time. In conclusion, Algorithm 8 brings a significant speed improvement and makes the PQ kernel practical for vocabularies of more than 1000 words.

### 5.5.4 PQ Kernel Results on Birds Experiment

The second experiment with the PQ kernel is on the Birds data set. As there is no validation set provided with this data set, the regularization parameter of SVM is cross-validated on the training set. Table 5.4 presents the classification accuracy of various BOVW models obtained on the test set, by combining different vocabulary dimensions, representations, and kernels.

The results of the PQ kernel on this experiment are consistent with the previous experiment, in that the PQ kernel outperforms the other kernels again. For the standard histogram representation, the accuracy rate of the PQ kernel is almost always 4–6 % above the accuracy rates of the other state of the art kernels. For the spatial pyramid representation, the PQ kernel accuracy rate stays 3–6 % above the accuracy rates of the state-of-the-art kernels. The results of each kernel improve by roughly 3–4 % when spatial pyramids are used instead of standard histograms. All the evaluated kernels obtain increasingly better results from 500 visual words to 1000 words, and from 1000 to 3000 words.

Overall, the results of the Hellinger's kernel are comparable to those of the JS kernel, while the results of the intersection kernel are considerably lower than those of the Hellinger's and the JS kernels, especially for larger vocabularies of visual words. The best results are obtained by the PQ kernel based on a vocabulary of 3000

**Table 5.4** Classification accuracy on the Birds data set for SVM based on different kernels

| Representation | Vocabulary (words) | Lin. $L_2$ (%) | Hel. $L_1$ (%) | Int. $L_1$ (%) | JS $L_1$ (%) | PQ $L_2$ (%) |
|---|---|---|---|---|---|---|
| Histogram | 500 | 59.67 | 72.00 | 70.00 | 70.00 | **76.67** |
| Histogram | 1000 | 64.67 | 78.33 | 71.00 | 77.00 | **81.00** |
| Histogram | 3000 | 69.33 | 80.33 | 74.67 | 79.67 | **84.33** |
| Spatial pyramid | 500 | 62.67 | 75.67 | 74.00 | 78.67 | **81.67** |
| Spatial pyramid | 1000 | 66.67 | 79.33 | 74.33 | 79.33 | **84.00** |
| Spatial pyramid | 3000 | 69.67 | 81.00 | 77.00 | 82.00 | **88.33** |

Results of the kernels based on the standard histogram representation are given in the top half, and those of the kernels based on the spatial pyramid representation are given in the bottom half of the table. The best accuracy for each vocabulary dimension is highlighted in bold words. For the standard representation, the accuracy rate of the SVM based on the PQ kernel is 84.33 %, which represents an improvement of 4.00 % over the second-best accuracy rate, which is obtained by the SVM based on the Hellinger's kernel. For the spatial pyramid representation, the accuracy rate of the SVM based on the PQ kernel is 88.33 %, which represents an improvement of 6.33 % over the second-best score, which is obtained by the SVM based on the JS kernel. The empirical results on the Birds data set show that the PQ kernel is significantly better at fine-grained object class recognition than the other state-of-the-art kernels.

### *5.5.5 SNAK Parameter Tuning*

The SNAK framework takes both the average position and the standard deviation of each visual word into account. In a set of preliminary experiments performed on the Birds data set, the two statistics were used independently to determine which one brings a more significant improvement. The empirical results demonstrated that they roughly achieve similar accuracy improvements, having an almost equal contribution to the proposed framework. Consequently, a decision was made to use the same value for the two constants $c_1$ and $c_2$ from Eq. (5.1). Only five values in the range 1–100 were chosen for preliminary evaluation. The best results were obtained with $c_1 = c_2 = 10$, while choices like 5 or 50 were only 2–3 % behind. Finally, a decision was made to use $c_1 = c_2 = 10$ in the experiments reported next, but it is very likely that better results can be obtained by fine-tuning the parameters $c_1$ and $c_2$ on each data set. An important remark is that $c_1$ and $c_2$ were tuned on the Birds data set, but the same choice was used on the Pascal VOC data set, without testing other values. Good results on Pascal VOC might indicate that $c_1$ and $c_2$ do not necessarily depend on the data set, but rather on the normalization procedure used for the spatial coordinate system. It is interesting to note that the two coordinates are independently normalized as described in Sect. 5.4.1, resulting in small distortions along the axes. Two other methods of size-normalizing the coordinate space without introducing distortions were also evaluated. One is based on dividing both coordinates by the diagonal of the image, and the other by the mean of the width and height of the image. Perhaps surprisingly, these have produced lower average precision scores on a subset of the Pascal VOC data set. For instance, size-normalizing by the mean of the width and height gives a mAP score that is roughly 0.5 % lower than normalizing each axis independently by the width and height.

In the Pascal VOC experiment, the validation set is used to validate the regularization parameter $C$ of the SVM algorithm. In the Birds experiment, the parameter $C$ was adjusted by cross-validation on the training set.

### 5.5.6 SNAK Results on Pascal VOC Experiment

The SNAK framework is first evaluated on the Pascal VOC 2007 data set. For each of the 20 classes, the data set provides a training set, a validation set and a test set. After validating the regularization parameter of the SVM algorithm on the validation set, the classifier is trained one more time on both the training and the validation sets, that have roughly 5000 images together.

Table 5.5 presents the mean AP of various BOVW models obtained on the test set, by combining different spatial representations, vocabulary dimensions, and kernels. For each model, the reported mAP represents the average score on all the 20 classes of the Pascal VOC data set. The results presented in Table 5.5 clearly indicate that spatial information significantly improves the performance of the BOVW model. This observation holds for every kernel and every vocabulary dimension. Indeed, the spatial pyramid based on two levels shows a performance increase that ranges between 3 % (for the linear kernel) and 6 % (for intersection kernel). As expected, the spatial pyramid based on three levels further improves the performance, especially for the linear kernel. When the $4 \times 4$ bins are added into the spatial pyramid, the mAP of the linear kernel grows by roughly 7–8 %, while the mAP scores of the other two kernels increase by 1–2 %. Among the three kernels based on spatial pyramids,

**Table 5.5** Mean AP on Pascal VOC 2007 data set for different representations that encode spatial information into the BOVW model

| Representation | Vocabulary (words) | Lin. $L_2$ (%) | Hel. $L_1$ (%) | Int. $L_1$ (%) |
|---|---|---|---|---|
| Histogram | 500 | 28.59 | 39.06 | 39.11 |
| Histogram | 1000 | 28.71 | 42.28 | 42.99 |
| Histogram | 3000 | 28.96 | 45.23 | 46.97 |
| Spatial pyramid (2 levels) | 500 | 31.17 | 44.21 | 45.17 |
| Spatial pyramid (2 levels) | 1000 | 31.38 | 46.94 | 48.27 |
| Spatial pyramid (2 levels) | 3000 | 31.85 | 49.21 | 50.78 |
| Spatial pyramid (3 levels) | 500 | 38.49 | 45.20 | 47.66 |
| Spatial pyramid (3 levels) | 1000 | 39.59 | 47.87 | 49.85 |
| Spatial pyramid (3 levels) | 3000 | 40.97 | 50.37 | 51.87 |
| SNAK | 500 | **42.56** | **47.39** | **49.75** |
| SNAK | 1000 | **44.69** | **49.54** | **51.99** |
| SNAK | 3000 | **45.95** | **52.49** | **54.05** |

For each representation, results are reported using several kernels and vocabulary dimensions. The best AP for each vocabulary dimension and each kernel is highlighted in bold the best mAP scores are obtained by the intersection kernel, which was previously reported to work best in combination with the spatial pyramid (Lazebnik et al. 2006).

The best results on the Pascal VOC data set are obtained by the SNAK framework. Indeed, the results are even better than the spatial pyramid based on three levels, which uses a representation that is more than four times greater than the SNAK representation. The mAP scores of the Hellinger's and the intersection kernels based on SNAK are roughly 2 % better than the mAP scores of the same kernels combined with the spatial pyramid based on three levels. On the other hand, a 4–5 % growth of the mAP score can be observed in case of the linear kernel. Among the three kernels, the best results are obtained by the intersection kernel. When the intersection kernel is combined with SNAK, the best overall mAP score is obtained, that is 54.05 %. This is 2.18 % better than the intersection kernel combined with the spatial pyramid based on three levels.

Overall, the empirical results indicate that the SNAK approach is significantly better than the state-of-the-art spatial pyramid framework, in terms of recognition accuracy. Perhaps this comes as a surprising result given that the images from the Pascal VOC data set usually contain multiple objects, and that SNAK implicitly assumes that there is a single relevant object in the scene, due to the use of the objecteness measure. The SNAK framework also provides a more compact representation, which brings improvements in terms of space and time over a spatial pyramid based on three levels, for example.

### *5.5.7 SNAK Results on Birds Experiment*

The SNAK framework is next evaluated on the Birds data set. Table 5.6 presents the classification accuracy of the BOVW model based on various representations that include spatial information. The results are reported on the test set, by combining different vocabulary dimensions and kernels.

The results of the SNAK framework on this data set are consistent with the results reported in the previous experiment, in that the SNAK framework yields better performance than the spatial pyramid representation. The spatial pyramid based on two levels improves the classification accuracy of the standard BOVW model by 3–4 %. On top of this, the spatial pyramid based on three levels further improves the performance. Significant improvements can be observed for the linear kernel and for the intersection kernel.

The spatial pyramid based on two levels shows little improvements over the histogram representation for the vocabulary of 3000 words, and more significant improvements for the vocabulary of 500 words. The certain fact is that the spatial information helps to improve the classification accuracy on this data set, but the best approach seems to be the SNAK framework. With only two exceptions, the SNAK framework gives better results than the spatial pyramid based on three levels. Compared to the spatial pyramid based on two levels, which has the same number of features, the SNAK approach is roughly 3–5 % better. An interesting observa-

**Table 5.6** Classification accuracy on the Birds data set for different representations that encode spatial information into the BOVW model

| Representation | Vocabulary (words) | Lin. $L_2$ (%) | Hel. $L_1$ (%) | Int. $L_1$ (%) |
|---|---|---|---|---|
| Histogram | 500 | 59.67 | 72.00 | 70.00 |
| Histogram | 1000 | 64.67 | 78.33 | 71.00 |
| Histogram | 3000 | 69.33 | 80.33 | 74.67 |
| Spatial pyramid (2 levels) | 500 | 62.67 | 75.67 | 74.00 |
| Spatial pyramid (2 levels) | 1000 | 66.67 | 79.33 | 74.33 |
| Spatial pyramid (2 levels) | 3000 | 69.67 | 81.00 | 77.00 |
| Spatial Pyramid (3 levels) | 500 | 68.33 | 76.67 | 76.00 |
| Spatial Pyramid (3 levels) | 1000 | 70.33 | **80.67** | 78.00 |
| Spatial Pyramid (3 levels) | 3000 | **73.00** | 82.67 | 79.67 |
| SNAK | 500 | **69.33** | **79.00** | **76.33** |
| SNAK | 1000 | **71.67** | 80.33 | **78.67** |
| SNAK | 3000 | 72.33 | **83.67** | **81.33** |

For each representation, results are reported using several kernels and vocabulary dimensions. The best accuracy for each vocabulary dimension and each kernel is highlighted in bold

tion is that the intersection kernel does not yield the best overall results as in the previous experiment, but it seems to gain a lot from the spatial information. For instance, the accuracy of the intersection kernel grows from 71.00 % with histograms to 78.67 % with SNAK, when the underlying vocabulary has 1000 words. The best accuracy (83.67 %) is obtained by the Hellinger's kernel combined with SNAK, using a vocabulary of 3000 visual words. When it comes to fine-grained object class recognition, the overall empirical results on the Birds data set indicate that the SNAK framework is more accurate than the spatial pyramid approach.

## 5.6 Bag of Visual Words for Facial Expression Recognition

In this section, several ways of improving and adapting the BOVW model for facial expression recognition are discussed. The improved BOVW model is evaluated on the data set used in the FER Challenge of the ICML 2013 WREPL Workshop. The experiments and the results obtained on the FER Challenge data set are presented in Sect. 5.8.

The bag of visual words model builds a vocabulary by vector quantizing local image features into visual words. For a particular image, the frequency of each visual word contained in the image is usually recorded in a histogram of visual words. For facial expression recognition, it seems that, according to common intuitions, the presence of a visual word is more important than its frequency. For example, it should be enough to detect the presence of a single cheek dimple to recognize a smiling face. Thus, instead of recording occurrence counts in a histogram, it is enough to record visual words presence in a presence vector. An important remark is that the presence vector should be normalized not to favor faces with more visual words.

The BOVW model proposed for facial expression recognition has two stages, one for training and one for testing. Each stage is divided into two major steps. The first step in both stages is for feature detection and representation. The second step is to train a kernel method (in the training stage) in order to predict the class label of a new image (in the testing stage). For each test image, a local classification problem is constructed by selecting only the nearest neighbors from the kernel feature space. The local learning part of the framework is further described in Sect. 5.7. The entire process, that involves both training and testing stages, is summarized in Fig. 5.5. It is interesting to notice the different steps from the model presented in Fig. 5.1.

The feature detection and representation step in the training stage is a slightly modified version of the approach described in Sect. 5.2, which uses k-means clustering (Leung and Malik 2001) to quantize dense SIFT descriptors (Bosch et al. 2007; Dalal and Triggs 2005) into visual words that are stored in a randomized forest of k-d trees (Philbin et al. 2007). The novelty consists of a semi-supervised approach, in the sense that the use of the SIFT descriptors from the unlabeled test set is leveraged by including these descriptors in the k-means clustering process as well, together with those from the training data set. For larger test sets this has led to a better representation of the faces manifold and to increased classification performance. The presence of each visual word is recorded in a presence vector which represents the final feature vector for the image. Normalized presence vectors of visual words can now enter the learning step. Figure 5.6 presents a sample of 30 SIFT descriptors extracted from two images of the FER data set.

Feature detection and representation is similar during the testing stage. The presence vector of visual words that represents the test image is compared with the training presence vectors, through the implicit distance defined by the kernel, to select a number of nearest neighbors. For a certain test image, only its nearest neighbors actually enter the learning step. In other words, a local recognition problem is built for each test image. A kernel method is employed to learn the local recognition problem and finally predict a class label for the test image. Classifiers such as the SVM, the KDA or the KRR are good choices to perform the local learning task.

The model described so far ignores spatial relationships between image features, but the performance improves when spatial information is included. This can be achieved by dividing the image into spatial bins. The presence of each visual word is then recorded in a presence vector for each bin. The final feature vector for the image is a concatenation of these presence vectors. A more robust approach is to use a spatial pyramid, as the work of Lazebnik et al. (2006) suggests. The spatial

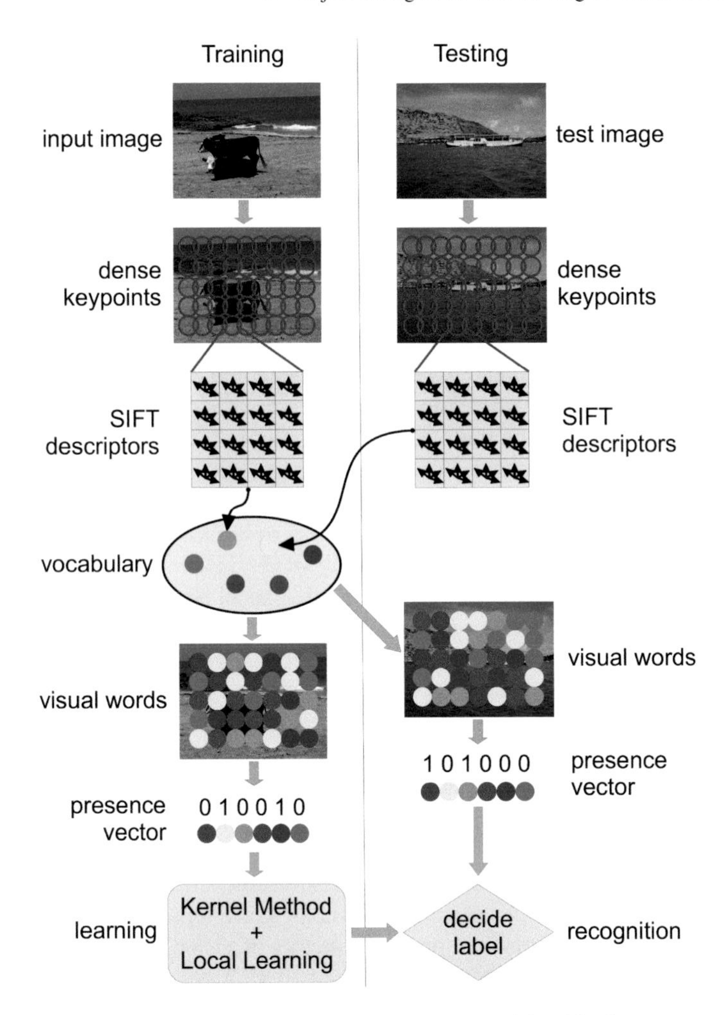


**Fig. 5.5** The BOVW learning model for facial expression recognition. The feature vector consists of SIFT features computed on a regular grid across the image (dense SIFT) and vector quantized into visual words. The presence of each visual word is then recorded in a presence vector. Normalized presence vectors enter the training stage. Learning is done by a local kernel method

pyramid is usually obtained by dividing the image into increasingly fine subregions (bins) and computing histograms of visual words found inside each bin.

The framework proposed in this section makes use of the spatial information by computing a spatial pyramid from presence vectors. It is reasonable to think that dividing an image representing a face into bins is a good choice, since most features, such as the contraction of the muscles at the corner of the eyes, are only visible in a certain region of the face. In other words, one does not expect to find raised eyebrows on the cheek, or cheek dimples on the forehead.

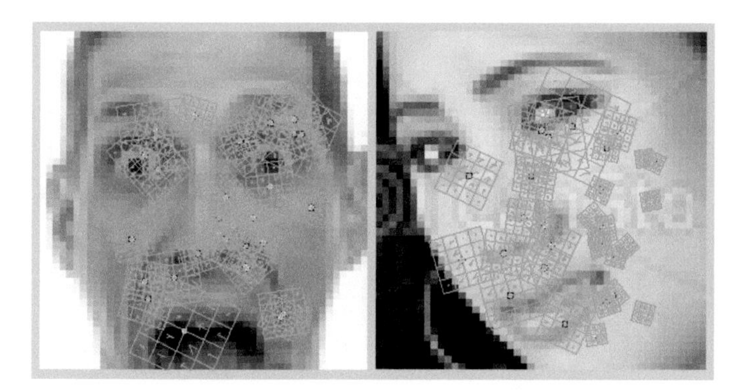

**Fig. 5.6** An example of SIFT features extracted from two images representing distinct emotions: fear (*left*) and disgust (*right*)

## 5.7 Local Learning

Local learning methods attempt to locally adjust the performance of the training system to the properties of the training set in each area of the input space. A simple local learning algorithm works as follows: for each test sample, the approach is to select a few training samples located in the vicinity of the test sample, train a classifier with only these few examples and apply the resulting classifier to the test sample. The generic local learning algorithm is presented in detail in Chap. 2.

The learning step of the BOVW framework adapted to facial expression recognition is based on a local learning algorithm that uses the presence kernel to select nearest neighbors in the vicinity of a test image. Local learning has a few advantages over standard learning methods. First, it divides a hard classification problem into more simple subproblems. Second, it reduces the variety of images in the training set, by selecting images that are most similar to the test one. Third, it improves accuracy for data sets affected by labeling noise. Considering all these advantages, a local learning algorithm is indeed suitable for the FER data set.

Figure 5.7 shows that the nearest neighbors selected from the vicinity of a particular test image are visually more relevant than a random selection of training images. It also gives a hint that local learning should perform better than a standard learning formulation.

## 5.8 Facial Expression Recognition Experiments

### *5.8.1 Data Set Description*

The data set of the FER Challenge consists of $48 \times 48$ pixel grayscale images of faces representing 7 categories of facial expressions: anger, disgust, fear, happiness,

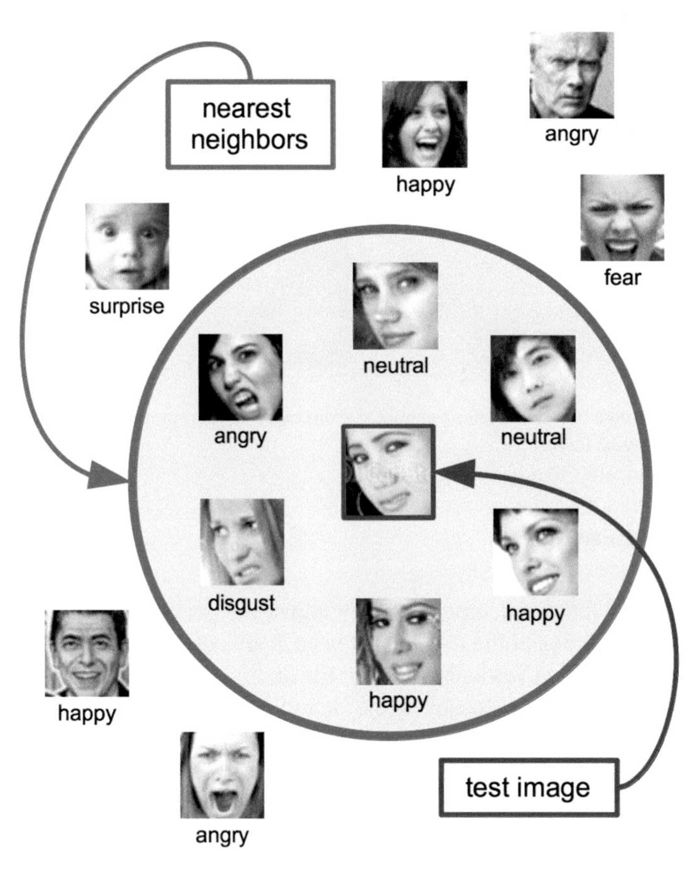


**Fig. 5.7** The six nearest neighbors selected with the presence kernel from the vicinity of the test image are visually more similar than the other six images randomly selected from the training set. Despite of this fact, the nearest neighbors do not adequately indicate the test label (disgust). Thus, a learning method needs to be trained on the selected neighbors to accurately predict the label of the test image

sadness, surprise, and neutral. There are 28709 examples for training, 3589 examples for testing, and another 3589 examples for private testing. The task is to categorize each face based on the emotion shown in the facial expression into one of the 7 categories. Images were collected from the web using a semi-automatic procedure. Therefore, the data set may contain images that do not represent faces. Another issue is that the training data may also contain labeling noise, meaning that the labels of some faces do not indicate the right facial expression.

### 5.8.2 *Implementation*

The framework described in Sect. 5.6 is used for facial expression recognition. Details about the implementation of the model are given next. In the feature detection and representation step, a variant of dense SIFT descriptors extracted at multiple scales is used (Bosch et al. 2007). The dense SIFT features are extracted using a grid step of 1 pixel at scales ranging from 2 to 8 pixels. These features are extracted from the entire FER data set. A random sample of 240,000 features is selected to compute the vocabulary. The number of visual words used in the experiments ranges from 5,000 to 20,000. A slight improvement in accuracy can be observed when the vocabulary dimension grows. Both histograms of visual words and presence vectors are tested in a set of preliminary experiments. The empirical results indicate that presence vectors are able to improve accuracy, by eliminating some of the noise encoded by the histogram representation.

Different spatial presence vectors are combined to obtain several spatial pyramid versions. Images are divided into $2 \times 2$ bins, $3 \times 3$ bins, and $4 \times 4$ bins to obtain spatial presence vectors. The basic presence vectors are also used in the computation of spatial pyramids.

The kernel trick is employed to obtain spatial pyramids. Kernel matrices are computed for each (spatial) presence vector representation. Then, kernel matrices are summed up to obtain the kernel matrix of a certain spatial pyramid. Some of the proposed models are based on a weighted sum of kernel matrices, with weights adjusted to match the accuracy level of each (spatial) presence vector. Summing up kernel matrices is equivalent to presence vector concatenation. But presence vectors are actually high-dimensional sparse vectors, and the concatenation of such vectors is not a viable solution in terms of space and time.

Several state of the art kernel methods are used to perform the local learning tasks, namely the SVM, the KDA and the KRR. The empirical results indicate that the SVM performs slightly better than the KDA, and much better than the KRR. The number of nearest neighbors selected to enter the local learning phase for each test image is 1000. However, an experiment is conducted to show the accuracy level as the number of nearest neighbors varies. The regularization parameter $C$ of the SVM was set to $10^3$. This choice is motivated by the fact that the data set is separable since there is a small number of training examples (1000 neighbors), in a high-dimensional feature space. Thus, the best working SVM is a hard margin SVM that can be obtained by setting the $C$ parameter of the SVM to a high value (Shawe-Taylor and Cristianini 2004).

### 5.8.3 *Parameter Tuning and Results*

For parameter tuning and validation, the training set was randomly split into two thirds kept for training and one third for validation. Preliminary experiments using

**Table 5.7** Accuracy levels for several BOVW models obtained on the FER validation, test, and private test sets

| Model | Neighbors | Validation (%) | Global SVM (%) | Test (%) | Private (%) |
|---|---|---|---|---|---|
| 8000 1 × 1 | 1000 | 59.07 | | | |
| 8000 2 × 2 | 1000 | 62.22 | | | |
| 8000 3 × 3 | 1000 | 62.27 | | | |
| 8000 4 × 4 | 1000 | 62.93 | | | |
| 8000 SUM | 1000 | 63.27 | | 65.73 | 66.73 |
| 17000 1 × 1 | 1000 | 60.86 | | | |
| 14000 2 × 2 | 1000 | 62.69 | | | |
| 11000 3 × 3 | 1000 | 62.36 | | | |
| MIX1 | 1000 | 63.03 | | 65.89 | |
| MIX2 | 1000 | 63.74 | | 66.42 | |
| MIX3 | 1000 | 63.61 | | 66.65 | |
| Rebuilt models | | | | | |
| MIX1 | 1000 | 62.91 | 59.35 | 66.59 | 66.73 |
| MIX2 | 1000 | 63.99 | 60.95 | 67.01 | 67.31 |
| MIX3 | 1000 | 64.30 | 62.27 | **67.32** | **67.48** |
| MIX3 | 3000 | 64.23 | 62.27 | | |
| MIX4 | 500 | 63.59 | 61.82 | | |
| MIX4 | 1000 | 63.90 | 61.82 | | |
| MIX4 | 1500 | 64.10 | 61.82 | 66.53 | 66.98 |
| MIX4 | 3000 | **64.45** | 61.82 | | |
| MIX4 | 5000 | 64.35 | 61.82 | | |

The best accuracy on each set is highlighted in bold

different models are performed on the validation set to assess the performance levels of the kernel methods. The obtained results indicate that the SVM is about 1–2 % better than the KRR, and about 0.5 % better than KDA. They also indicate that presence vectors are 0.5–1 % better than histograms. In the experiments presented in Table 5.7, only the results obtained with various SVM models based on presence vectors are included.

Several kernels based on different vocabularies and various ways of encoding spatial information were proposed for the FER Challenge. The model names of the form 8000 3 × 3 specify the size of the vocabulary, followed by the size of the grid used to partition the image into spatial bins. The kernel of 8000 *SUM* is the mean of the kernels based on 8000 visual words computed on spatial bins ranging from 1 × 1 bins to 4 × 4 bins. In other words, the 8000 SUM model is based on spatial pyramids. It performs better than each of its terms on the validation set. The kernel identified by MIX3 is the mean of 17000 1 × 1, 14000 2 × 2, 11000 3 × 3, and 8000 4 × 4. Again, it performs betterthan each of its individual components. The kernel identified by MIX4 is a variant of MIX3 built on even larger vocabularies, as it is given by the mean of 20000 $1 \times 1$, 20000 $2 \times 2$, 12000 $3 \times 3$, and 8000 $4 \times 4$. Using the same number of neighbors, it does not perform better than MIX3, neither on the validation data set nor on the actual test set. The kernel identified as MIX1 is the weighted mean of 7000 $1 \times 1$, 7000 $2 \times 2$, 7000 $3 \times 3$, and 5000 $4 \times 4$, with the weights 0.1, 0.2, 0.4, 0.3, respectively. Finally, the last proposed model is the kernel identified as MIX2, which is the weighted mean of 11000 $1 \times 1$, 9000 $2 \times 2$, 7000 $3 \times 3$, and 5000 $4 \times 4$, with the weights 0.2, 0.3, 0.3, 0.2, respectively. The vector quantization process of these MKL models initially included the descriptors extracted from the 28709 examples for training and the 3589 examples for preliminary testing. Some of the best performing models were rebuilt to add the 3589 examples from the private testing set in the vector quantization process.

The empirical results presented in Table 5.7 indicate that the performance of the global one-versus-all SVM is at least 2 % lower than that obtained with the local learning based on a one-versus-all SVM with the same parameters. Another interesting behavior that can be observed in this table is the effect on accuracy caused by dividing the image area into spatial bins, namely the accuracy increases as the image is divided into finer subregions. This table also shows the performance impact of the number of neighbors, another parameter that must be properly adjusted.

No model with more than 1500 neighbors was submitted to the FER Challenge, yet it may well be that using 3000 neighbors could have led to somewhat higher scores. The parameter tuning was limited both by the amount of RAM available in the machines (24 GB for the largest one) used to train the models, and by the speed of the CPUs (4-core Xeon E5540 at 2.53 GHz in the fastest one). Test cycles took therefore up to 9 hours. The best accuracy on the final test set is 67.484 %. An interesting remark is that the proposed model has roughly achieved human-level performance on this data set (Goodfellow et al. 2015).

## 5.9 Discussion

This chapter discussed several improvements of the BOVW model either for object recognition or for facial expression recognition. First, the work presented in this chapter showed that the results for image classification, image retrieval, or related tasks can be improved by using the PQ kernel. The PQ kernel comes from the idea of treating feature vectors of visual words as ordinal variables. Object recognition experiments compared the PQ kernel with other state of the art kernels on two benchmark data sets. The PQ kernel has the best accuracy on both data sets, being constantly better than the other methods. Nonetheless, the accuracy can be improved even further by increasing the number of visual words used for building the vocabulary. Because the size of the PQ feature map has a quadratic dependence on the number of visual words, the vocabulary size should be adjusted by taking into consideration the trade-off between accuracy and speed. Despite the fact that the kernel trick can be employed to use the PQ kernel with larger vocabularies, a greater importance should be given to the idea of selecting good visual words. A possible way of improving the results for the PQ kernel may be that of using a TF-IDF measure for visual words as suggested by Philbin et al. (2007). Indeed, eliminating visual words that have a low TF-IDF score can lead to an approximation of the PQ kernel that works faster and possibly better.

Another contribution described in this chapter is an approach to improve the BOVW model by encoding spatial information in a more efficient way than spatial pyramids, by using a kernel function termed SNAK. More precisely, SNAK includes the spatial distribution of the visual words in the similarity of two images. The empirical results indicate that the SNAK framework can improve the object recognition accuracy over the spatial pyramid representation. Considering that SNAK uses a more compact representation, the results become even more impressive. In conclusion, SNAK has all the ingredients to become a viable alternative to the spatial pyramid approach.

In this work, the objectness measure was used to add some level of translation invariance into the SNAK framework. In future work, the SNAK framework can be further improved by including ways of obtaining scale and rotation invariance. Ground truth information about an object's scale can be obtained from manually annotated bounding boxes. A first step would be to use such bounding boxes to determine if it helps to compare objects at the same scale with the SNAK kernel. Another direction, is to extend the SNAK framework to use the valuable information offered by objectness (Alexe et al. 2010), which is only barely used in the current framework.

Other improvements to the bag of visual words were proposed in order to make the model suitable for the FER Challenge data set of faces. Histograms of visual words were replaced with normalized presence vectors, then local learning was used to predict class labels of test images. The proposed model also includes spatial information in the form of spatial pyramids computed from presence vectors. Experiments were performed to validate the proposed model. By reserving one third of the training data set as validation set, the method's parameters were tuned without overfitting, as can be seen in Table 5.7. Empirical results showed that the proposed model has an almost 5 % improvement in accuracy over a classical BOVW model. Also, using multiple kernel learning (with sum or weighted sum kernels) led to accuracy levels higher than that of the individual kernels involved. Finally, the proposed model ranked fourth in the FER Challenge, with an accuracy of 67.484 % on the final test. A different approach to local learning, that of clustering train images to divide the learning task on each cluster separately, can be studied in future work.

## References

Alexe B, Deselaers T, Ferrari V (2010) What is an object? In: Proceedings of CVPR. pp 73–80
Alexe B, Deselaers T, Ferrari V (2012) Measuring the objectness of image windows. IEEE Trans Pattern Anal Mach Intell 34(11):2189–2202

Bosch A, Zisserman A, Munoz X (2007) Image classification using random forests and ferns. In: Proceedings of ICCV. pp 1–8
Christensen D (2005) Fast algorithms for the calculation of Kendall's $\tau$. Comput Stat 20(1):51–62
Dalal N, Triggs B (2005) Histograms of oriented gradients for human detection. In: Proceedings of CVPR, vol 1. pp 886–893
Dinu LP, Manea F (2006) An efficient approach for the rank aggregation problem. Theoret Comput Sci 359(1–3):455–461
Dinu LP, Popescu M (2009) Comparing statistical similarity measures for stylistic multivariate analysis. In: Proceedings of RANLP
Dinu LP, Ionescu RT, Popescu M (2012) Local Patch Dissimilarity for images. In: Proceedings of ICONIP, vol 7663. pp 117–126
Everingham M, van Gool L, Williams CK, Winn J, Zisserman A (2010) The Pascal Visual Object Classes (VOC) challenge. Int J Comput Vision 88(2):303–338
Fei-Fei L, Fergus R, Perona P (2007) Learning generative visual models from few training examples: an incremental Bayesian approach tested on 101 object categories. Comput Vis Image Underst 106(1):59–70
Felzenszwalb PF, Girshick RB, McAllester D, Ramanan D (2010) Object detection with discriminatively trained part-based models. IEEE Trans Pattern Anal Mach Intell 32(9):1627–1645
Goodfellow IJ, Erhan D, Carrier PL, Courville A, Mirza M, Hamner B, Cukierski W, Tang Y, Thaler D, Lee DH, Zhou Y, Ramaiah C, Feng F, Li R, Wang X, Athanasakis D, Shawe-Taylor J, Milakov M, Park J, Ionescu RT, Popescu M, Grozea C, Bergstra J, Xie J, Romaszko L, Xu B, Chuang Z, Bengio Y (2015) Challenges in representation learning: a report on three machine learning contests. Neural Netw 64:59–63
Ionescu RT (2013) Local Rank Distance. In: Proceedings of SYNASC. pp 221–228
Ionescu RT, Popescu M (2013) Kernels for visual words histograms. In: Proceedings of ICIAP, vol 8156. pp 81–90
Ionescu RT, Popescu M (2015a) PQ kernel: a rank correlation kernel for visual word histograms. Pattern Recogn Lett 55:51–57
Ionescu RT, Popescu M (2015b) Have a SNAK. Encoding Spatial Information with the Spatial Non-Alignment Kernel. In: Proceedings of ICIAP, vol 9279. pp 97–108
Ionescu RT, Popescu M, Grozea C (2013) Local learning to improve bag of visual words model for facial expression recognition. In: Workshop on challenges in representation learning, ICML
Kendall MG (1948) Rank correlation methods. Griffin, London
Knight WR (1966) A computer method for calculating Kendall's tau with ungrouped data. J Am Stat Assoc 61(314). ISSN 01621459
Krapac J, Verbeek J, Jurie F (2011) Modeling spatial layout with Fisher vectors for image categorization. In: Proceedings of ICCV. pp 1487–1494
Lazebnik S, Schmid C, Ponce J (2005) A maximum entropy framework for part-based texture and object recognition. In: Proceedings of ICCV, vol 1. pp 832–838
Lazebnik S, Schmid C, Ponce J (2006) Beyond bags of features: spatial pyramid matching for recognizing natural scene categories. In: Proceedings of CVPR, vol 2. pp 2169–2178
Leung T, Malik J (2001) Representing and recognizing the visual appearance of materials using three-dimensional textons. Int J Comput Vision 43(1):29–44
Lowe DG (1999) Object recognition from local scale-invariant features. In: Proceedings of ICCV, vol 2. pp 1150–1157
Lowe DG (2004) Distinctive image features from scale-invariant keypoints. Int J Comput Vision 60(2):91–110
Philbin J, Chum O, Isard M, Sivic J, Zisserman A (2007) Object retrieval with large vocabularies and fast spatial matching. In: Proceedings of CVPR. pp 1–8
Sánchez J, Perronnin F, de Campos T (2012) Modeling the spatial layout of images beyond spatial pyramids. Pattern Recogn Lett 33(16):2216–2223. ISSN 0167–8655
Shawe-Taylor J, Cristianini N (2004) Kernel methods for pattern analysis. Cambridge University Press

Sivic J, Russell BC, Efros AA, Zisserman A, Freeman WT (2005) Discovering objects and their localization in images. In: Proceedings of ICCV. pp 370–377
Uijlings JRR, Smeulders AWM, Scha RJH (2009) What is the spatial extent of an object? In: Proceedings of CVPR. pp 770–777
Upton G, Cook I (2004) A dictionary of statistics. Oxford University Press, Oxford
Vedaldi A, Fulkerson B (2008) VLFeat: an open and portable library of computer vision algorithms. http://www.vlfeat.org/
Vedaldi A, Zisserman A (2010) Efficient additive kernels via explicit feature maps. In: Proceedings of CVPR. pp 3539–3546
Yagnik J, Strelow D, Ross DA, Lin RS (2011) The power of comparative reasoning. In: Proceedings of ICCV. pp 2431–2438
Zhang J, Marszalek M, Lazebnik S, Schmid C (2007) Local features and kernels for classification of texture and object categories: a comprehensive study. Int J Comput Vision 73(2):213–238
Ziegler A, Christiansen EM, Kriegman DJ, Belongie SJ (2012) Locally uniform comparison image descriptor. In: Proceedings of NIPS. pp 1–9

# Part II
# Knowledge Transfer from Computer Vision to Text Mining

# Chapter 6
# State-of-the-Art Approaches for String and Text Analysis

## 6.1 Introduction

Researchers have developed a wide variety of methods for string data that can be applied with success in different fields such as computational biology, text mining, information retrieval, and so on. Such methods range from clustering techniques used to analyze the phylogenetic trees of different organisms, to kernel methods used to identify authorship or native language from text. This chapter gives a concise overview of the state of the art methods used in two major fields of study, namely computational biology and text mining. The main goal of this chapter is to provide some context for the models that are used throughout this work for various string processing tasks.

String distance measures are briefly discussed in Sect. 6.2. Particular consideration is given to rank distance, since it represents the cornerstone concept for many of the developments presented in this book, namely the Local Patch Dissimilarity, the Local Texton Dissimilarity, the Local Rank Distance, and the Spatial Non-Alignment Kernel.

Two intensively studied problems in computational biology are DNA sequencing and phylogenetic analysis. A state of the art of the methods used for sequencing and comparing DNA is given in Sect. 6.3.1. Phylogenetic analysis, one of the first problems in computational biology, is discussed in Sect. 6.3.2. Methods that provide solutions to these computational biology problems are presented in Chap. 7.

*Text mining* refers to the process of acquiring relevant information from text using natural language processing (NLP) tools and machine learning methods. Text mining is a broad domain that studies document classification by topic, authorship identification, sentiment analysis, native language identification, among others. An overview of the state-of-the-art approaches in text mining is given in Sect. 6.4. Special consideration is given to the approach based on string kernels, that works at the character level. The string kernel method has various applications that are discussed in Sect. 6.4.1.



R.T. Ionescu and M. Popescu, *Knowledge Transfer between Computer Vision and Text Mining*, Advances in Computer Vision and Pattern Recognition,
DOI 10.1007/978-3-319-30367-3_6

One of the recently studied problems in NLP is native language identification (Brooke and Hirst 2012; Jarvis and Crossley 2012; Tetreault et al. 2012). A machine learning solution for native language identification based on string kernels is presented in Chap. 8. Finally, text categorization by topic using the bag of words model is addressed in Chap. 9. Remarkably, the performance of the bag of words is enhanced by including spatial information in the representation, an idea inspired by the research in the computer vision community presented in Chap. 3.

## 6.2 String Distance Measures

Perhaps the most widely known distance measures for strings are the Hamming distance (Hamming 1950) and the edit distance (Levenshtein 1966). Rank distance is a rather more recently introduced measure (Dinu and Manea 2006) that has demonstrated its effectiveness in biology (Dinu and Ionescu 2012b; Dinu and Sgarro 2006), natural language processing (Dinu and Dinu 2005; Dinu et al. 2008), computer science, and many other fields. Rank distance can be computed fast and benefits from some features of the edit distance. The previously mentioned string distance measures are formally described next. However, it is important to know that other distance measures used to compare strings also exist, such as the Dice's coefficient, the Jaro–Winkler distance, or the Jaccard distance, among others.

### 6.2.1 *Hamming Distance*

The Hamming distance (Hamming 1950) between two strings is given by the number of positions at which the symbols are different. It can also be interpreted as the minimum number of errors that, if eliminated, could transform one string into the other. Hamming distance is also used to analyze bit codes in several other fields such as information theory, coding theory, and cryptography. Example 5 shows how to compute the Hamming distance between two strings over a binary alphabet.

*Example 5* Let $x = ABABA$, $y = BABBA$ be two strings from $\{A, B\}^*$. For each position in which the symbols are different among $x$ and $y$, the procedure is to add 1 to the total distance:

$$\Delta_H(x, y) = 1 + 1 + 1 + 0 + 0 = 3.$$

It can be easily observed that the first three letters in $x$ are different from the first three letters in $y$, so the Hamming distance is 3.

### 6.2.2 Edit Distance

The edit distance, also known as the Levenshtein Distance (Levenshtein 1966), is given by the minimum number of operations required to transform a string into another. Usually, the allowed operations are substitution, deletion, and insertion. The edit distance finds its applications in natural language processing, where it can be used for automatic spelling correction, and in computational biology, where it can be used to measure the similarity of DNA sequences. Example 6 shows how to compute the edit distance between two strings over a binary alphabet.

*Example 6* Let $x = ABABA$, $y = BABBA$ be two strings from $\{A, B\}^*$. By deleting the first $A$ from $x$ the string $x' = BABA$ is obtained. Now, by inserting a $B$ in the middle of $x'$, the string $x'' = BABBA$ is obtained. Since $x''$ is identical to $y$, it implies that the edit distance between $x$ and $y$ is equal to the number of operations used to transform $x$ into $x''$. As such, the edit distance is 2.

Although this example points out that it is fairly easy to compute the edit distance, it is worth mentioning that the edit distance is traditionally computed by dynamic programming.

### 6.2.3 Rank Distance

A ranking is an ordered list that represents the result of applying an ordering criterion to a set of objects. A formal definition is given next.

**Definition 9** Let $\mathcal{U} = \{x_1, x_2, \ldots, x_m\}$ be a finite set of objects, named universe, where $m = |\mathcal{U}|$ denotes the cardinality of $\mathcal{U}$. A *ranking* over $\mathcal{U}$ is an ordered list: $\tau = (x_1 > x_2 > \cdots > x_d)$, where $x_i \in \mathcal{U}$ for all $1 \leq i \leq d$, $x_i \neq x_j$ for all $1 \leq i \neq j \leq d$, and $>$ is a strict ordering relation on the set $\{x_1, x_2, \ldots, x_d\}$.

A ranking defines a partial function on $\mathcal{U}$, where for each object $x \in \mathcal{U}$, $\tau(x)$ represents the position of the object $x$ in the ranking $\tau$.

The rankings that contain all the objects of a universe $\mathcal{U}$ are termed *full rankings*, while the others are *partial rankings*. The order of an object $x \in \mathcal{U}$ in a ranking $\sigma$ of length $d$ is defined by $ord(\sigma, x) = |d + 1 - \sigma(x)|$. By convention, if $x \in \mathcal{U} \setminus \{\sigma\}$, then $ord(\sigma, x) = 0$.

**Definition 10** Given two partial rankings $\sigma$ and $\tau$ over the same universe $\mathcal{U}$, the rank distance between them is defined as

$$\Delta_{RD}(\sigma, \tau) = \sum_{x \in \sigma \cup \tau} |ord(\sigma, x) - ord(\tau, x)|.$$

Rank distance is an extension to partial rankings of the Spearman's footrule distance (Diaconis and Graham 1977), defined below.

**Definition 11** If $\sigma$ and $\tau$ are two permutations of the same length, then $\Delta(\sigma, \tau)$ is named the Spearman's footrule distance.

The rank distance (Dinu and Manea 2006) can naturally be extended to strings using the following observation. If a string does not contain identical symbols, it can be transformed directly into a ranking (the rank of each symbol is its position in the string). Conversely, each ranking can be viewed as a string over an alphabet equal to the universe of the objects in the ranking. Given a string $w = a_1 \ldots a_n$ that contains multiple occurrences of a symbols $a_j$, each occurrence of $a_j$ is indexed using the number of occurrences of $a_j$ in the string $a_1 a_2 \ldots a_j$. The next definition formalizes the transformation of strings, that can have multiple occurrences of identical symbols, into rankings.

**Definition 12** Let $n$ be an integer and let $w = a_1 \ldots a_n$ be a finite word of length $n$ over an alphabet $\Sigma$. The extension to rankings of $w$, is defined as $\bar{w} = a_{1,i(1)} \ldots a_{n,i(n)}$, where $i(j) = |a_1 \ldots a_j|_{a_j}$ for all $j = 1, \ldots, n$.

An immediate observation is that given $\bar{w}$, the string $w$ can be obtained by simply removing all the index annotations. Rank distance can be extended to arbitrary strings as follows.

**Definition 13** Given $w_1, w_2 \in \Sigma^*$:

$$\Delta_{RD}(w_1, w_2) = \Delta_{RD}(\bar{w}_1, \bar{w}_2).$$

*Example 7* Given two strings $x = ABABA$ and $y = BABBA$ from $\{A, B\}^*$, the annotated versions of $x$ and $y$ are $\bar{x} = A_1 B_1 A_2 B_2 A_3$ and $\bar{y} = B_1 A_1 B_2 B_3 A_2$, respectively. Thus, the rank distance between $x$ and $y$ is the sum of the absolute differences between the orders of the characters in $\bar{x}$ and $\bar{y}$:

$$\Delta_{RD}(x, y) = |1-2| + |2-1| + |3-5| + |4-3| = 5.$$

For the missing characters $A_3$ and $B_3$, the maximum possible offset 4 can be considered in practice, so the distance becomes:

$$\Delta_{RD}(x, y) = |1-2| + |2-1| + |3-5| + |4-3| + 4 + 4 = 13.$$

It is important to note that the transformation of a string into a ranking can be done in linear time, by memorizing for each symbol, in an array, how many times it appears in the string. The computation of the rank distance between two rankings can also be done in linear time in the cardinality of the universe (Dinu and Sgarro 2006).

# 6.3 Computational Biology

## *6.3.1 Sequencing and Comparing DNA*

In many important problems in computational biology a common task is to compare a new DNA sequence with sequences that are already well studied and annotated. DNA sequence comparison was ranked in the top of two lists with major open problems in bioinformatics (Koonin 1999; Wooley 1999) over a decade ago, but it still receives the attention of researchers nowadays. Sequences that are similar would probably have the same function, or, if two sequences from different organisms are similar, there may be a common ancestor sequence (Liew et al. 2005). Another important problem with practical motivations for biologists is finding of motifs or common patterns in a set of given DNA sequences. A typical case where the last-mentioned problem occurs is, for example, when one needs to design genetic drugs with structure similar to a set of existing sequences of RNA (Lanctot et al. 2003). Other applications in computational biology which involve this task are (from a rich literature): PCR primer design (Gramm et al. 2002; Lanctot et al. 2003), genetic probe design (Lanctot et al. 2003), antisense drug design (Deng et al. 2003), finding unbiased consensus of a protein family (Ben-Dor et al. 1997), motif finding (Li et al. 2002; Wang and Dong 2005), and many others.

The standard method used in computational biology for sequence comparison is *sequence alignment*. Sequence alignment is a procedure of comparing DNA sequences, that aims at identifying regions of similarity that may be a consequence of functional, structural, or evolutionary relationships between the sequences. Algorithmically, the standard pairwise alignment method is based on dynamic programming (Smith and Waterman 1981). The method compares every pair of characters of the two sequences and generates an alignment and a score, which is dependent on the scoring scheme used, for example, a scoring matrix for the different base pair combinations, match and mismatch scores, or a scheme for insertion or deletion (gap) penalties. Although dynamic programming for sequence alignment is mathematically optimal, in practice, it is far too slow for comparing a large number of bases, and too slow to be performed in a reasonable time.

Recently, many tools designed to align short reads have been proposed (Li and Homer 2010). The main efforts in the design of such tools are on improving the speed and correctness. Fast tools are needed to keep the pace with data production, while the number of correctly placed reads is maximized. Usually, tools sacrifice correctness over speed, allowing only few mismatches between the reads and the reference genome. Tools that maximize such trade-off are BOWTIE (Langmead et al. 2009) and BWA (Li and Durbin 2009). They make use of the seed-and-extend heuristic: in order to align a read $r$, an almost exact match of the first $l < |r|$ bases of the read is a necessary condition. The BFAST (Homer et al. 2009) tool moves towards favoring correctness over speed, allowing alignments with a high number of mismatches and indels.

Another highly accurate tool able to align reads in the presence of extensive polymorphisms, high error rates, and small indels is rNA (Vezzi et al. 2012). It achieves an accuracy greater than other tools in a feasible amount of time.

Most of the techniques for comparing and aligning DNA need to compare DNA strings based on a distance measure. Thus, several distance measures for strings have been proposed and developed. Since most variations between organisms of the same species consist of point mutations like single nucleotide polymorphisms, or small insertions or deletions, edit distance is the standard string measure in many biomedical analyses, such as the detection of genomic variation, genome assembly (Zerbino and Birney 2008), identification and quantification of RNA transcripts (Tomescu et al. 2013; Trapnell et al. 2009, 2010), identification of transcription factor binding sites (Levy and Hannenhalli 2002), or methylation patterns (Prezza et al. 2012).

In the case of genomic sequences coming from different related species, other mutations are present, such as reversals (Bader et al. 2001), transpositions (Bafna and Pevzner 1998), translocations (Hannenhalli 1996), fissions and fusions (Hannenhalli and Pevzner 1995). For this reason, there have been a series of different proposals of similarity between entire genomes, including rearrangement distance (Belda et al. 2005), $k$-break rearrangements (Alekseyev and Pevzner 2008), edit distance with block operations (Shapira and Storer 2003). The study of genome rearrangement (Palmer and Herbon 1988) was also investigated under Kendall's tau distance. Other choices of distance metrics in recent techniques are the Hamming distance (Chimani et al. 2011; Vezzi et al. 2012), Kendall's tau distance (Popov 2007), and many others (Felsenstein 2004).

Rank distance (Dinu 2003) is another such measure of similarity, having low computational complexity, but high significance in phylogenetic analysis (Dinu and Ionescu 2012a; Dinu and Sgarro 2006). More recently, an improved version of rank distance, termed Local Rank Distance, has demonstrated its application in genome sequence alignment (Dinu et al. 2014). The Local Rank Distance sequence aligner is also presented in Chap. 7.

### 6.3.2 *Phylogenetic Analysis*

Biologists have spent many years creating a taxonomy (hierarchical classification) of all living things: kingdom, phylum, class, order, family, genus, and species. Thus, it is perhaps not surprising that much of the early work in cluster analysis sought to create a discipline of mathematical taxonomy that could automatically find such classification structures. More recently, biologists have applied clustering to analyze the large amounts of genetic information that are now available. For example, clustering has been used to find groups of genes that have similar functions.

The phylogenetic analysis of organisms remains one of the most important problems in biology. When a new organism is discovered, it needs to be placed in a phylogenetic tree in order to determine its class, order, family, and species. Yet, the phylogenetic trees of already known organisms, obtained with different methods, are still disputed by researchers. For example, there is no definitive agreement on either the correct branching order or differential rates of evolution among the higher primates, despite the amount of research in this area. Joining human with chimpanzee and the gorilla with the orangutan is currently favored, but the alternatives that group humans with either gorillas or the orangutan rather than with chimpanzees also have support (Holmquist et al. 1988). Others have tried to find the right place of an entire order of organisms in the evolutionary tree of species. One such example is the work of Reyes et al. (2000), which finds that the position of Rodents in the mammalian tree remains an open question.

While distance methods are commonly utilized (for example, the neighbor-joining method due to Saitou and Nei (1987) uses only distances), the standard method of phylogenetic analysis is probably the maximum likelihood for the evolution of the strings under a biologically motivated model of evolution, for example the Markov model (also known as General Time Reversible Model) (Saccone et al. 1990) with various supplements such as some invariant states. Similarly, the now commonly used Bayesian methods (Munch et al. 2008; Yang and Rannala 1997) are based on the aligned strings themselves, not on some distances between the strings. Many trees are also found using parsimony methods. Some researchers have also proposed to examine the phylogenetic evolution using only proteins encoded by mitochondrial DNA (Cao et al. 1998), instead of entire mtDNA sequences.

There are many standard methods of phylogenetic inference from distances, such as the Unweighted Pair Group Method with Arithmetic Mean (UPGMA) of Sneath and Sokal (1973), least square methods (Bryant and Waddell 1998; Fitch and Margoliash 1967), minimum evolution methods (Rzhetsky and Nei 1992), or neighbor-joining (Saitou and Nei 1987). These have been studied considerably in the literature. Broad overviews may be found in the works of Nei and Kumar (2000) and Felsenstein (2004). The phylogenetic analysis approach proposed in Chap. 7 is also distance based. More precisely, Chap. 7 presents a novel distance measure for strings, termed Local Rank Distance (Ionescu 2013), with applications in phylogeny.

## 6.4 Text Mining

Natural language processing is an active area of research with many applications. Researchers have developed a broad variety of techniques that aim at solving different NLP tasks. Extensive overviews of such techniques may be found in the works of Jurafsky and Martin (2000), Manning and Schütze (1999), Manning et al. (2008). Most NLP tasks, such as machine translation, require the understanding of the syntactical and semantic structure of text. Some researchers have developed part-of-speech tagging techniques to identify the part of speech of words in a given text. Others have studied word sense disambiguation (WSD) techniques that try to resolve the ambiguity of words in a given context (Agirre and Edmonds 2006; Hristea et al. 2008). All these techniques, that try to automatically extract syntactic or semantic information from text, have applications beyond machine translation. For example, unsupervised

WSD has been used to improve the precision of an information retrieval system for difficult queries (Chifu and Ionescu 2012). To summarize, the basic approach is to develop methods based on linguistic knowledge, that comes from the study of morphology, syntax, semantics, and so on. An alternative approach, mainly studied in text mining, is based on finding machine learning methods that can be used as good approximate solutions for certain problems such as text categorization by topic, sentiment analysis, or native language identification, to name only a few. The two approaches are a result of the conjunction of researchers with different backgrounds. While some of them dedicated their time to the study of language, the others developed machine learning methods that can be applied with success in several domains, including NLP. The study presented in this book falls in the second category, as the book studies machine learning approaches for text categorization by topic and native language identification. Therefore, this state-of-the-art section is only focused on the machine learning approach.

In the recent years, massive amounts of unstructured and semi-structured data, including text documents, have become available with the help of the Internet. Data mining and machine learning techniques can be employed to find patterns in this data, but they rely on other methods for structuring text. Indeed, machine learning methods usually work with features or pairwise similarities, that must be obtained from the unstructured data. In order to treat text as a set of features, researchers have proposed several representations that also keep the relevant information from text. One of the most popular representations is the bag of words model. In this model, text is represented as an unordered collection of words, disregarding grammar and even word order. However, Chap. 9 shows that the performance of the bag of words can be improved for the task of text categorization by topic, by including spatial information. Other popular representations are based on word $n$-grams (Tan et al. 2002). An $n$-gram is defined as a contiguous sequence of $n$ items from a given sequence of text. An $n$-gram of size 1 is referred to as *unigram*, one of size 2 is referred to as *bigram*, and one of size 3 is referred to as *trigram*. While the use of word $n$-grams seems natural for text analysis, another approach is to use character $n$-grams (also known as character $p$-grams). String kernels based on character $p$-grams have achieved state-of-the-art performance in text categorization (by topic), authorship identification, plagiarism detection. A recent application of string kernels, that of identifying the native language from text, is presented in Chap. 8.

### 6.4.1 *String Kernels*

Using words as basic units is natural in text analysis tasks such as text categorization, authorship identification, or plagiarism detection. Perhaps surprisingly, recent results have proved that methods handling the text at character level can also be very effective in text analysis tasks (Grozea et al. 2009; Lodhi et al. 2002; Popescu 2011; Popescu and Dinu 2007; Popescu and Grozea 2012; Sanderson and Guenter 2006).

Lodhi et al. (2002) used string kernels for document categorization with very good results. Trying to explain why treating documents as symbol sequences and why using string kernels led to such good results, the authors suppose that: "the [string] kernel is performing something similar to stemming, hence providing semantic links between words that the word kernel must view as distinct".

String kernels were also successfully used in authorship identification (Popescu and Dinu 2007; Popescu and Grozea 2012; Sanderson and Guenter 2006). For example, the system described by Popescu and Grozea (2012) ranked first in most problems and overall in the PAN 2012 Traditional Authorship Attribution tasks. A possible reason for the success of string kernels in authorship identification is given by Popescu and Dinu (2007): "the similarity of two strings as it is measured by string kernels reflects the similarity of the two texts as it is given by the short words (2-5 characters) which usually are function words, but also takes into account other morphemes like suffixes ('ing' for example) which also can be good indicators of the author's style".

Even more interesting is the fact that two methods, that are essentially the same, obtained very good results for text categorization (by topic) (Lodhi et al. 2002) and authorship identification (Popescu and Dinu 2007). Both are based on SVM and a string kernel of length 5. Traditionally, the two tasks, text categorization by topic and authorship identification, are viewed as opposite. How is it possible to obtain good results with the same technique? When words are considered as features, for text categorization the (stemmed) content words are used (the stop words being eliminated), while for authorship identification the function words (stop words) are used as features, the other words (content words) being eliminated. Then, why did the same string kernel (of length 5) work well in both cases? It seems that the key factor is the kernel-based learning algorithm. The string kernel implicitly embeds the texts in a high-dimensional feature space, in this case the space of all (sub)strings of length 5. The kernel-based learning algorithm (SVM or any other kernel method), aided by regularization, implicitly assigns a weight to each feature, thus selecting the features that are important for the discrimination task. In this fashion, in the case of text categorization the learning algorithm enhances the features (substrings) representing stems of content words, while in the case of authorship identification the same learning algorithm enhances the features representing function words.

String kernels are used for native language identification in Chap. 8. Some new string kernels are also presented in the same chapter. Using string kernels makes the corresponding learning method completely language independent, because the texts will be treated as sequences of symbols (strings). Methods working at the word level or above very often restrict their feature space according to theoretical or empirical principles. For instance, they select only features that reflect various types of spelling errors or only some type of words, such as function words. These features prove to be very effective for specific tasks, but it is possible that other good features also exist. String kernels embed the texts in a very large feature space, given by all the substrings of length $p$, and leave it to the learning algorithm to select the important features for a specific task, by highly weighting these features. Since it does not restrict the feature space according to any linguistic theory, the string kernels approach is linguistic theory neutral. Furthermore, the method does not explicitly consider any features of natural language such as words, phrases, or meaning, contrary to the usual NLP approach. However, an interesting remark is that such features can implicitly be discovered within the extracted $p$-grams. On the other hand, explicitly extracting such features could get very difficult for less-studied or low-resource languages, and the methods that rely on linguistic features become inherently language dependent. Even a method that considers words as features cannot be completely language independent, since the definition of a word is necessarily language specific. A method that uses only function words as features is also not completely language independent because it needs a list of function words which is specific to a language. When features such as part-of-speech tags are used, as in the work of Jarvis et al. (2013), the method relies on a part-of-speech tagger which might not be available (yet) for some languages. Furthermore, a way to segment a text into words is not an easy task for some languages such as Chinese.

## References

Agirre E, Edmonds PG (2006) Word Sense Disambiguation: Algorithms and Application. Springer

Alekseyev MA, Pevzner PA (2008) Multi-break rearrangements and chromosomal evolution. Theor Comput Sci 395(2–3):193–202

Bader DA, Moret BME, Yan M (2001) A linear-time algorithm for computing inversion distance between signed permutations with an experimental study. In: Proceedings of the 7th international workshop on algorithms and data structures, pp 365–376

Bafna V, Pevzner PA (1998) Sorting by transpositions. SIAM J Discrete Math 11(2):224–240

Belda E, Moya A, Silva FJ (2005) Genome rearrangement distances and gene order phylogeny in gamma-proteobacteria. Mol Biol Evol 22(6):1456–1467

Ben-Dor A, Lancia G, Perone J, Ravi R (1997) Banishing bias from consensus sequences. In: Proceedings of CPM 1264:247–261

Brooke J, Hirst G (2012) Robust, Lexicalized native language identification. In: Proceedings of COLING 2012:391–408

Bryant D, Waddell P (1998) Rapid evaluation of least squares and minimum evolution criteria on phylogenetic trees. Mol Biol Evol 15(10):1346–1359

Cao Y, Janke A, Waddell PJ, Westerman M, Takenaka O, Murata S, Okada N, Paabo S, Hasegawa M (1998) Conflict among individual mitochondrial proteins in resolving the phylogeny of Eutherian orders. J Mol Evol 47:307–322

Chifu A-G, Ionescu RT (2012) Word sense disambiguation to improve precision for ambiguous queries. Cent Eur J Comput Sci 2(4):398–411

Chimani M, Woste M, Bocker S (2011) A closer look at the closest string and closest substring problem. In: Proceedings of ALENEX, pp 13–24

Deng X, Li G, Li Z, Ma B, Wang L (2003) Genetic design of drugs without side-effects. SIAM J Comput 32(4):1073–1090

Diaconis P, Graham RL (1997) Spearman footrule as a measure of disarray. J Roy Stat Soc B (Methodological), 39(2):262–268

Dinu A, Dinu LP (2005) On the syllabic similarities of romance languages. In: Proceedings of CICLing 3406:785–788

Dinu LP (2003) On the classification and aggregation of hierarchies with different constitutive elements. Fundam Informaticae 55(1):39–50

Dinu LP, Ionescu RT (2012a) Clustering based on rank distance with applications on DNA. In: Proceedings of ICONIP 7667:722–729

Dinu LP, Ionescu RT (2012b) An efficient rank based approach for closest string and closest substring. PLoS ONE 7(6):e37576
Dinu LP, Manea F (2006) An efficient approach for the rank aggregation problem. Theor Comput Sci 359(1–3):455–461
Dinu LP, Sgarro A (2006) A low-complexity distance for DNA strings. Fundam Informaticae 73(3):361–372
Dinu LP, Popescu M, Dinu A (2008) Authorship identification of romanian texts with controversial paternity. In: Proceedings of LREC
Dinu LP, Ionescu RT, Tomescu AI (2014) A rank-based sequence aligner with applications in phylogenetic analysis. PLoS ONE, 9(8):e104006. doi:10.1371/journal.pone.0104006
Felsenstein J (2004) Inferring phylogenies. Sinauer Associates, Sunderland
Fitch WM, Margoliash E (1967) Construction of phylogenetic trees. Science 155(760):279–284
Gramm J, Huffner F, Niedermeier R (2002) Closest strings, primer design, and motif search. Presented at RECOMB 2002 poster session, pp 74–75
Grozea C, Gehl C, Popescu M (2009) ENCOPLOT: pairwise sequence matching in linear time applied to plagiarism detection. In: 3rd PAN Workshop. Uncovering Plagiarism, Authorship and Social Software Misuse, pp 10
Hamming RW (1950) Error detecting and error correcting codes. Bell Syst Tech J 26(2):147–160
Hannenhalli S (1996) Polynomial-time algorithm for computing translocation distance between genomes. Discrete Appl Math 71(1–3):137–151
Hannenhalli S, Pevzner PA (1995) Transforming men into mice (polynomial algorithm for genomic distance problem. In: Proceedings of FOCS, pp 581–592
Holmquist R, Miyamoto MM, Goodman M (1988) Higher-primate phylogeny—why can't we decide? Mol Biol Evol 3(5):201–216
Homer N, Merriman B, Nelson SF (2009) BFAST: an alignment tool for large scale genome resequencing. PLoS ONE 4(11):e7767+
Hristea F, Popescu M, Dumitrescu M (2008) Performing word sense disambiguation at the border between unsupervised and knowledge-based techniques. Artif Intell Rev 30(1–4):67–86
Ionescu RT (2013) Local Rank Distance. In: Proceedings of SYNASC, pp 221–228
Jarvis S, Crossley S (eds) (2012) Approaching language transfer through text classification: explorations in the detection-based approach, vol 64. Multilingual Matters Limited, Bristol
Jarvis S, Bestgen Y, Pepper S (2013) Maximizing classification accuracy in native language identification. In: Proceedings of the Eighth Workshop on Innovative Use of NLP for Building Educational Applications, pp 111–118
Jurafsky D, Martin JH (2000) Speech and language processing: an introduction to natural language processing, computational linguistics, and speech recognition, 1st edn. Prentice Hall PTR, Upper Saddle River
Koonin EV (1999) The emerging paradigm and open problems in comparative genomics. Bioinformatics 15:265–266
Lanctot KJ, Li M, Ma B, Wang Shaojiu, Zhang L (2003) Distinguishing string selection problems. Inf Comput 185(1):41–55
Langmead B, Trapnell C, Pop M, Salzberg S (2009) Ultrafast and memory-efficient alignment of short DNA sequences to the human genome. Genome Biol 10(3):R25–10
Levenshtein VI (1966) Binary codes capable of correcting deletions, insertions and reverseals. Cybern Control Theory 10(8):707–710
Levy S, Hannenhalli S (2002) Identification of transcription factor binding sites in the human genome sequence. Mamm Genome 13(9):510–514
Li H, Durbin R (2009) Fast and accurate short read alignment with Burrows-Wheeler transform. Bioinformatics 25(14):1754–1760
Li H, Homer N (2010) A survey of sequence alignment algorithms for next-generation sequencing. Briefings Bioinf 11(5):473–483
Li M, Ma B, Wang L (2002) Finding similar regions in many sequences. J Comput Syst Sci 65(1): 73–96

Liew AW, Yan H, Yang M (2005) Pattern recognition techniques for the emerging field of bioinformatics: a review. Pattern Recogn 38(11):2055–2073
Lodhi H, Saunders C, Shawe-Taylor J, Cristianini N, Watkins CJCH (2002) Text classification using string kernels. J Mach Learn Res 2:419–444
Manning CD, Schütze H (1999) Foundations of statistical natural language processing. MIT Press, Cambridge
Manning CD, Raghavan P, Schütze H (2008) Introduction to information retrieval. Cambridge University Press, New York
Munch K, Boomsma W, Huelsenbeck JP, Willerslev E, Nielsen Rasmus (2008) Statistical assignment of DNA sequences using Bayesian phylogenetics. Syst Biol 57(5):750–757
Nei M, Kumar S (2000) Molecular evolution and phylogenetics, 1 edn. Oxford University Press, USA. ISBN 0195135857
Palmer J, Herbon L (1988) Plant mitochondrial DNA evolves rapidly in structure, but slowly in sequence. J Mol Evol 28:87–89
Popescu M (2011) Studying translationese at the character level. In: Proceedings of RANLP, pp 634–639
Popescu M, Dinu LP (2007) Kernel methods and string kernels for authorship identification: the federalist papers case. In: Proceedings of RANLP
Popescu M, Grozea C (2012) Kernel methods and string kernels for authorship analysis. CLEF (Online Working Notes/Labs/Workshop)
Popov YV (2007) Multiple genome rearrangement by swaps and by element duplications. Theor Comput Sci 385(1–3):115–126
Prezza N, Fabbro CD, Vezzi F, De Paoli E, Policriti A (2012) ERNE-BS5: aligning BS-treated sequences by multiple hits on a 5-letters alphabet. In: Proceedings of BCB, pp 12–19
Reyes A, Gissi C, Pesole G, Catzeflis FM, Saccone C (2000) Where do rodents fit? Evidence from the complete mitochondrial genome of Sciurus vulgaris. Mol Biol Evol 17(6):979–983
Rzhetsky A, Nei M (1992) A simple method for estimating and testing minimum-evolution trees. Mol Biol Evol 9(5):945–967
Saccone C, Lanave C, Pesole G, Preperata G (1990) Influence of base composition on quantitative estimates of gene evolution. In: Doolittle RF (ed) Molecular evolution: computer analysis of protein and nucleic acid sequences, vol 183 of methods in enzymology, chapter 35. Academic Press, New York, pp 570–583
Saitou N, Nei M (1987) The neighbor-joining method: a new method for reconstructing phylogenetic trees. Mol Biol Evol 4(4):406–425. ISSN 1537–1719
Sanderson C, Guenter S (2006) Short text authorship attribution via sequence kernels, markov chains and author unmasking: an investigation. In: Proceedings of EMNLP, pp 482–491
Shapira D, Storer JA (2003) Large edit distance with multiple block operations. In: Proceedings of SPIRE 2857:369–377
Smith T, Waterman M (1981) Comparison of biosequences. Adv Appl Math 2(4):482–489
Sneath P, Sokal R (1973) Numerical taxonomy. W. H. Freeman and Company, San Francisco
Tan C-M, Wang Y-F, Lee C-D (2002) The use of bigrams to enhance text categorization. Inf Process Manage 38(4):529–546
Tetreault J, Blanchard D, Cahill A, Chodorow M (2012) Native tongues, lost and found: resources and empirical evaluations in native language identification. In: Proceedings of COLING 2012:2585–2602
Tomescu AI, Kuosmanen A, Rizzi R, Mäkinen VA (2013) Novel min-cost flow method for estimating transcript expression with RNA-Seq. BMC Bioinformatics 14(Suppl 5):S15. Presented at RECOMB-Seq 2013
Trapnell C, Pachter L, Salzberg SL (2009) TopHat: discovering splice junctions with RNA-Seq. Bioinformatics 25(9):1105–1111
Trapnell C, Williams BA, Pertea G, Mortazavi A, Kwan G, van Baren MJ, Salzberg SL, Wold BJ, Pachter L (2010) Transcript assembly and quantification by RNA-Seq reveals unannotated transcripts and isoform switching during cell differentiation. Nat Biotechnol 28:511–515

Vezzi F, Del Fabbro C, Tomescu AI, Policriti A (2012) rNA: a fast and accurate short reads numerical aligner. Bioinformatics 28(1):123–124
Wang L, Dong L (2005) Randomized algorithms for motif detection. J Bioinform Comput Biol 3(5):1039–1052
Wooley JC (1999) Trends in computational biology: a summary based on a RECOMB plenary lecture. J Comput Biol 6:459–474
Yang Z, Rannala B (1997) Bayesian phylogenetic inference using DNA sequences: a Markov chain Monte Carlo method. Mol Biol Evol 14(7):717–724
Zerbino DR, Birney E (2008) Velvet: algorithms for de Novo short read assembly using de Bruijn graphs. Genome Res 18(5):821–829

# Chapter 7
# Local Rank Distance

## 7.1 Introduction

Computer science researchers have developed a wide variety of methods that can be applied with success in computational biology. Such methods range from clustering techniques used to analyze the phylogenetic trees of different organisms, to genetic algorithms used to find motifs or common patterns in a set of given DNA sequences. The results of many state-of-the-art techniques for phylogenetic analysis or DNA sequence comparison are inaccurate from a biological point of view, and can always be improved. Some of these methods are based on a distance measure for strings. Popular choices for recent techniques are the Hamming distance (Chimani et al. 2011; Vezzi et al. 2012), edit distance (Shapira and Storer 2003), Kendall's tau distance (Popov 2007), rank distance (Dinu and Ionescu 2012a, b), and many others (Felsenstein 2004). Some of these distance measures are presented in detail in Chap. 6.

In this context, the present chapter aims to introduce a new distance measure, termed Local Rank Distance (LRD) (Ionescu 2013), inspired from the recently introduced Local Patch Dissimilarity for images (Dinu et al. 2012), which in turn, is inspired by rank distance (Dinu and Manea 2006). Unlike rank distance, LRD does not require for strings to be annotated. Furthermore, it uses substrings instead of single characters, just as Local Patch Dissimilarity uses patches instead of single pixels. Local Rank Distance can be viewed as a generalization of rank distance that has various applications from phylogenetic analysis (Ionescu 2013) and sequence alignment (Dinu et al. 2014) to native language identification (Ionescu et al. 2014; Popescu and Ionescu 2013). The computational biology applications of LRD are discussed in this chapter, while native language identification is addressed in Chap. 8.

A series of phylogenetic analysis and DNA comparison experiments are conducted to compare LRD with other distance measures. The empirical results show that the use of LRD enables significant improvements over several state-of-the-art methods.



R.T. Ionescu and M. Popescu, *Knowledge Transfer between Computer Vision and Text Mining*, Advances in Computer Vision and Pattern Recognition,
DOI 10.1007/978-3-319-30367-3_7

Furthermore, this chapter presents a method for assigning a set of short DNA reads to a reference genome, under Local Rank Distance. The rank-based sequence aligner works as follows. Given a set of reads that need to be aligned against a reference genome, the aligner determines the position of each read in the reference genome that gives the minimum Local Rank Distance. The proposed aligner will be referred to as the *LRD aligner* through the rest of this chapter. Some strategies of optimizing the search for the best positions of reads are also proposed and investigated. The LRD aligner is improved in terms of speed by storing $k$-mer positions in a hash table for each read. An approximate LRD aligner that works even faster is obtained through the following strategy. The approximate aligner considers only the positions in the reference that are likely to give the minimum distance, by previously counting the number of $k$-mers from the read that can be found at every position in the reference genome.

The LRD sequence aligners are designed to work with genomic data produced by Next-Generation Sequencing technologies. These high-throughput technologies are able to produce up to 200 million DNA reads of length between 30 and 400 base pairs in a single experiment. Despite this abundance of reads, their short length makes the problem of assembling them into the originating genome a difficult one in practice. Therefore, methods for finding the class, the order, the family, or even the species of an unknown organism, given only a set of short Next-Generation Sequencing DNA reads originating from its genome, are of interest. A method that can be used to solve this phylogenetic analysis task is presented in this chapter. The method works as follows: given a collection $\mathcal{R}$ of short DNA reads, and a collection $\mathcal{G}$ of genomes, it finds the genome $G \in \mathcal{G}$ that gives a minimum score with resptect to $\mathcal{R}$. This method serves two purposes. First, the method can be used to determine the place of an individual in a phylogenetic tree, by finding the most similar organism in the phylogenetic tree. This can be achieved by using only a set of short DNA reads originating from the genome of the new individual. Second, the method is used to evaluate the performance level of the rank-based aligners and to compare them with other state-of-the-art alignment tools, such as BWA (Li and Durbin 2009), BOWTIE (Langmead et al. 2009), or BLAST (Altschul et al. 1990). Experimental results on simulated reads are obtained under two scenarios: low and high error rates. In the former scenario, all the aligners besides BWA have full precision. In the latter scenario, the LRD aligners are the only ones that attain full precision. It seems that the LRD aligners give the most accurate results, while being more computationally expensive than the other aligners.

A set of experiments are conducted to determine the precision and the recall of the proposed LRD aligners, in the presence of contaminated reads. The task is to align reads sampled from several mammals on the human mitochondrial DNA sequence genome. The goal is to maximize the number of aligned reads sampled from the human genome (true positives), and to minimize the number of aligned reads sampled from the other mammals (false positives). Again, the LRD aligners seem to have the best performance, followed closely by BOWTIE and BLAST.

The fast LRD aligner is also tested on three human vibrio pathogens with results that point towards the same conclusion of Chen et al. (2003) and Lin et al. (2005).

In all the experiments presented in this chapter, the rank-based aligners show results that are better than the state-of-the-art alignment tools, in terms of accuracy. The results obtained in this work can be considered as a strong argument in favor of using rank-based distance measures for computational biology tasks, in order to obtain results that are more accurate from a biological point of view.

It is important to point out that the main focus of the sequence alignment experiments is on the alignment accuracy of the aligners based on LRD. Therefore, the simple strategy of assigning each read to the genomic sequence with the best LRD distance was used. However, in other biological problems, these alignments can be fed to other more elaborate methods. For example, in profiling bacterial species from a metagenomics sample, various tools, such as the MG-RAST server (Meyer et al. 2008), MEGAN (Huson et al. 2007), and metaBEETL (Ander et al. 2013), align the reads to a reference taxonomy, but report as hit the Lowest Common Ancestor node of a set of significant hits in this taxonomic tree.

The chapter is organized as follows. LRD and its principles are described in Sect. 7.2. Section 7.3 presents the LRD definition. An efficient algorithm to compute LRD is presented in Sect. 7.4. Some theoretical properties of LRD are given in Sect. 7.5. The rank-based sequence aligners are described in Sect. 7.6. Phylogenetic analysis and sequence alignment experiments are presented in Sect. 7.7. The final remarks are given in Sect. 7.8.

## 7.2 Approach

The development of Local Rank Distance (Ionescu 2013) is based on Local Patch Dissimilarity (Dinu et al. 2012), which is a generalization of rank distance for two-dimensional input (digital images), as mentioned in Chap. 4. Rank distance (Dinu and Manea 2006) has applications in biology (Dinu and Ionescu 2012b; Dinu and Sgarro 2006), natural language processing (Dinu and Dinu 2005; Dinu et al. 2008), computer science, and many other fields. Despite of having such broad applications, rank distance might sometimes be not fully adequate for specific data types, such as DNA strings. This probably happens because rank distance is a rather trivial extension for strings of the Spearman's footrule which was developed to measure the distance between permutations. Local Rank Distance comes from the idea of better adapting rank distance to string data, in order to comprise a better similarity (or dissimilarity) between strings, such as DNA sequences or text documents.

As described in Chap. 6, the distance between two strings can be measured with rank distance by scanning (from left to right) both strings. First, characters need to be annotated with indexes in order to eliminate duplicates. For each annotated letter, rank distance measures the offset between its position in the first string and its position in the second string. Finally, all these offsets are summed up to obtain the rank distance. In other words, the rank distance measures the offset between the positions of a letter in the two given strings, and then sums up these values. Intuitively, the rank distance computes the total non-alignment score between two sequences.

Rank distance is based on mathematical principles that are not suited for specific input data, such as images or DNA sequences. The reason is that rank distance was initially designed to work with rankings (ordered sets of objects). It is useful to consider the following example to understand how rank distance works on text and how it can be adapted to specific input data. For two strings $s_1$ and $s_2$, the characters are first annotated. The annotation step is necessary to transform the strings into (full or partial) rankings. Now, rank distance can be computed between annotated strings $\bar{s_1}$ and $\bar{s_2}$.

*Example 8* If $s_1 = CCGAATACG$ and $s_2 = TGACTCA$, the annotated strings are $\bar{s_1} = C_1C_2G_1A_1A_2T_1A_3C_3G_2$ and $\bar{s_2} = T_1G_1A_1C_1T_2C_2A_2$. The rank distance between $s_1$ and $s_2$ is

$$\begin{aligned}\Delta_{RD}(s_1, s_2) = \Delta_{RD}(\bar{s_1}, \bar{s_2}) = &|1-4| + |2-6| + |3-2| + |4-3| \\ &+ |5-7| + |6-1| + x + x + x + x,\end{aligned}$$

where $x$ represents the offset of characters that cannot be matched (because they are missing in one or the other string), such as $A_3$ or $C_3$.

In order to compute rank distance on strings, a global order is introduced by the annotation step. The drawback of the annotation procedure is that there are annotated characters in $\bar{s_1}$ (such as $A_3$ or $C_3$) that have no matching characters in $\bar{s_2}$. Hence, conventions are made in practice to replace $x$ (the offset of unmatched characters) with the maximum possible offset between two characters, or with the average offset. To reduce the effect of missing characters, strings can be annotated both from left to right and from right to left. Some of these approaches are studied by Dinu and Sgarro (2006). Another idea with results similar to previous approaches is to consider circular DNA sequences (Dinu and Ghetu 2011). However, these mathematical tricks are not natural from a biological point of view nor from a practical point of view. Thus, one may ask whether this global order is really necessary or whether the global order is defined in the right way (for example, should strings be annotated from right to left instead of left to right). It can be argued that strings can be annotated both from left to right and from right to left, and the two distances obtained after annotation can be summed up. However, in order to define rank distance for specific input data, answering such questions becomes difficult. In the case of digital images, Local Patch Dissimilarity solves this problem by simply dropping the annotation step and by replicating only the local behavior of rank distance. Therefore, pixels or patches have no global order to conform to. This will also be the case of Local Rank Distance, for which the annotation step is dropped, and characters in one string are simply matched with the nearest similar characters in the other string. The advantage of using no annotation immediately becomes clear. In Example 8, $A_3$ remains unmatched in the case of rank distance, while in the case of LRD, $A_3$ from $\bar{s_1}$ will be matched with $A_2$ from $\bar{s_2}$, as shown in Example 9, since there are no annotations.

If long DNA strings (as found in nature) are considered, that contain only characters in the alphabet $\Sigma = \{A, C, G, T\}$, it can be naturally observed that measuring the non-alignment score between characters that might be randomly distributed (with a uniformly distributed frequency) is not too relevant, because scores between different strings will almost be the same. Therefore, important information encoded in the DNA strings might be lost or not completely captured by a distance measure that works at the character level (this is the case of Hamming or rank distance, for example). There is a similar situation in image processing. Instead of analyzing images at the pixel level, computer vision researchers have developed techniques that work with patches (Barnes et al. 2011; Cho et al. 2010). To extract meaning from image, computer vision techniques, including Local Patch Dissimilarity, look at certain features such as contour, contrast, or other primitive shapes. It is clear that these features cannot be captured in single pixels, but rather in small, overlapping rectangles of fixed size (e.g., $10 \times 10$ pixels), called patches. A similar idea was introduced in the early years of bioinformatics, where $k$-mers (also known as $p$-grams or substrings) are used instead of single characters. There are recent studies that use $k$-mers for the phylogenetic analysis of organisms (Li et al. 2004), or for sequence alignment (Melsted and Pritchard 2011). Analyzing DNA at substring level is also more suited from a biological point of view, because DNA substrings may contain meaningful information. For example, genes are encoded by a number close to 100 base pairs, or codons that encode the 20 standard amino acids are formed of 3-mers. Hence, LRD should also be designed to compare DNA strings based on $k$-mers and not on single characters.

It is interesting to note that LRD does not require an alignment of the strings by the addition of insertions and deletions, unlike other commonly used phylogenetic analysis methods. The most common methods in phylogeny deal with aligned strings. Removing this step greatly speeds up the calculation of the distance between two strings.

Both rank distance (Dinu and Manea 2006) and LRD (Ionescu 2013) are related to the rearrangement distance (Amir et al. 2006). The rearrangement distance works with indexed $k$-mers and is based on a process of converting a string into another, while LRD does not impose such global rules. LRD focuses only on the local phenomenon present in DNA: strings may contain similar $k$-mers that are not perfectly aligned.

## 7.3 Local Rank Distance Definition

In order to describe LRD, the following notations are defined. Given a string $x$ over an alphabet $\Sigma$, the length of $x$ is denoted by $|x|$. Strings are considered to be indexed starting from position 1, that is $x = x[1]x[2] \dots x[|x|]$. Moreover, $x[i : j]$ denotes the substring $x[i]x[i+1] \dots x[j-1]$ of $x$.

Local Rank Distance is inspired by rank distance (Dinu and Manea 2006), the main differences being that it uses $p$-grams instead of single characters, and that it matches each $p$-gram in the first string with the nearest equal $p$-gram in the second string. Given a fixed integer $p \geq 1$, a threshold $m \geq 1$, and two strings $x$

and $y$ over an alphabet $\Sigma$, the *Local Rank Distance* between $x$ and $y$, denoted by $\Delta_{LRD}(x, y)$, is defined through the following algorithmic process. For each position $i$ in $x$ ($1 \leq i \leq |x| - p + 1$), the algorithm searches for a certain position $j$ in $y$ ($1 \leq j \leq |y| - p + 1$) such that $x[i : i + p] = y[j : j + p]$ and $|i - j|$ is minimized. If $j$ exists and $|i - j| < m$, then the offset $|i - j|$ is added to the Local Rank Distance. Otherwise, the maximal offset $m$ is added to the Local Rank Distance. An important remark is that LRD does not impose any mathematically developed global constraints, such as matching the $i$-th occurrence of a $p$-gram in $x$ with the $i$-th occurrence of that same $p$-gram in $y$. Instead, it is focused on the local phenomenon, and tries to pair equal $p$-grams at a minimum offset. To ensure that LRD is a (symmetric) distance function, the algorithm also has to sum up the offsets obtained from the above process by exchanging $x$ and $y$. LRD is formally defined next.

**Definition 14** Let $x, y \in \Sigma^*$ be two strings, and let $p \geq 1$ and $m \geq 1$ be two fixed integer values. The Local Rank Distance between $x$ and $y$ is defined as

$$\Delta_{LRD}(x, y) = \Delta_{\text{left}}(x, y) + \Delta_{\text{right}}(x, y),$$

where $\Delta_{\text{left}}(x, y)$ and $\Delta_{\text{right}}(x, y)$ are defined as follows:

$$\Delta_{\text{left}}(x, y) = \sum_{i=1}^{|x|-p+1} \min\{|i - j| \text{ such that}$$
$$1 \leq j \leq |y| - p + 1 \text{ and } x[i : i + p] = y[j : j + p]\} \cup \{m\},$$
$$\Delta_{\text{right}}(x, y) = \sum_{j=1}^{|y|-p+1} \min\{|j - i| \text{ such that}$$
$$1 \leq i \leq |x| - p + 1 \text{ and } y[j : j + p] = x[i : i + p]\} \cup \{m\}.$$

A string may contain multiple occurrences of a $p$-gram $z \in \Sigma^p$. LRD matches each occurrence of the $p$-gram $z$ from a string, with the nearest occurrence of the $p$-gram $z$ in the other string. Overlapping $p$-grams are also permitted in the computation of LRD. This makes perfect sense in the biological context, since there is no restriction that specifies where $k$-mers should start or end in a DNA string. An important notice is that, in order to be a symmetric distance measure, LRD must consider every $p$-gram in both strings. The symmetric property of LRD is ensured by computing both $\Delta_{\text{left}}$ and $\Delta_{\text{right}}$.

To gain some additional insights on how LRD actually works, it is useful to consider Example 9 where LRD is computed between strings $s_1$ and $s_2$ defined as in Example 8, using 1-mers (single characters).

*Example 9* Given $s_1 = CCGAATACG$ and $s_2 = TGACTCA$, defined as in Example 8, and the maximum offset $m = 10$, the LRD between $s_1$ and $s_2$ can be computed as follows:

$$\Delta_{LRD}(s_1, s_2) = \Delta_{\text{left}} + \Delta_{\text{right}},$$

where the two sums $\Delta_{\text{left}}$ and $\Delta_{\text{right}}$ are computed as follows:

$$\begin{aligned}\Delta_{\text{left}} = &|1-4| + |2-4| + |3-2| + |4-3| + |5-3| \\ &+ |6-5| + |7-7| + |8-6| + |9-2| = 19,\end{aligned}$$

$$\begin{aligned}\Delta_{\text{right}} = &|1-6| + |2-3| + |3-4| + |4-2| + |5-6| \\ &+ |6-8| + |7-7| = 12.\end{aligned}$$

In other words, $\Delta_{\text{left}}$ considers every 1-mer from $s_1$, while $\Delta_{\text{right}}$ considers every 1-mer from $s_2$. It is easy to observe that $\Delta_{LRD}(s_1, s_2) = \Delta_{LRD}(s_2, s_1)$.

## 7.4 Local Rank Distance Algorithm

A brute force search algorithm to compute LRD is given by Ionescu (2013). It is based on performing an extensive search within a window of fixed size to find the spatial displacement of equal $p$-grams between two strings, which is computationally expensive. This section presents an efficient algorithm that aims to improve the computational time of LRD by avoiding this costly search step. The efficient LRD algorithm, which was recently introduced by Ionescu (2015), is more than two orders of magnitude faster than the original brute force algorithm presented in the work of Ionescu (2013).

Given two strings $x$ and $y$, the efficient algorithm to compute $\Delta_{LRD}(x, y)$ works as described next. First of all, a hash table is used to store all the $p$-grams and their positions in $y$. More precisely, each $p$-gram that occurs at least once in $y$ will correspond to a key in the hash table. The value associated to a specific key ($p$-gram) is a set that will contain all the positions of the respective $p$-gram in the string $y$. The hash table $h$ can also be described as a positional inverted index structure. The next step is to consider every $p$-gram in $x$ and to look it up in the hash table $h$. Let $i$ denote the position of the currently considered $p$-gram in $x$. If the $p$-gram is found in $h$, the next step is to carry out a binary search in the set stored at $h(x[i : i + p])$, in order to find the nearest position $j$ that minimizes $|i - j|$. The spatial offset $|i - j|$ is added to the distance sum only if $|i - j|$ is less than the maximal offset $m$, otherwise, $m$ is added. If the $p$-gram $x[i : i + p]$ is not found in $h$, $m$ is added to the distance sum. The final sum obtained by this algorithm is the $\Delta_{\text{left}}(x, y)$ partial sum from Definition 14. It can be easily observed that $\Delta_{\text{right}}(x, y) = \Delta_{\text{left}}(y, x)$. Therefore, by switching the roles of $x$ and $y$ in the algorithm described so far, the $\Delta_{\text{right}}(x, y)$ partial sum can be obtained.

The efficient algorithm to compute either one of the partial sums is formally described in Algorithm 9. The $[i]$ operator is used in the algorithm to identify the

**Algorithm 9**: Efficient LRD Algorithm

```
1  Input:
2  x, y – the input strings;
3  p – the length of the p-grams to be compared;
4  m – the maximum spatial offset.
5  Notations:
6  h – the positional inverted index table;
7  ⌊r⌉ – the rounding function that returns the nearest integer value to r.
8  Computation:
9  for i ∈ {1, ..., |y| − p + 1} do
10     if h(y[i : i + p]) ≠ ∅ then
11         h(y[i : i + p]) ← h(y[i : i + p]) ∪ {i};
12     else
13         h(y[i : i + p]) ← {i};
14 Δ ← 0;
15 for i ∈ {1, ..., |x| − p + 1} do
16     if h(x[i : i + p]) ≠ ∅ then
17         o ← m;
18         beg ← 1;
19         end ← |h(x[i : i + p])|;
20         while end − beg > 1 do
21             mid ← ⌊(beg + end)/2⌉;
22             if i = h(x[i : i + p])[mid] then
23                 beg ← mid;
24                 end ← mid;
25             else
26                 if i < h(x[i : i + p])[mid] then
27                     end ← mid;
28                 else
29                     beg ← mid;
30         for j ∈ {beg, ..., end} do
31             if |i − h(x[i : i + p])[j]| < o then
32                 o ← |i − h(x[i : i + p])[j]|;
33         Δ ← Δ + o;
34     else
35         Δ ← Δ + m;
36 Output:
37 Δ – the Δ_left partial sum between x and y.
```

$i$-th element of a set, such as the set of positions stored in $h$ for a certain $p$-gram. The algorithm can be divided into two stages. In the first stage (steps 9–13), the positional inverted index table $h$ is computed from the input string $y$. In the second stage (steps 14–35), a binary search algorithm is employed for each $p$-gram that $x$ and $y$ have in common, in order to find the position in $y$ that minimizes the spatial offset from the position in $x$ of each common $p$-gram. In order to compute LRD, the algorithm has to be executed twice, once to compute $\Delta_{\text{left}}(x, y)$ and once to compute $\Delta_{\text{right}}(x, y)$. The analysis of the computational complexity of Algorithm 9 is straightforward. Let $l_1 = |x|$ and $l_2 = |y|$ denote the lengths of the two strings. Searching and storing keys in the hash table can be done in $O(1)$ time. The binary search takes $O(log\, n)$. The first stage of the algorithm can be computed in $O(l_2)$ time. The second phase is more computationally expensive, requiring $O(l_1 \cdot log\, l_2)$ time. Consequently, the overall complexity of Algorithm 9 is $O(l_1 \cdot log\, l_2)$. Since Algorithm 9 has to be executed twice to compute each partial sum, LRD can be computed in $O(l_1 \cdot log\, l_2 + l_2 \cdot log\, l_1)$ time complexity. Unlike the brute force search LRD algorithm (Ionescu 2013), the computational time of the efficient LRD algorithm is no longer limited by the maximal offset $m$ of LRD. This is a clear advantage of this faster implementation. It is worth noting that an open source Java implementation of the efficient LRD algorithm is provided at http://lrd.herokuapp.com.

In practice, the input parameters of Algorithm 9 should be carefully adjusted with respect to the kind of strings (texts or DNA sequences) and the approached task. For example, setting $p = 10$ to use 10-mers for DNA strings of 100 or 200 bases is probably not reasonable, since finding similar 10-mers in such short DNA strings is rare, if not almost impossible. But 3 to 5-mers are probably more suitable for such short DNA strings. In the case of authorship identification, it is worthwhile using 2 to 5-grams that correspond to function words which have been found to provide a good indication of the author style (Popescu and Grozea 2012). Thus, good knowledge of the domain and the approached task is very useful, but the parameters can be easily tuned through a validation procedure. An interesting note on improving the efficiency when computing LRD for multiple strings is to build the hash table for each string only once. For this, the two stages of Algorithm 9 need to be completely separated. Once the hash table for a certain string $y$ has been built, it can be reused to compute all partial sums $\Delta_{\text{left}}(x, y), \forall x \in X \setminus \{y\}$, where $X$ is the set of strings. Furthermore, it is sufficient to keep only one hash table in memory at a time, considerably reducing the memory footprint. Nevertheless, the algorithm can also be easily executed in a parallel computing environment.

Example 10 illustrates how Algorithm 9 works in a particular case. The example shows that binary search is required only when a $p$-gram occurs several times in one of the strings.

*Example 10* Let $s_1 = ABAB$, $s_2 = BABBABA$ be two strings from $\{A, B\}^*$, and the maximum offset $m = 5$. In order to compute $\Delta_{\text{left}}(s_1, s_2)$, the inverted index table $h$ with 3-grams from $s_2$ needs to be constructed first:

$$h = \{BAB => [1, 4], ABB => [2], BBA => [3], ABA => [5]\}$$

The next step is to take the 3-grams from $s_1$ and look them up in $h$:

- The first 3-gram from $s_1$, namely $ABA$, is found at position 5 in $s_2$ according to $h$, so the offset $|1 - 5|$ is added to the distance.

- The second 3-gram from $s_1$, namely $BAB$, is found at positions [1, 4] in $s_2$ according to $h$. In this case, a binary search in the array [1, 4] is employed to find the closest position to 2. The closest position is 1, so the offset $|2 - 1|$ is added to the distance.

In conclusion, $\Delta_{\text{left}}(s_1, s_2)$ is the sum of $|1 - 5|$ and $|2 - 1|$, namely 5.

## 7.5 Properties of Local Rank Distance

Local Rank Distance replicates the behavior of Local Patch Dissimilarity. As Local Patch Dissimilarity is adapted to images, LRD is based on principles that make it more suitable for DNA sequences and strings, in general. LRD essentially measures the non-alignment score between $p$-grams. Both the Local Rank Distance and the Local Patch Dissimilarity can be considered as extended versions of rank distance (Dinu and Manea 2006), which are developed under practical motivations, specifically to provide a good measure of similarity between certain kinds of objects, such as strings or images. Although LRD was designed with a practical motivation in mind, it still has a few interesting theoretical properties. A proof that LRD is a semi-metric is first given. After that, a proof of the expected distance between two random strings is presented.

The definition of a semi-metric is considered next, as the first concern of this section is to prove that LRD is a semi-metric.

**Definition 15** A semi-metric on a set $X$ is a function $d : X \times X \to \mathbb{R}$ that satisfies the following conditions for all $x, y, z \in X$:

$$(i)\ d(x, y) \geqslant 0 \text{ (non-negativity, or separation axiom);}$$
$$(ii)\ d(x, y) = 0 \text{ if and only if } x = y \text{ (coincidence axiom);}$$
$$(iii)\ d(x, y) = d(y, x) \text{ (symmetry).}$$

It is worth noting that axioms $(i)$ and $(ii)$ from Definition 15 produce positive definiteness.

**Theorem 3** *Local Rank Distance is a semi-metric.*

*Proof* The proof is based on showing that $\Delta_{LRD}$ given by Definition 14 satisfies the conditions in Definition 15.

Let $x, y \in \Sigma^*$ and let $p \geq 1$ and $m \geq 1$ be two fixed integer values.

$(i)$ **Proof of non-negativity**

$\Delta_{LRD}(x, y)$ is a sum of positive or zero values that represent offsets between $p$-grams in $x$ and $y$. This sum can only be greater or equal to 0. This ensures the non-negativity condition in Definition 15.

$(ii)$ **Proof of coincidence axiom**

($\Rightarrow$): Let $\Delta_{LRD}(x, y) \neq 0$. There is at least one position $i$ in $x$ such that $\min\{|i - j|, \forall 1 \leq j \leq |y| - p + 1 \text{ and } x[i : i + p] = y[j : j + p]\} \cup \{m\} > 0$, or at least one position $j$ in $y$ such that $\min\{|j - i|, \forall 1 \leq i \leq |x| - p + 1 \text{ and } y[j : j + p] = x[i : i + p]\} \cup \{m\} > 0$, otherwise $\Delta_{LRD}(x, y)$ would be 0.

Case 1: If $\min\{|i - j|, \forall 1 \leq j \leq |y| - p + 1 \text{ and } x[i : i + p] = y[j : j + p]\} \cup \{m\} > 0$, then the $p$-gram $x[i : i + p]$ does not occur in the same position in $y$ as in $x$. In other words, $x[i : i + p] \neq y[i : i + p]$. The strings $x$ and $y$ have at least one different $p$-gram, therefore $x \neq y$.

Case 2: If $\min\{|j - i|, \forall 1 \leq i \leq |x| - p + 1 \text{ and } y[j : j + p] = x[i : i + p]\} \cup \{m\} > 0$, the proof is analogous to Case 1. The implication is proven.

($\Leftarrow$): Let $x \neq y$. There is at least one position $i$ in $x$, or $j$ in $y$ with different $p$-grams.

Case 1: The $p$-gram $x[i : i + p]$ is considered. Since $y$ does not contain the same $p$-gram at position $i$, or in other words $x[i : i + p] \neq y[i : i + p]$, it means that $\min\{|i - j|, \forall 1 \leq j \leq |y| - p + 1 \text{ and } x[i : i + p] = y[j : j + p]\} \cup \{m\} > 0$. Hence, there is at least one term in the sum $\Delta_{\text{left}}$ that is positive. Consequently, $\Delta_{LRD}(x, y) \neq 0$.

Case 2: The proof for the $p$-gram $y[j : j + p]$ is analogous to Case 1. The coincidence axiom is proven.

$(iii)$ **Proof of symmetry**

Using the commutative property of the $+$ operation and that of the euclidean distance, the proof of symmetry for $\Delta_{LRD}$ is immediate:

$$
\begin{aligned}
\Delta_{LRD}(x, y) = & \sum_{i=1}^{|x|-p+1} \min\{|i - j| \text{ such that} \\
& 1 \leq j \leq |y| - p + 1 \text{ and } x[i : i + p] = y[j : j + p]\} \cup \{m\} \\
& + \sum_{j=1}^{|y|-p+1} \min\{|j - i| \text{ such that} \\
& 1 \leq i \leq |x| - p + 1 \text{ and } y[j : j + p] = x[i : i + p]\} \cup \{m\} \\
= & \sum_{j=1}^{|y|-p+1} \min\{|j - i| \text{ such that} \\
& 1 \leq i \leq |x| - p + 1 \text{ and } y[j : j + p] = x[i : i + p]\} \cup \{m\} \\
& + \sum_{i=1}^{|x|-p+1} \min\{|i - j| \text{ such that} \\
& 1 \leq j \leq |y| - p + 1 \text{ and } x[i : i + p] = y[j : j + p]\} \cup \{m\} \\
= & \Delta_{LRD}(y, x).
\end{aligned}
$$

This ends the proof of Theorem 3. □

It is worth noting that the coincidence axiom is not longer verified if approximate matches are allowed between $p$-grams. The use of Hamming distance to compare $k$-mers (and match them under a similarity threshold) may seem appropriate from a biological point of view, but from a mathematical point of view, LRD can no longer be considered a semi-metric. In practice, better results can probably be obtained using approximate matches, but this subject is not addressed in this book.

Another important note is that it can be demonstrated that LRD is a distance function for a given $p \geq 1$, given that all the $p$-grams occur only a single time in each of the two strings. This is rarely the case in practice, so the proof that LRD satisfies the triangle inequality only when $p$-grams are unique is not of great interest from a practical point of view. Usually, a $p$-gram from one string can be associated with two or more identical $p$-grams from the other string, and LRD is no longer a distance function, but it still remains a semi-metric, as proven in Theorem 3.

An upper bound for LRD can be computed as the product between the maximum offset $m$ and the number of pairs of compared $p$-grams. Thus, LRD can be normalized to a value in the $[0, 1]$ interval. By normalizing, LRD can also be transformed into a similarity or dissimilarity measure.

The following theorem gives the expected value of LRD between two random strings.

**Theorem 4** *Given $x, y \in \Sigma^*$, $p \geq 1$ and $m \geq 1$ two fixed integer values, and the set of p-grams $Z = \{z \mid z \in \Sigma^p, \exists i \in \{1, 2, \ldots, |x| - p + 1\}$ such that $z = x[i : i + p] \vee \exists j \in \{1, 2, \ldots, |y| - p + 1\}$ such that $z = y[j : j + p]\}$, the expected Local Rank Distance between x and y can be approximated by*

$$\Delta_{LRD}(x, y) \approx \left[ \left( 1 - \left( \frac{|Z| - 1}{|Z|} \right)^{2m-p+1} \right) \frac{m}{2} + \left( \frac{|Z| - 1}{|Z|} \right)^{2m-p+1} m \right] (|x| + |y|), \tag{7.1}$$

*where $|Z|$ gives the number of p-grams in Z, $|x|$ is the length of x, and $|y|$ is the length of y, respectively.*

*Proof* For every $p$-gram that occurs in $x$, LRD adds the offset to the nearest occurrence of the same $p$-gram in $y$. In the same manner, for every $p$-gram that occurs in $y$, LRD adds the offset to the nearest occurrence in $x$. If the expected offset is defined by $d \in [0, m]$, then the following equation approximates LRD:

$$\Delta_{LRD}(S_1, S_2) \approx d(|x| + |y|). \tag{7.2}$$

To determine the expected offset $d$ of a certain $p$-gram, the following discussion is made. Let $z$ be a $p$-gram from $Z$ that appears in either one of the strings. LRD searches for the $p$-gram $z$ in a window of at most $2m + 1$ characters in the other string. However, the $p$-gram $z$ may or may not occur in the search window. Let $A$ be the event defined as "there is at least one occurrence of $p$-gram $z$ in a string of $2m + 1$ symbols". Let $A^C$ denote the complementary event of $A$.

The probability of event $A^C$ is given by

$$P(A^C) = \left(\frac{|Z|-1}{|Z|}\right)^{2m-p+1}. \tag{7.3}$$

The probability of event $A$ can be determined using Eq. (7.3) as follows:

$$P(A) = 1 - P(A^C) = 1 - \left(\frac{|Z|-1}{|Z|}\right)^{2m-p+1}. \tag{7.4}$$

When event $A$ occurs, the average offset of $p$-gram $z$ is $m/2$, since each position in the search window is equally likely to be the occurrence of $z$. When the complementary event $A^C$ occurs, LRD adds the maximum offset, since the $p$-gram $z$ is not found in the search window. Thus, the expected offset is given by

$$d = \left(1 - \left(\frac{|Z|-1}{|Z|}\right)^{2m-p+1}\right)\frac{m}{2} + \left(\frac{|Z|-1}{|Z|}\right)^{2m-p+1} m \tag{7.5}$$

By replacing $d$ in Eq. (7.2) with the value determined in Eq. (7.5), the result in Eq. (7.1) is obtained. This ends the proof. □

Theorem 4 approximates the expected distance between two strings. This approximation is useful in practice, especially when one wants to determine the similarity of two strings without having additional knowledge about the strings. If the obtained distance is lower than the expected value given in Eq. (7.1), then the strings can be considered as being similar. Otherwise, the strings can be considered as being dissimilar.

## 7.6 Local Rank Distance Sequence Aligners

This section introduces two sequence aligners that work under Local Rank Distance. The first one is based on a basic algorithm that aligns a read of length $l$ against a reference DNA sequence of length $n$. For efficiency reasons, it actually computes only $\Delta_{\text{left}}$ from Definition 14 between the read and a certain substring from the reference genome. It is perfectly reasonable to use only one of the two partial sums, $\Delta_{\text{left}}$ or $\Delta_{\text{right}}$, since the symmetric property of LRD is no longer needed in the context of sequence alignment.

The basic alignment algorithm compares the read with the first substring in the reference and remembers the offset of each $k$-mer in the read. As it continues to compare the read with the following substrings at position $2, 3, \ldots, n - l + 1$ in the reference genome, respectively, the algorithm only needs to update the offset of each $k$-mer to obtain the new $\Delta_{\text{left}}$ distance at a certain position. The read is aligned in the position that gives the minimum $\Delta_{\text{left}}$ distance, but only if the obtained distance is less

**Algorithm 10**: LRD Sequence Aligner

1 **Input**:
2 $r$ – a short DNA string of length $l$;
3 $s$ – a reference DNA sequence of length $n$;
4 $k$ – the size of the $k$-mers to be compared;
5 $m$ – the maximum offset;
6 $th$ – the maximum distance threshold accepted for the aligned read;
7 $d$ – the threshold that can be adjusted to skip the alignment at some positions.
8 **Initialization**:
9 $\Delta_{min} \leftarrow th$;
10 $bestPos \leftarrow 0$;
11 **Computation**:
12 **for** $i \in \{1, ..., l - k + 1\}$ **do**
13     add $i$ in the array stored at $h(r[i : i + k])$;
14 **for** $i \in \{1, ..., n - k + 1\}$ **do**
15     **if** $|h(s[i : i + k])| > 0$ **then**
16         $f[i] \leftarrow$ true;
17     **else**
18         $f[i] \leftarrow$ false;
19 $count \leftarrow 0$;
20 **for** $i \in \{1, ..., l - k + 1\}$ **do**
21     **if** $f[i] = true$ **then**
22         $count \leftarrow count + 1$;
23     $c[1] \leftarrow count$;
24 **for** $i \in \{2, ..., n - l + 1\}$ **do**
25     **if** $f[i - 1] = true$ **then**
26         $count \leftarrow count - 1$;
27     **if** $f[i + l - k + 1] = true$ **then**
28         $count \leftarrow count + 1$;
29     $c[i] \leftarrow count$;
30 **for** $i \in \{1, 2, ..., n - l + 1\}$ **do**
31     **if** $c[i] \geq \max\{c\} - d$ ***and*** $(|r| - k - c[i]) \cdot m < \Delta_{min}$ **then**
32         $\Delta \leftarrow 0$;
33         **for** $j \in \{1, ..., l - k + 1\}$ **do**
34             **if** $\Delta > \Delta_{min}$ **then**
35                 abort and proceed to the next value of $i$ in the loop from step 30;
36             **else**
37                 **if** $f[i + j - 1] = true$ **then**
38                     do a binary search in the array stored at $h(s[i + j - 1 : i + j - 1 + k])$ to obtain the position $p$ that minimizes $|j - p|$;
39                     $\Delta \leftarrow \Delta + \min\{|j - p|, m\}$;
40                 **else**
41                     $\Delta \leftarrow \Delta + m$;
42         **if** $\Delta < \Delta_{min}$ **then**
43             $\Delta_{min} \leftarrow \Delta$;
44             $bestPos \leftarrow i$;
45 **Output**:
46 $bestPos$ – the position where the read $r$ was aligned;
47 $\Delta_{min}$ – the minimum LRD (or $\Delta_{\text{right}}$ to be more precise) obtained at position $bestPos$.

than a certain threshold. It is worth mentioning that the algorithm described above needs to be applied for the short DNA string obtained by reverse complementing the original read. Several efficiency improvements are described next. In the end, they lead to the development of the faster LRD aligner presented in Algorithm 10. Both the basic and the efficient LRD aligners are provided in a software package, which is freely available for future development and use at http://lrd.herokuapp.com/aligners. The software is implemented in C++ and Java, being supported on UNIX and MS Windows.

### *7.6.1 Indexing Strategies and Efficiency Improvements*

The main efficiency improvement brought to the LRD aligner is to store $k$-mer positions in a hash table for each read. More precisely, the hash table $h$ constructed from a short DNA read $r$ will contain an array for each $k$-mer in the read. The array will contain all the positions of that $k$-mer in the read $r$. This hash table is actually a positional inverted index structure that is very popular in information retrieval. When LRD is computed for the read at a certain position $i$ in the reference genome $s$, it is no longer necessary to do an extensive search within a window of fixed size to find equal $k$-mers between the read and the substring $s[i : i + |r|]$. The alternative solution is to take every $k$-mer in $s[i : i + |r|]$ and to look it up in the hash table $h$. Let $j$ denote the position of the currently considered $k$-mer in $s[i : i + |r|]$. If the $k$-mer is found in $h$, the next step is to try a binary search in the positional array that is stored in $h(s[i + j - 1 : i + j - 1 + k])$, in order to find the nearest position $p$ that minimizes $|j - p|$. The offset $|j - p|$ is added to the distance sum only if $|j - p|$ is less than the maximal offset $m$, otherwise, $m$ is added. If the $k$-mer is not found in $h$, $m$ is added to the distance sum. The final sum obtained by this algorithm is the $\Delta_{\text{right}}$ partial sum from Definition 14. As mentioned before, one of the two partial sums can be left out for efficiency reasons, without affecting the accuracy. Thus, the hash LRD implementation described in Algorithm 10 is based only on $\Delta_{\text{right}}$, as opposed to the basic implementation that uses only $\Delta_{\text{left}}$. Consequently, there are some minor differences in the results obtained by the two implementations, but the accuracy levels are very similar and in some cases almost the same. For instance, in the experiments performed to evaluate the aligners in the presence of contaminated reads, the hash LRD aligner is only slightly better, while for the task of clustering unknown organisms, the results of the two LRD aligners are exactly the same.

The following strategies are designed to further improve the hash LRD aligner, in terms of speed. First of all, a Boolean array $f$ of size $|s| - k + 1$ is used. Each element $f[i]$ indicates if the $k$-mer $s[i : i + k]$ is in the hash table $h$. When the algorithm tries to align the read at every position $i$ in the reference sequence $s$, by computing the distance from the read $r$ to the substring $s[i : i + |r|]$, it will have to look up some of the $k$-mers in $h$, several times (more precisely, $|r|$ times). Despite the fact that the hash table look up takes $O(1)$ time in theory, it is still faster to check the value of $f[i]$ instead of doing a hash table look up. Another improvement is to stop the alignment at a certain position $i$, if the distance sum computed between $r$ and $s[i : i + |r|]$ becomes greater than the minimum $\Delta_{\text{right}}$ obtained so far.

The next efficiency improvement is to count the number of $k$-mers that are found in $h$, for every substring $s[i : i + |r|]$ in the reference genome. These counts are stored in an array $c$ of length $|s| - |r|$. The algorithm can now consider the alignment only in the positions in the reference that are more likely to give the minimum $\Delta_{\text{right}}$ distance. It is fairly easy to observe that the more equal $k$-mers $r$ and $s[i : i + |r|]$ have in common, the lower $\Delta_{\text{right}}(r, s[i : i + |r|])$ should be, since LRD is first based on finding equal $k$-mers between the two strings and, then, on minimizing the offsets between these $k$-mers. However, there is no guarantee that this is always the case. Therefore, the approach to skip the alignment for some positions $i$ with low $c[i]$, in order to speed up the hash LRD aligner, gives approximate alignment results. More precisely, lower distances can probably be obtained for some of the disregarded positions. These positions are disregarded by the two rules described next. The first rule is to eliminate the position $i$ if $c[i] < \max\{c\} - d$, where $d$ is a new input parameter of the aligner. This parameter can take values in the interval $[0, max\{c\}]$. When $d = 0$, more positions are disregarded. When $d = \max\{c\}$, no positions are disregarded at all, since $c[i]$ is always greater than 0. If the parameter $d$ is set to eliminate more positions during the alignment, the algorithm will be faster, but it will also give less accurate alignment results. However, choosing $d = 5$ for reads of length 100 gives similar results to the basic LRD aligner in terms of accuracy, while drastically reducing the computational time, as the empirical results presented in this chapter show. In all the experiments, the results of the approximate hash LRD aligner are obtained with $d = 5$. The second rule used by the approximate aligner is to eliminate the position $i$, if $(|r| - k - c[i]) \cdot m$ is greater than the minimum $\Delta_{\text{right}}$ distance obtained until position $i$. The difference $|r| - k - c[i]$ gives the number of $k$-mers in $s[i : i + |r|]$ that are not found in $h$. For each of these missing $k$-mers, the maximal offset $m$ is added to the $\Delta_{\text{right}}$ sum. Thus, $\Delta_{\text{right}}(r, s[i : i + |r|])$ is always greater than $(|r| - k - c[i]) \cdot m$. But, if $(|r| - k - c[i]) \cdot m$ is already greater than the minimum $\Delta_{\text{right}}$ distance obtained so far, there is no point in aligning the read at position $i$.

All the improvements described above are actually combined together to obtain an efficient LRD aligner. It is fairly obvious that these efficiency improvements and indexing strategies produce a different yet more efficient algorithm than the basic LRD aligner. The approximate hash LRD aligner is described in Algorithm 10. As in the basic LRD aligner, a read is only aligned by the hash LRD aligner if the minimum LRD (or $\Delta_{\text{right}}$, to be more precise) obtained by the algorithm is less than a certain threshold.

Algorithm 10 is also applied for the short DNA string obtained by reverse complementing the original read. But, another speed improvement is considered here. The alignment tool tries to align the reverse complement only if the minimum distance for the original read is not acceptable. An internal threshold is used to determine if the minimum $\Delta_{\text{right}}$ is acceptable (lower than the threshold) or not. This threshold is computed as follows:

$$t = \min\{th, \min\{k, m\} \cdot (|r| - k + 1)\}.$$

The threshold $t$ is low enough to ensure, with a certain probability, that if $\Delta_{\text{right}} < t$ then the read is aligned in the right place. This parameter speeds up the alignment tool especially when the reads and the reference genome belong to the same species. If the reads belong to other species (as in the case of contaminated reads, for example), the aligner will most likely try to align the reverse complements too.

In the end, the computational complexity of Algorithm 10 is $O(n \cdot l \cdot \log \frac{l}{|\Sigma|^k})$. Unlike the basic LRD aligner, the computational time of the approximate hash LRD aligner is no longer limited by the maximal offset $m$ of LRD, which represents an advantage of this faster implementation. However, in the experiments, the results of both the basic LRD aligner and the approximate hash LRD aligner are obtained with $m = 36$ in order to compare the results of the two aligners and to show that they produce almost the same results.

In practice, the input parameters of Algorithm 10 should be carefully adjusted with respect to length of the DNA reads and to the amount of mutations and errors in DNA. For example, setting $k = 10$ to use 10-mers for reads of 100 or 200 bases is not reasonable, since finding similar 10-mers in such short DNA strings is rare, if not almost impossible. But 3 to 5-mers are probably more suitable for aligning short DNA reads. Notice that the maximum offset parameter $m$ should be adjusted accordingly. Using 5-mers and a maximum offset that is too small (less than 10, for example) might result in finding almost no similar 5-mers in the search window. The best practice to choose the parameters of the aligner is to tune them on a validation data set first.

## 7.7 Experiments and Results

### 7.7.1 Data Sets Description

LRD is evaluated on two important problems in bioinformatics: the phylogenetic analysis of mammals and the finding of common substrings in a set of given DNA strings. In the experiments presented in this chapter, mitochondrial DNA sequence genome of 27 mammals is used. The 27 mammals along with their associated accession numbers are given in Table 7.1. The mtDNA sequences are available for download in the EMBL database (http://www.ebi.ac.uk/ena/). The mammals selected for the experiments belong to one of the following 8 biological orders: Carnivora, Cetartiodactylae, Chiroptera, Metatheria, Monotremata, Perissodactylae, Primates, and Rodentia.

**Table 7.1** The 27 mammals from the EMBL database used in the phylogenetic experiments

| Mammal | Latin name | Accession |
|---|---|---|
| Human | *Homo sapiens* | V00662 |
| Common chimpanzee | *Pan troglodytes* | D38116 |
| Pigmy chimpanzee | *Pan paniscus* | D38113 |
| Gorilla | *Gorilla gorilla* | D38114 |
| Orangutan | *Pongo pygmaeus* | D38115 |
| Sumatran orangutan | *Pongo pygmaeus abelii* | X97707 |
| Gibbon | *Hylobates lar* | X99256 |
| Hamadryas baboon | *Papio hamadryas* | Y18001 |
| Horse | *Equus caballus* | X79547 |
| Donkey | *Equus asinus* | X97337 |
| Indian rhinoceros | *Rhinoceros unicornis* | X97336 |
| White rhinoceros | *Ceratotherium simum* | Y07726 |
| Harbor seal | *Phoca vitulina* | X63726 |
| Gray seal | *Halichoerus grypus* | X72004 |
| Cat | *Felis catus* | U20753 |
| Fin whale | *Balaenoptera physalus* | X61145 |
| Blue whale | *Balaenoptera musculus* | X72204 |
| Cow | *Bos taurus* | V00654 |
| Sheep | *Ovis aries* | AF010406 |
| Pig | *Sus scrofa* | AF034253 |
| Rat | *Rattus norvegicus* | X14848 |
| Mouse | *Mus musculus* | V00711 |
| Fat dormouse | *Myoxus glis* | AJ001562 |
| Jamaican fruit-eating bat | *Artibeus jamaicensis* | AF061340 |
| North American opossum | *Didelphis virginiana* | Z29573 |
| Wallaroo | *Macropus robustus* | Y10524 |
| Platypus | *Ornithorhynchus anatinus* | X83427 |

The accession number is given on the last column

To evaluate the LRD aligners, several experiments are conducted on two data sets of genome sequences. The first data set is a subset of 20 mitochondrial DNA sequence genomes presented in Table 7.1. They belong to the following biological orders: Carnivora, Cetartiodactylae, Perissodactylae, Primates, Rodentia.

Mitochondrial DNA (mtDNA) is the DNA located in organelles called mitochondria. The DNA sequence of mtDNA has been determined from a large number of organisms and individuals, and the comparison of those DNA sequences represents a mainstay of phylogenetics, in that it allows biologists to elucidate the evolutionary relationships among species. In mammals, each double-stranded circular mtDNA molecule consists of 15,000–17,000 base pairs. DNA from two individuals of the same species differs by only 0.1 %. This means, for example, that mtDNA from two

**Table 7.2** The genomic sequence information of three vibrio pathogens consisting of two circular chromosomes

| Species | Chromosome | Accession no. | Size (Mbp) |
|---|---|---|---|
| *V. vulnificus* YJ016 | I (VV1) | NC_005139 | 3.4 |
| *V. vulnificus* YJ016 | II (VV2) | NC_005140 | 1.9 |
| *V. parahaemolyticus* RIMD 2210633 | I (VP1) | NC_004603 | 3.3 |
| *V. parahaemolyticus* RIMD 2210633 | II (VP2) | NC_004605 | 1.9 |
| *V. cholerae* El Tor N16961 | I (VC1) | NC_002505 | 3.0 |
| *V. cholerae* El Tor N16961 | II (VC2) | NC_002506 | 1.0 |

different humans differs by less than 20 base pairs. Because this small difference cannot affect the study, the experiments are conducted using a single mtDNA sequence for each mammal.

The second data set used for sequence alignment contains chromosomal DNA sequence genomes of three vibrio pathogens available in the NCBI database (http://www.ncbi.nlm.nih.gov): *Vibrio vulnificus* YJ106, *Vibrio parahaemolyticus* RIMD 2210633, and *Vibrio cholerae* El Tor N16961. The genomes of these three organisms consist of two circular chromosomes. Additional information about these chromosomes, including accession number and size (given in Megabase pairs), is given in Table 7.2. The genomic sequences of these vibrio species have been revealed by different studies (Chen et al. 2003; Heidelberg et al. 2000; Makino et al. 2003). Several studies report that Vibrio vulnificus shares morphological and biochemical characteristics with other human vibrio pathogens, including Vibrio cholerae and Vibrio parahaemolyticus (Chen et al. 2003; Lin et al. 2005).

### *7.7.2 Phylogenetic Analysis*

Two experiments are performed on the phylogenetic analysis of mammals. The first one includes mitochondrial DNA sequence genome of 22 mammals from the EMBL database. Results on this data set of 22 mammals are also reported by Dinu and Ionescu (2012a, 2013), Dinu and Sgarro (2006). Similar studies were also performed by Cao et al. (1998), Li et al. (2004), Reyes et al. (2000).

In this work, a hierarchical clustering technique based on LRD, using the average linkage criterion, is tested on the 22 mammals data set. Single or complete linkage criteria show similar results, but the average linkage seems to work best in combination with LRD. Figures 7.1, 7.2, 7.3, 7.4, and 7.5 show the results obtained with the hierarchical clustering based on LRD using different $k$-mer sizes. As the size of the $k$-mers grows, the dendrograms look better and better. For LRD with 8-mers (Fig. 7.4) and 10-mers (Fig. 7.5), perfect phylogenetic trees are obtained. In other words, mammals are clustered according to their biological orders. The maximum offset of LRD ranges from 640 to 1000, and it is adjusted proportional to the size of

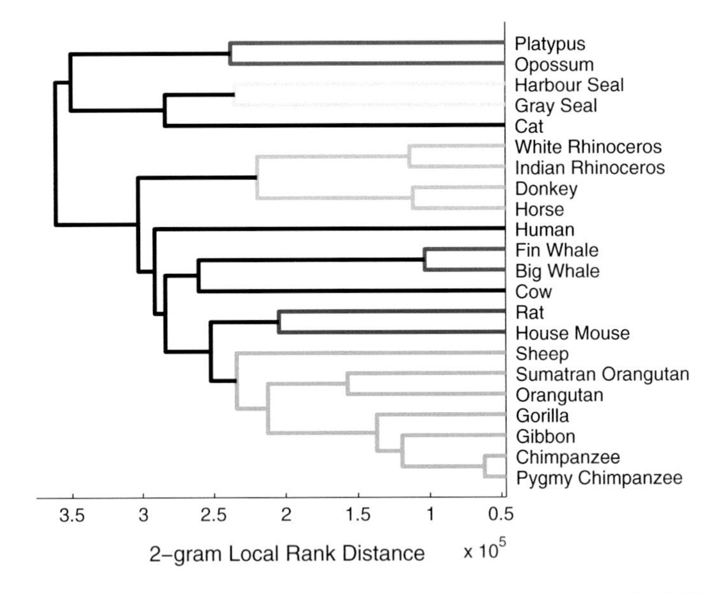


**Fig. 7.1** Phylogenetic tree obtained for 22 mammalian mtDNA sequences using LRD based on 2-mers

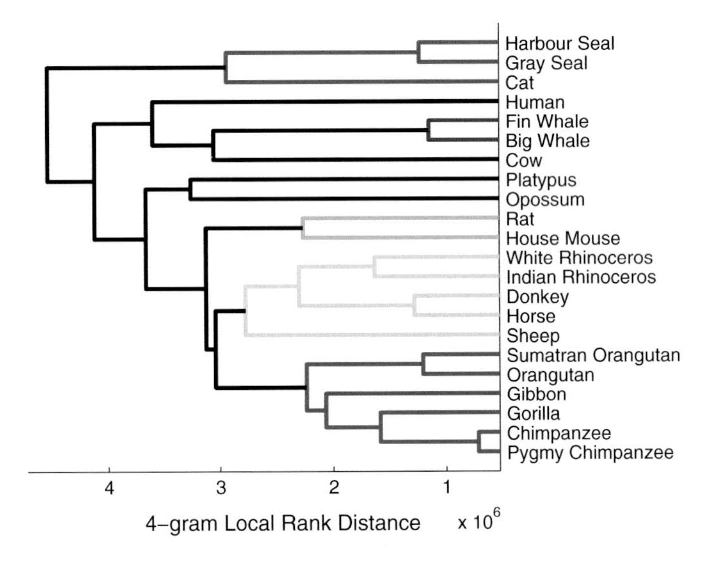


**Fig. 7.2** Phylogenetic tree obtained for 22 mammalian mtDNA sequences using LRD based on 4-mers

$k$-mers used. Another dendrogram presented in Fig. 7.6 is obtained by summing up the Local Rank Distances with 6, 7, 8, 9, and 10-mers, respectively. Again, mammals are perfectly clustered according to their orders. The idea of summing up distances obtained with different $k$-mers makes the hierarchical clustering method more robust.

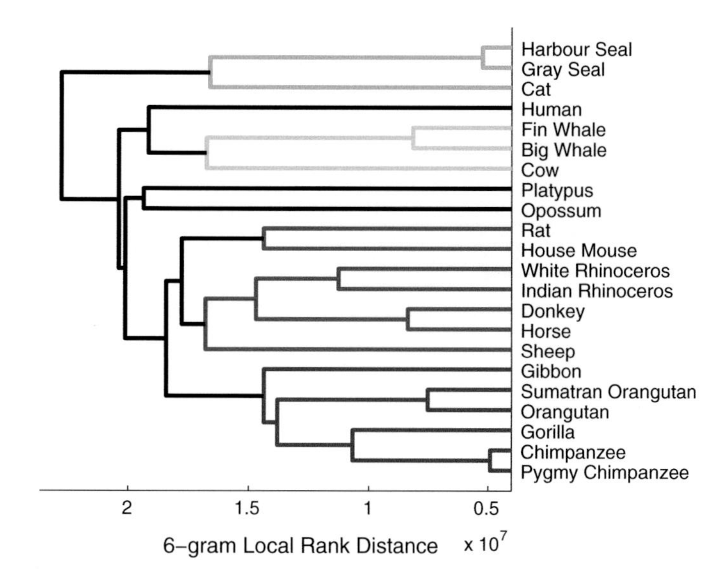


**Fig. 7.3** Phylogenetic tree obtained for 22 mammalian mtDNA sequences using LRD based on 6-mers

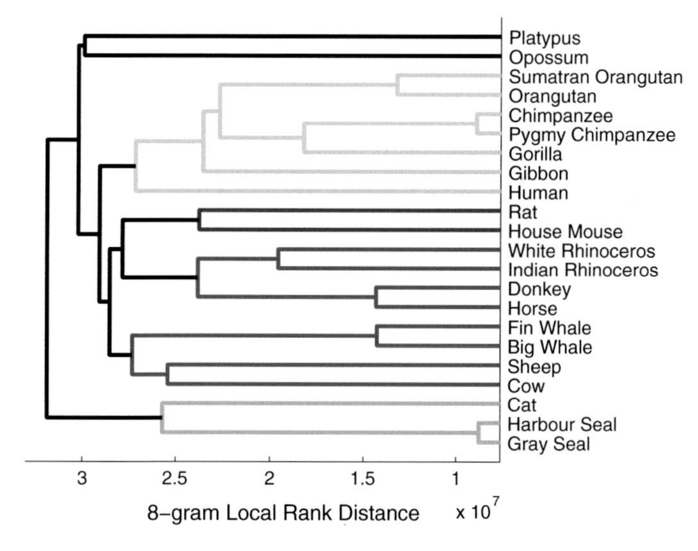


**Fig. 7.4** Phylogenetic tree obtained for 22 mammalian mtDNA sequences using LRD based on 8-mers

Table 7.3 shows the number of misclustered mammals of previously proposed techniques and that of the hierarchical clustering based on LRD with sum of $k$-mers. The best result with 100 % accuracy is obtained by the hierarchical clustering based on LRD.

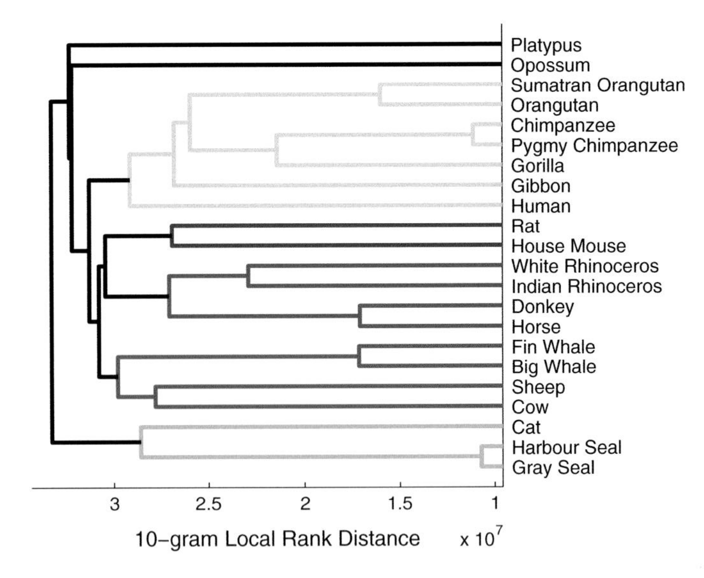


**Fig. 7.5** Phylogenetic tree obtained for 22 mammalian mtDNA sequences using LRD based on 10-mers

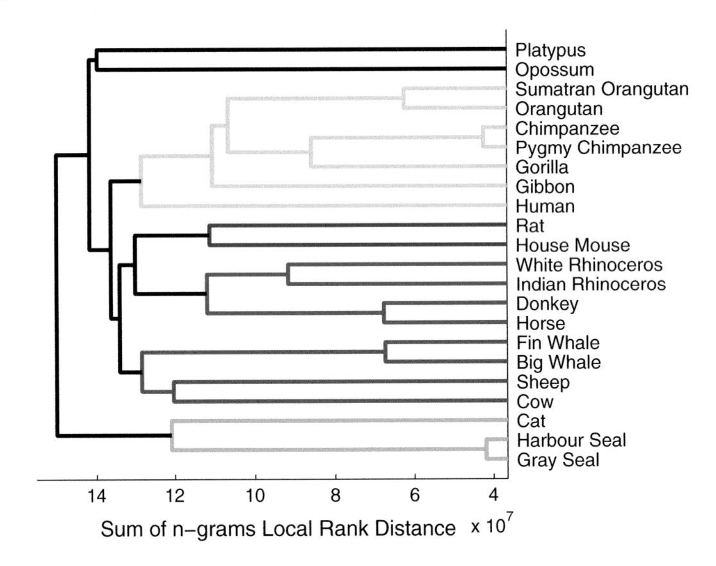


**Fig. 7.6** Phylogenetic tree obtained for 22 mammalian mtDNA sequences using LRD based on sum of $k$-mers

Since the hierarchical clustering based on LRD obtains perfect accuracy, another clustering experiment with more DNA sequences is performed. This second experiment includes mtDNA sequences from all the 27 mammals obtained from the EMBL database. If 8-mers are enough to obtain a perfect phylogenetic tree in the first experiment, longer $k$-mers are considered for the second experiment. Figure 7.7 shows the

**Table 7.3** The number of misclustered mammals for different clustering techniques on the 22 mammals data set

| Method | Misclustered | Accuracy (%) |
|---|---|---|
| Proposed by Dinu and Sgarro (2006) | 3/22 | 86.36 |
| Proposed by Dinu and Ionescu (2012a) | 3/22 | 86.36 |
| LRD + sum of $k$-mers | 0/22 | 100.00 |

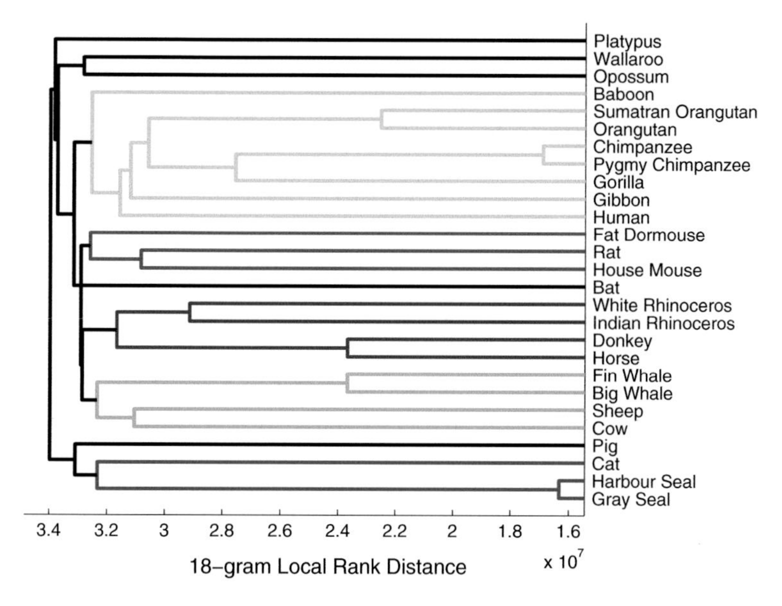


**Fig. 7.7** Phylogenetic tree obtained for 27 mammalian mtDNA sequences using LRD based on 18-mers

dendrogram obtained by the hierarchical clustering based on LRD using 18-mers. Again the average linkage criterion gives the best results. The only mistake of the proposed method is that it clusters the pig together with members of the Carnivora order instead of the Cetartiodactylae order. Having 1 out of 27 misclustered mammals, the accuracy is 96.29 % this time.

Overall, the accuracy level achieved by the clustering method based on LRD is better or at least comparable with state-of-the-art methods proposed in similar studies (Cao et al. 1998; Dinu and Sgarro 2006; Li et al. 2004; Reyes et al. 2000).

### *7.7.3 DNA Comparison*

In this section an experiment is performed to show that LRD can also be used to find the closest string (or closest substring) for a set of DNA strings, using a genetic

**Table 7.4** Closest string results for the genetic algorithm based on LRD with 3-mers

| LRD results | Rat–house mouse | Rat–fat dormouse | Rat–cow |
|---|---|---|---|
| Distance | 1524 | 2379 | 4169 |
| Average time | 12 s | 14 s | 17 s |

algorithm based on LRD. Here, the genetic algorithm proposed by Dinu and Ionescu (2012b) is combined with LRD.

Only the rat, house mouse, fat dormouse, and cow genomes are used in the experiment. The task is to find the closest string between the rat and house mouse DNA strings, between the rat and fat dormouse DNA strings, and between the rat and cow DNA strings. The goal of this experiment is to compare the distances associated to the three closest strings. The cow belongs to the Cetartiodactylae order, while the rat, the house mouse, and the fat dormouse belong to the Rodentia order. Expected results should show that the rat–house mouse distance and the rat–fat dormouse distance are smaller than the rat–cow distance. The same experiment was also conducted in the work of Dinu and Ionescu (2012b) using three genetic algorithms based on Hamming distance, edit distance, and rank distance, respectively. To compare the results of these algorithms with the results obtained by the genetic algorithm based on LRD, the same experiment setting is used. The genetic algorithm parameters used to obtain the presented results are given next. The population size is 600 chromosomes, the number of generations is 200, the crossover probability is 0.36, the mutation probability is 0.005, and size of each DNA sequence is 150 bases. Details regarding experiment organization and parameters of the genetic algorithm can also be found in the paper of Dinu and Ionescu (2012b). The results presented in Table 7.4 are obtained using LRD with 3-mers and a maximum offset of 48. The reported time is computed by measuring the average time of 10 runs of the genetic algorithm on a computer with Intel Core i7 2.3 GHz processor and 8 GB of RAM memory using a single Core.

Figure 7.8 shows the distance evolution of the best chromosome at each generation for Local Rank Distance, rank distance, Hamming distance, and edit (Levenshtein) distance. The use of LRD enables the genetic algorithm to achieve better results than previously studied genetic algorithms (Dinu and Ionescu 2012b), but with a smaller population of chromosomes (600 instead of 1800) and a lower number of generations (200 instead of 300). In terms of speed, the proposed method is similar to the other algorithms based on low-complexity distances (Hamming and rank distance).

### *7.7.4 Alignment in the Presence of Contaminated Reads*

In this experiment, reads sampled from the genomes of several mammals are aligned on the human mtDNA sequence genome. The reads were simulated with the *wgsim*

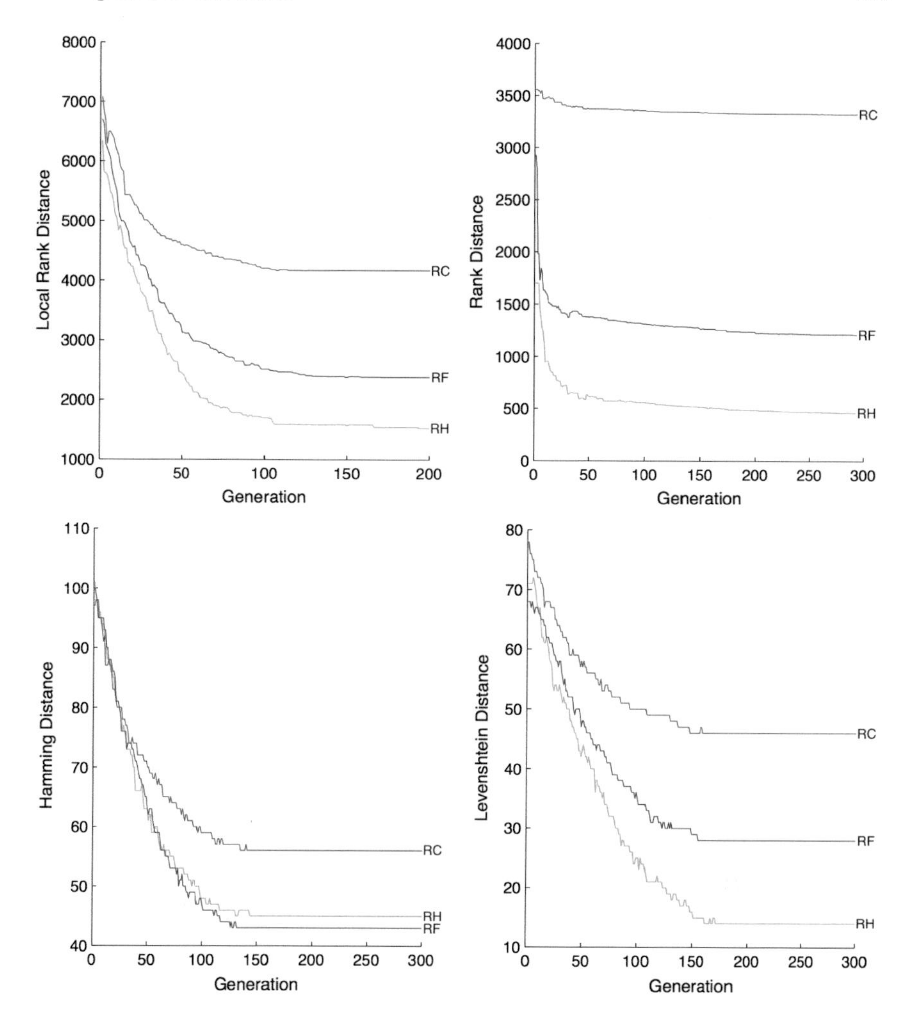


**Fig. 7.8** The distance evolution of the best chromosome at each generation for the rat–mouse–cow experiment. The *green line* represents the rat–house mouse (RH) distance, the *blue line* represents the rat–fat dormouse (RF) distance, and the *red line* represents the rat–cow (RC) distance

tool (Li 2011), using the default parameters. More precisely, the reads were generated using an error rate of 0.02, a mutation rate of 0.001, a fraction of indels of 0.15 (out of the total number of mutations) and a probability of extending an indel of 0.30.

The LRD aligners are compared to the BWA, the BOWTIE2, and the BLAST aligners, under two different scenarios. In the first scenario, 10,000 contaminated reads are sampled from the orangutan genome. In the second scenario, 50,000 contaminated reads are sampled from 5 mammals, namely the orangutan, the blue whale, the harbor seal, the donkey, and the house mouse. There are actually 10,000 reads sampled from each of the 5 mammals. In both scenarios 10,000 reads simulated from the human genome are included. The simulated reads are always 100 bases long. The goal is to maximize the number of aligned reads sampled from the human genome (true positives), and to minimize the number of aligned reads from the other mammals (false positives). Unlike the other sequence alignment experiments, reverse complement reads were not included in this experiment. However, it is important to mention that the aligners are dealing with a hard task, since the contaminated reads were sampled only from organisms that are in the same class as the human. It may be that contaminated reads from other species that are not in the Mammalia class (such as viruses, for example) can be identified and discarded more easily.

The parameters of the aligners were adjusted as described next. For the BOWTIE2 aligner, two variants are evaluated. The first one uses the *local* and the *very-sensitive-local* options. The second variant uses the *end-to-end* and the *very-sensitive* options. For the BLAST aligner, the *megablast* option is used. Two variants of the LRD aligner based on 3-mers and a maximum offset between paired 3-mers of 36 are also evaluated. One is based on the exact search algorithm, while the other one uses the approximate algorithm based on hash tables that runs much faster.

To evaluate and compare the aligners, the precision and recall curve is used. Note that the precision is given by the proportion of aligned reads that are positive, while the recall is given by the proportion of true positive reads that are aligned. In order to obtain the precision–recall curve for each aligner, the idea is to vary the threshold that gives the maximum distance allowed for an aligned read. In the case of the BWA and the BOWTIE aligners, the edit distance threshold takes values from 0 to 30. The score of the BLAST aligner ranges from 185 to 100. The LRD threshold takes values from 50 to 600, for both variants of the LRD aligner. Higher precision is obtained for lower distance thresholds, while higher recall is obtained for higher distance thresholds. The only aligner that works the other way around, and gives higher precision for higher scores, and higher recall for lower scores, is the BLAST aligner.

Several statistical measures, such as the Area Under the ROC Curve (AUC), the $F_1$ measure, and the $F_2$ measure, are also presented in order to better compare the aligners. The ROC curve plots the fraction of true positive reads versus the fraction of false positive reads, at various threshold settings. The AUC score represents the area under the ROC curve. The $F_1$ measure (also known as the $F_1$ score) can be interpreted as a weighted average of the precision and recall at a certain distance threshold. The $F_2$ measure is similar to the $F_1$ measure, only that it weights recall higher than precision. For each aligner, the highest $F_1$ and $F_2$ scores can indicate the thresholds that give a good trade-off between precision and recall. The $F_\beta$ measure is computed as follows:

$$F_\beta = (1 + \beta^2) \frac{\text{precision} \cdot \text{recall}}{\beta^2 \cdot \text{precision} + \text{recall}}. \tag{7.6}$$

The $F_1$ and the $F_2$ scores are immediately obtained from Eq. (7.6), by replacing $\beta$ with 1 and 2, respectively.

In the first scenario, there are 20,000 reads to be aligned on the human mtDNA sequence. Half of them are sampled from the same human mitochondrial genome,

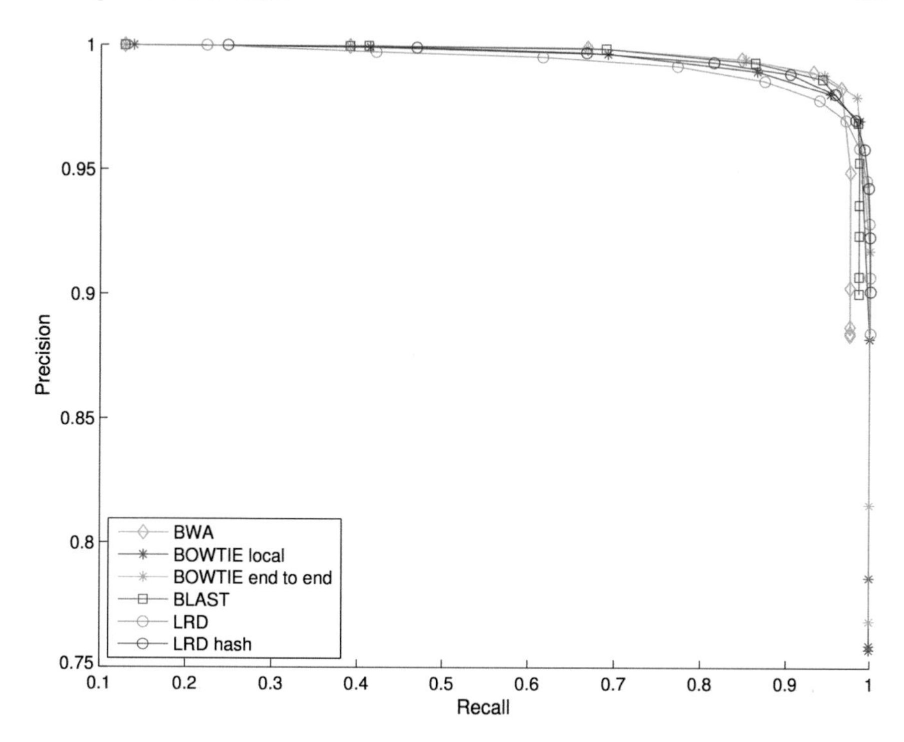


**Fig. 7.9** The precision–recall curves of the state-of-the-art aligners versus the precision–recall curves of the two LRD aligners, when 10,000 contaminated reads of length 100 from the orangutan are included. The two variants of the BOWTIE aligner are based on local and global alignment, respectively. The LRD aligner based on hash tables is a fast approximate version of the original LRD aligner

while the other half are sampled from the orangutan mitochondrial genome. Thus, the contamination rate is 50 %.

The precision–recall curves of the BWA, the BOWTIE, and the BLAST aligners together with the precision–recall curves of the two variants of the LRD aligner are presented in Fig. 7.9. By analyzing Fig. 7.9, it can be observed that the aligners obtain roughly similar results in terms of precision and recall. To better assess the performance of the evaluated aligners, the AUC measure and the best $F_1$ and $F_2$ scores for each aligner are presented in Table 7.5. In terms of the AUC, the BOWTIE and the LRD aligners attain the best results, while the other aligners fall behind. In terms of the $F_1$ measure, the BOWTIE aligner seems to be slightly better than the LRD aligner, while in terms of the $F_2$ measure, the LRD aligner achieves the best score, followed closely by the BOWTIE aligner. The BLAST aligner comes in third place after the LRD and the BOWTIE aligners. The results of the BWA aligner are also not too far from the other top scoring aligners.

The results presented in Fig. 7.9 indicate that all the aligners obtain a good trade-off between precision and recall. Indeed, all of them are able to align more than 90 % of

**Table 7.5** Several statistics of the state-of-the-art aligners versus the LRD aligner, when 10,000 contaminated reads of length 100 sampled from the orangutan genome are included

| Aligner | AUC (%) | Best $F_1$ score (%) | Best $F_2$ score (%) |
|---|---|---|---|
| BWA | 97.37 | 97.38 | 97.03 |
| BOWTIE local | 99.46 | 97.80 | 98.30 |
| BOWTIE end-to-end | **99.63** | **98.13** | 98.24 |
| BLAST | 98.38 | 97.67 | 98.15 |
| LRD aligner | 99.46 | 97.25 | 98.48 |
| Hash LRD aligner | **99.63** | 97.58 | **98.61** |

The AUC is computed from the ROC curve, while the best $F_1$ and $F_2$ measures were computed using different points on the precision–recall curve. The $F_2$ measure puts a higher weight on recall

**Table 7.6** Metrics of the human reads mapped to the human mitochondrial genome (true positives) by the hash LRD aligner versus the human reads that are not mapped to the genome (false negatives)

| LRD | Precision (%) | Recall (%) | TP | FN | TP edit | FN edit |
|---|---|---|---|---|---|---|
| 51 | 100 | 24.79 | 2479 | 7521 | 30.66 | 31.17 |
| 100 | 99.91 | 46.92 | 4692 | 5308 | 30.29 | 31.72 |
| 150 | 99.72 | 66.70 | 6670 | 3330 | 30.37 | 32.41 |
| 200 | 99.34 | 81.43 | 8143 | 1857 | 30.66 | 32.74 |
| 250 | 98.87 | 90.36 | 9036 | 964 | 30.81 | 33.26 |
| 300 | 98.06 | 95.74 | 9574 | 426 | 30.97 | 32.76 |
| 350 | 97.02 | 98.16 | 9816 | 184 | 31.02 | 32.61 |
| 400 | 95.85 | 99.24 | 9924 | 76 | 31.04 | 31.92 |
| 539 | 90.18 | 100 | 10000 | 0 | 31.05 | – |

The average edit distance is reported for true positive (TP) and false negative (FN) reads, respectively. The average edit distance is given for several points on the precision–recall curve of the hash LRD aligner, going from 100 % precision to 100 % recall. The points are obtained by varying the LRD threshold from 51 to 539

the human reads with a precision that is higher than 90 %. For instance, the hash LRD aligner is able to align 98.6 % of the human reads with 97.02 % precision. However, it would be interesting to observe how the LRD aligner behaves at the sequence level. For this purpose, some metrics of the reads simulated from the human mitochondrial genome are provided in Table 7.6. More precisely, the average edit distance of the human reads that are mapped to the human genome (true positives) is reported at different precision and recall levels. In the same time, the average edit distance of the human reads that are not mapped to the human genome (false negatives) is also reported. Perhaps it would be more interesting to give the average number of errors and mutations in the true positive reads versus the average number of errors and mutations in the false negative reads. Unfortunately, the *wgsim* tool does not output these values for the simulated reads. Nevertheless, the simulation tool does output the exact location from which each read was simulated. Therefore, a standard distance can be computed between a simulated read and its corresponding original substring (of 100 bases) from the human genome, that was used by *wgsim* to generate the read. The edit distance should give some indication of the number of changes in the human reads that are not mapped to the human genome. It can be observed that for each LRD threshold presented in Table 7.6, the average edit distance of the mapped reads is always less than the average edit distance of the false negative reads. Nonetheless, the difference between the average distance of true positives and the average distance of false negatives is not very high. Basically, only a few more bases are different from the source substring for the false negatives compared to the true positives. The highest difference is reported for the LRD threshold of 250. Table 7.6 shows that, on average, the reads that are mapped to the genome have less errors and mutations than the reads that are not mapped. However, the difference is not significant, since the false negative reads have at most 3 more errors, on average, than the mapped reads. An interesting remark is that the LRD aligner accepts more and more errors and mutations in the aligned reads as the LRD threshold increases, but even with the highest threshold of 539 that gives 100 % recall rate, the precision of the hash LRD aligner is still very high (90.18 %). In other words, the LRD aligner does a good job in discarding most of the reads simulated from the orangutan genome (true negatives), while mapping all the human reads, even those with higher error rates.

In the second scenario, there are 60,000 reads to be aligned on the human mtDNA sequence. Only 10,000 reads are actually sampled from the same human genome. The 50,000 contaminated reads were sampled from 5 different mammals. The mammals were chosen to represent 5 orders available in the first data set: Primates, Perissodactylae, Cetartiodactylae, Rodentia, and Carnivora. The contamination rate of 83.33 % is much higher than in the previous scenario.

The precision–recall curves of the BWA, the BOWTIE, and the BLAST aligners versus the precision–recall curve of the two variants of the LRD aligner are presented in Fig. 7.10. Among the evaluated aligners, the BOWTIE local aligner has the lowest results in terms of precision and recall. Figure 7.10 seems to indicate that the LRD, the BLAST, and the BOWTIE end-to-end aligner give fairly similar results.

To make a better distinction between the aligners, the AUC measure and the best $F_1$ and $F_2$ scores for each aligner are presented in Table 7.7. The results presented in Table 7.7 indicate that the LRD aligner achieves the best AUC score, followed closely by the BOWTIE end-to-end aligner. As in the previous experiment, the BOWTIE aligner attains the highest $F_1$ score, while the LRD aligner attains the highest $F_2$ score. The BLAST aligner falls in third place.

An advantage of the LRD aligners is that they are the most flexible aligners in terms of precision and recall. The aligners proposed in this work are the only ones that can be adjusted to go from 100 % precision to 100 % recall. Even if the other state-of-the-art aligners do reach full recall, it is interesting to show the best recall that can be obtained by each one. The BWA aligner reaches a maximum recall of 97.57 %, while the BLAST aligner reaches a maximum recall of 98.63 %. Both variants of the BOWTIE aligner go up to 99.91 % recall. As mentioned before, the maximum recall obtained by the LRD aligner is 100 %.

Another interesting statistics is the recall when 100 % precision is achieved. The recall at best precision is recorded in two scenarios. In the first scenario, only the

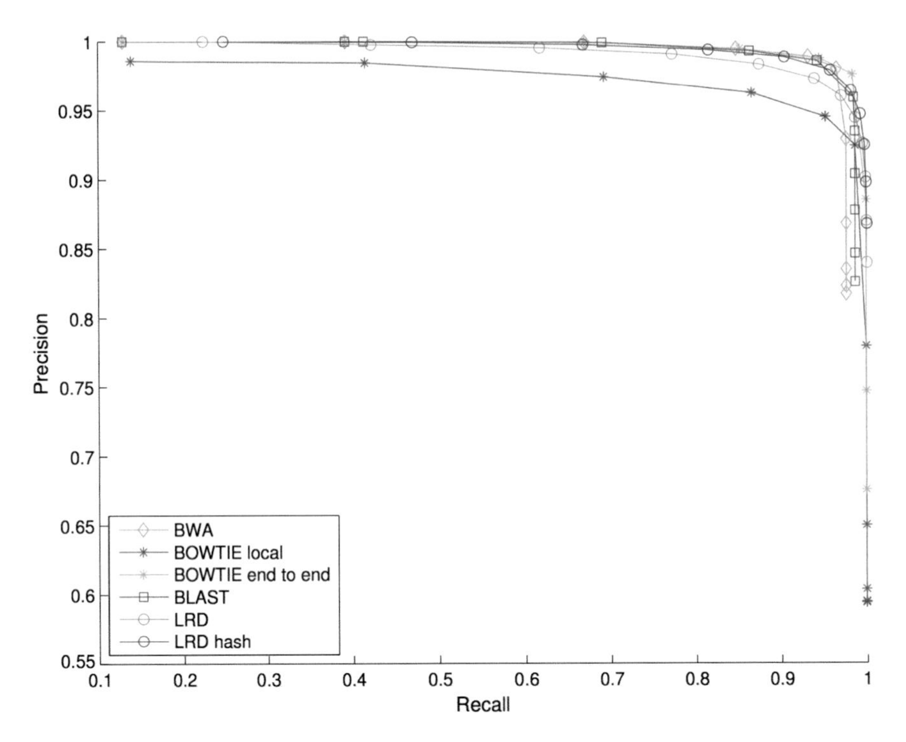


**Fig. 7.10** The precision–recall curves of the state-of-the-art aligners versus the precision–recall curves of the two LRD aligners, when 50,000 contaminated reads of length 100 from 5 mammals are included. The two variants of the BOWTIE aligner are based on local and global alignment, respectively. The LRD aligner based on hash tables is a fast approximate version of the original LRD aligner

**Table 7.7** Several statistics of the state-of-the-art aligners versus the LRD aligner, when 50,000 contaminated reads of length 100 sampled from the genomes of five mammals are included

| Aligner | AUC (%) | Best $F_1$ score (%) | Best $F_2$ score (%) |
|---|---|---|---|
| BWA | 97.52 | 97.20 | 96.75 |
| BOWTIE local | 99.55 | 95.41 | 97.32 |
| BOWTIE end-to-end | 99.84 | **97.93** | 98.16 |
| BLAST | 98.57 | 97.15 | 97.93 |
| LRD aligner | 99.86 | 96.49 | 98.04 |
| Hash LRD aligner | **99.92** | 97.25 | **98.29** |

The AUC is computed from the ROC curve, while the best $F_1$ and $F_2$ measures were computed using different points on the precision–recall curve. The $F_2$ measure puts a higher weight on recall

contaminated reads from the orangutan are included, while in the second scenario, the rest of 40,000 contaminated reads from all the other mammals, besides the orangutan, are included. Since the orangutan and the human belong to the Primates order, the first scenario is more difficult.

**Table 7.8** The recall at best precision of the state-of-the-art aligners versus the LRD aligner, when 10,000 contaminated reads of length 100 sampled from the orangutan genome are included

| Aligner | Recall at Best Precision (%) | Best Precision (%) |
|---|---|---|
| BLAST | 12.83 | 100.0 |
| BWA | 12.84 | 100.0 |
| BOWTIE end-to-end | 12.84 | 100.0 |
| BOWTIE local | 13.87 | 100.0 |
| LRD aligner | 22.36 | 100.0 |
| Hash LRD aligner | **24.79** | 100.0 |

**Table 7.9** The recall at best precision of the state-of-the-art aligners versus the LRD aligner, when 40,000 contaminated reads of length 100 sampled from the blue whale, the harbor seal, the donkey, and the house mouse genomes are included, respectively

| Aligner | Recall at Best Precision (%) | Best Precision (%) |
|---|---|---|
| BLAST | 68.95 | 100.0 |
| BWA | 66.84 | 100.0 |
| BOWTIE end-to-end | 66.84 | 100.0 |
| BOWTIE local | 13.87 | 98.58 |
| LRD aligner | 52.25 | 100.0 |
| Hash LRD aligner | **81.43** | 100.0 |

The recall at best precision for each aligner evaluated in the first scenario is given in Table 7.8. When 10,000 contaminated reads sampled from the orangutan genome are used, it seems that the LRD aligners obtain the highest recall at 100 % precision. The LRD aligners are roughly 10 % better (in terms of recall) than the state-of-the-art aligners, which all give similar recall values.

The recall at best precision for each aligner evaluated in the second scenario is given in Table 7.9. This time, the recall at 100 % precision for each aligner is much higher than in the first scenario. This indicates that if contaminated reads do not belong to an organism that is closely related to the human, the tools are able to align most of the true positive reads with 100 % precision. The best aligner is the LRD aligner based on the hash tables implementation. It attains a recall of 81.43 %, being roughly 13 % better than most of the state-of-the-art aligners. In the second scenario, it seems that the BOWTIE local aligner falls very far behind the other alignment tools.

Overall, the hash LRD aligner seems to be the best tool among all the evaluated aligners, in the presence of contaminated reads. It is closely followed by the BOWTIE end-to-end aligner. The high accuracy of the LRD aligners also comes with the cost of being the slowest ones among the evaluated aligners.

### 7.7.5 Clustering an Unknown Organism

The rank-based aligners are evaluated in the context of finding a solution for the task of clustering a new (or unknown) organism, given only a set of short Next-Generation Sequencing DNA reads. More precisely, the task is to find the order, the family, or the species of the unknown organism, without having to sequence its genome first, by aligning its reads into several genomes in order to obtain the nearest neighbor species (or the most similar species). The LRD aligners are compared to the BWA, the BOWTIE2, and the BLAST aligners. In the case of the BOWTIE2 aligner, two variants are evaluated, one based on local alignment and the other based on global alignment. The LRD aligners are based on 3-mers with a maximum offset between paired 3-mers of 36. A maximum distance threshold of 1000 was used for the LRD aligners. The distance threshold for the LRD aligners was adjusted in order to allow more reads to be aligned, especially for the mammals that are more distantly related, more precisely, that are not from the same order. The approximate hash LRD aligner achieves similar results to the basic LRD aligner, when it aligns reads only in the positions that have at most 5 similar $k$-mers less than the maximum number of $k$-mers from the read that can be found at any given position in the reference sequence. For this reason, only the results of the approximate LRD aligner are reported in the following experiments.

One by one, each of the 20 mammalian genomes from the EMBL database will be considered to be unknown for the purpose of this experiment. The unknown individual will be represented by a set $\mathcal{R}$ of short DNA reads randomly sampled from its genome. The task is to find the most similar individual (or species) from the remaining 19 individuals, for each unknown individual. In order to solve the task, the collection $\mathcal{R}$ of reads (that represents an unknown individual) is aligned on each of the 19 genomes from the collection $\mathcal{G}$ of genomes. Reads are aligned under a maximum distance threshold. Thus, only a subset $\mathcal{S} \subseteq \mathcal{R}$ of reads is aligned on each genome. An alignment score is computed for each genome in order to obtain the most similar individual. The score is given by the average minimum distances of the reads in $\mathcal{S}$ divided by the number of aligned reads. The minimum distance for a specific read is given by the best positional match in the reference genome. Lower scores indicate greater similarity between species, and higher scores indicate a greater dissimilarity between species. The individual (or the species) with the lowest score is considered to be the most similar one. Finally, the unknown organism is considered to be part of the same order as its most similar individual. The unknown individual is correctly clustered if it is indeed a member of the order predicted by the aligner. Thus, the performance of each aligner on this task is determined by the number of correctly clustered unknown individuals. The evaluation procedure can also be described as the leave-one-out cross-validation procedure. It is important to notice that the procedure described above does not generate a partitioning of the data set, but rather assigns a newly discovered (or unknown) organism to a specific cluster in an existing phylogenetic tree. An evaluation tool to obtain this score has also been added to the software package provided at http://lrd.herokuapp.com/aligners.

An interesting remark is that the tools evaluated on this task align reads under a given maximum distance threshold and, hence, many reads remain unaligned. The distance measure depends on the aligner. While the BWA and the BOWTIE aligners are based on the edit distance, the BLAST aligner uses a score of its own. The rank-based aligner is based on Local Rank Distance. Therefore, the alignment score is obtained by the average distance divided by the number of aligned reads. In other words, a genome with more aligned reads is more likely to be similar to the unknown individual.

The aligners are evaluated and compared under two different scenarios. In both scenarios, reads of 100 bases long were simulated using the *wgsim* tool (Li 2011). In the first scenario, 20,000 short DNA reads per mitochondrial genome are sampled using the default parameters of the simulation tool. More precisely, the reads were generated using an error rate of 0.02, a mutation rate of 0.001, a fraction of indels of 0.15 (out of the total number of mutations), and a probability of extending an indel of 0.30. With an average base coverage of 100, the number of reads should be far than enough to correctly determine the order of unknown organisms. This scenario is designed to simulate a real-world setting where a high number of Next-Generation Sequencing reads is usually available. In the second scenario, only 200 simulated short DNA reads per genome are used in order to make the task harder to solve. The alignment methods should be challenged by the small amount of available reads. The generated reads also have more errors. More precisely, the reads for this second test case were simulated using an error rate of 0.08, a mutation rate of 0.008, a fraction of indels of 0.15 (out of the total number of mutations), and a probability of extending an indel of 0.30. In both test cases, half of the simulated reads from each genome are reverse complements.

Table 7.10 compares the results of the LRD aligner with the other state-of-the-art aligners in the first scenario with 20,000 simulated DNA reads of length 100 per genome. The BWA aligns only the reads that fall under a certain edit distance threshold. The BWA aligner based on the default threshold 5 is listed in Table 7.10 under the name of *BWA edit 5*. Another BWA aligner with a threshold of 10 was used in the experiments. As the latter one aligns more reads, it should be able to give more accurate results than the default BWA aligner. The LRD aligner is based on a distance threshold of 1000.

In the first scenario, it seems that the BLAST, the BOWTIE, and the LRD aligners achieve perfect results. More precisely, they are all able to identify the most similar individual as being part of the same order as the unknown organism, for the entire set of 20 mammals. On the other hand, the BWA edit 5 aligner is only able to predict the correct order for 17 out of 20 mammals. It clusters the cat as Primates, and the fin whale and the gorilla as part of the Carnivora order. The BWA edit 10 aligner works even worse, correctly predicting the order for 14 mammals.

It is interesting to observe that all the methods are usually able to determine not only the correct order, but also the most similar species in the group. For example, the horse is always clustered near the donkey, rather than the Indian or the white rhinoceros, despite the fact that they are all members of the same order, namely Perissodactylae. The same situation can be observed in the case of the gray seal,

**Table 7.10** The results for the real-world setting experiment on mammals

| Mammal | Class (Label) | BWA edit 5 | BWA edit 10 | BLAST | BOWTIE local | BOWTIE global | LRDa 1000 |
|---|---|---|---|---|---|---|---|
| Cow | Cet (1) | 4 | **9*** | 2 | 2 | 2 | 2 |
| Sheep | Cet (2) | 1 | **9*** | 1 | 1 | 1 | 1 |
| Blue whale | Cet (3) | 4 | 4 | 4 | 4 | 4 | 4 |
| Fin whale | Cet (4) | **6*** | 3 | 3 | 3 | 3 | 3 |
| Cat | Car (5) | **9*** | **9*** | 7 | 7 | 7 | 6 |
| Gray seal | Car (6) | 7 | 7 | 7 | 7 | 7 | 7 |
| Harbor seal | Car (7) | 6 | 6 | 6 | 6 | 6 | 6 |
| Human | Pri (8) | 11 | 11 | 11 | 11 | 11 | 11 |
| Gibbon | Pri (9) | 8 | **2*** | 8 | 11 | 13 | 11 |
| Gorilla | Pri (10) | **5*** | 13 | 13 | 11 | 11 | 11 |
| P. chimpanzee | Pri (11) | 13 | 13 | 13 | 13 | 13 | 13 |
| Orangutan | Pri (12) | 14 | 14 | 14 | 14 | 14 | 14 |
| Chimpanzee | Pri (13) | 11 | 11 | 11 | 11 | 11 | 11 |
| S. orangutan | Pri (14) | 12 | 12 | 12 | 12 | 12 | 12 |
| Horse | Per (15) | 16 | 16 | 16 | 16 | 16 | 16 |
| Donkey | Per (16) | 15 | 15 | 15 | 15 | 15 | 15 |
| I. rhinoceros | Per (17) | 18 | 18 | 18 | 18 | 18 | 18 |
| w. rhinoceros | Per (18) | 17 | 17 | 17 | 17 | 17 | 17 |
| Mouse | Rod (19) | 20 | **17*** | 20 | 20 | 20 | 20 |
| Rat | Rod (20) | 19 | **17*** | 19 | 19 | 19 | 19 |
| Accuracy | | 17/20 | 14/20 | 20/20 | 20/20 | 20/20 | 20/20 |

The results of clustering unknown organisms using the BWA aligner, the BLAST aligner, the BOWTIE aligner, and the LRD aligner are presented on columns, respectively. Mammals are labeled with numbers from 1 to 20, given on the second column. The label of the closest species obtained by each aligner is reported for each mammal. Incorrectly clustered mammals are marked in bold and with an asterisk. Classes are actually 3-letter prefixes of order names. Unknown organisms are represented by 20,000 reads of length 100 simulated from the original genomes. Half of the reads are reverse complements

which is always considered to be most similar with the harbor seal rather than the other member of the Carnivora order, namely the cat.

The empirical results show that, with the exception of the BWA aligner, all the other methods work very well. This also demonstrates that the evaluation procedure gives a relevant measure of similarity between a set of reads and a reference genome, that can be used for solving the task of clustering unknown organisms.

The first scenario is not enough to make a clear distinction between the compared methods, with respect to the accuracy and the biological relevance. To better assess the performance levels of these aligners, another experiment is conducted using only 200 short DNA reads of length 100 per genome. As described above, the reads also contain more errors and mutations than in the previous test case.

**Table 7.11** The results for the hard setting experiment on mammals

| Mammal | Class (Label) | BWA edit 5 | BWA edit 10 | BLAST | BOWTIE local | BOWTIE global | LRDa 1000 |
|---|---|---|---|---|---|---|---|
| Cow | Cet (1) | * | 2 | 2 | **19*** | 2 | 2 |
| Sheep | Cet (2) | * | **5*** | 1 | 1 | **12*** | 1 |
| Blue whale | Cet (3) | * | 4 | 4 | **12*** | 4 | 4 |
| Fin whale | Cet (4) | 3 | 1 | 3 | 3 | 3 | 3 |
| Cat | Car (5) | * | * | **1*** | **9*** | **19*** | 7 |
| Gray seal | Car (6) | 7 | 7 | 7 | 7 | 7 | 7 |
| Harbor seal | Car (7) | 6 | 6 | 6 | 6 | 6 | 6 |
| Human | Pri (8) | * | 13 | 11 | 11 | 13 | 11 |
| Gibbon | Pri (9) | * | 11 | 13 | **16*** | 14 | 13 |
| Gorilla | Pri (10) | 8 | 11 | 8 | 11 | 8 | 11 |
| P. chimpanzee | Pri (11) | 13 | 13 | 13 | 13 | 13 | 13 |
| Orangutan | Pri (12) | * | 14 | 14 | 14 | 14 | 14 |
| Chimpanzee | Pri (13) | 11 | 11 | 11 | 11 | 11 | 11 |
| S. orangutan | Pri (14) | 12 | 12 | 12 | 12 | 12 | 12 |
| Horse | Per (15) | 16 | 16 | 16 | 16 | 16 | 16 |
| Donkey | Per (16) | 15 | 15 | 15 | 15 | 15 | 15 |
| I. rhinoceros | Per (17) | 18 | 18 | 18 | 15 | **12*** | 18 |
| w. rhinoceros | Per (18) | * | 17 | 17 | **14*** | 17 | 17 |
| Mouse | Rod (19) | * | **6*** | **12*** | **14*** | **12*** | 20 |
| Rat | Rod (20) | * | * | **12*** | **8*** | **5*** | 19 |
| Accuracy | | 10/20 | 16/20 | 17/20 | 13/20 | 15/20 | 20/20 |

The results of clustering unknown organisms using the BWA aligner, the BLAST aligner, the BOWTIE aligner, and the LRD aligner are presented on columns, respectively. Mammals are labeled with numbers from 1 to 20, given on the second column. The label of the closest species obtained by each aligner is reported for each mammal. Incorrectly clustered mammals are marked in bold and with an asterisk. Classes are actually 3-letter prefixes of order names. Unknown organisms are represented by 200 reads of length 100 (half of them being reverse complements) simulated from the original genomes, using an error rate of 0.08 and a mutation rate of 0.008

The results of the state-of-the-art aligners together with the results of the LRD aligner are shown in Table 7.11. Compared to the previous scenario, the results of the state-of-the-art aligners are much lower this time. The BWA aligners predict the correct order for 10 and 16 mammals, respectively. Unlike the previous test case, the BWA edit 10 aligner works better than the BWA edit 5 aligner, probably because it is able to align more reads with high error and mutation rates. The BOWTIE aligners obtain results that are roughly similar to the results of the BWA aligners. The BOWTIE local aligner predicts the right order for 13 out of 20 mammals, while the BOWTIE end-to-end aligner is able to correctly cluster two more mammals, reaching a total of 15 correctly clustered mammals. The BLAST aligner works fairly well, predicting the correct order for 17 mammals. It wrongly predicts the order for the cat and for the two members of the Rodentia order, namely the house mouse and the rat. It seems that all the aligners, besides the LRD aligner, have trouble predicting the right order for the Rodentia members. On the other hand, it seems that the aligners find it very easy to predict the correct order for the Primates. Finally, the LRD aligner is able to predict the correct class for the entire set of mammals. The LRD aligner seems to be more robust to high error and mutation rates, as it achieves the best results among all the evaluated aligners.

It is interesting to observe that the BWA with an edit distance threshold of 10 is not able to align any reads at all, for two of the mammals. This is the reason why no similar mammal is found for the cat or for the rat. The same problem occurs in the case of the BWA edit 5 aligner, which is not able to find any similar genomes for 10 mammals, due to the lack of aligned reads. This problem is likely caused by the high error and mutation rates that were used to sample the reads from the original genomes. It may be concluded that the BWA aligner is the most fragile aligner with respect to high error and mutation rates.

### 7.7.6 *Time Evaluation of Sequence Aligners*

The time taken by each aligner to produce the results for the two test cases of the experiment on clustering unknown organisms is shown in Table 7.12. For both test cases, there are 20,200 short DNA reads that must be aligned for each mammal on the rest of 19 mammalian genomes. In total, each tool must align 7,676,000 short DNA reads of 100 bases long, on a reference mtDNA genome of roughly 15,000–17,000 bases. Note that the reference genome is not necessarily always the same, since the reads sampled from a genome are aligned into the remaining 19 genomes. The time

**Table 7.12** The running times of the BWA aligner, the BLAST aligner, the BOWTIE aligner, and the LRD aligner

| Method | Time |
|---|---|
| BWA edit 5 | 3 minutes 14 seconds |
| BWA edit 10 | 3 minutes 50 seconds |
| BOWTIE local | 9 minutes 43 seconds |
| BOWTIE end-to-end | 7 minutes 14 seconds |
| BLAST | 30 minutes |
| LRDa | 285 hours |
| LRDa + hash (C++ implementation) | 16 hours 33 minutes |
| LRDa + hash (Java implementation) | 326 minutes |

The aligners are compared with the task of aligning 7,676,000 short DNA reads of 100 bases long on a reference mtDNA genome of roughly 15,000–17,000 bases. The aligners were evaluated on a computer with Intel Core i7 2.3 GHz processor and 8 GB of RAM memory using a single Core

was measured on a computer with Intel Core i7 2.3 GHz processor and 8 GB of RAM memory using a single Core.

Among the evaluated aligners, the BWA aligner is the fastest one, taking just over 3 minutes to align all the reads. The BOWTIE2 aligner is also very fast. It takes roughly 7 minutes to align the reads when the *local* option is used, and 9–10 minutes for the *end-to-end* option. The BLAST aligner takes 30 minutes when the *megablast* option is turned on. Finally, the LRD aligner is the slowest one, but it also has the advantage of being the most accurate on all the test cases. The approximate LRD aligner based on the hash optimization implemented in C++ needs 16–17 hours to align all the reads. The Java implementation of LRD aligner based on hash tables is roughly three times faster, with a total time of 5–6 hours. The speed gain of the Java implementation is given by the optimized hash table implementation available in the Java API. It is important to mention that the parameters of the approximate LRD aligner are optimized for accuracy, not for speed. Even so, the approximate hash LRD aligner implemented in Java is roughly 50 times faster than the basic LRD aligner. The reported time of the approximate hash LRD aligner is comparable to that of the other tools that favor correctness over speed, such as BFAST (Homer et al. 2009). Parallel or GPU processing could be used to further reduce the running time of the LRD aligner and to make it run as fast as BOWTIE2 or BLAST.

An important advantage of the LRD aligner is that it obtains very accurate results even for a very low base coverage. For instance, the LRD aligner predicts the correct order for the entire set of 20 mammals by aligning 200 reads per genome (with high error and mutation rates), while the BWA edit 5 aligner is only able to predict the correct order for 17 mammals using 20,000 reads per genome (with low error and mutation rates). Considering this fact, the LRD aligner obtains better results than the fastest aligner (BWA) in the same amount of time (roughly 3 minutes). This being said, the LRD aligner can produce accurate results in an amount of time which is comparable to the other state-of-the-art aligners, simply by aligning considerably less reads than the other tools would require.

### 7.7.7 Experiment on Vibrio Species

Chen et al. (2003) conduct a comparative study of the *V. vulnificus* YJ106, *V. parahaemolyticus* RIMD 2210633, and *V. cholerae* El Tor N16961 genomes to compare relative positions of conserved genes and to investigate the movement of genetic materials within and between the two chromosomes of these vibrio species. The study shows that *V. vulnificus* has a higher degree of conservation in gene organization in the two chromosomes relative to *V. parahaemolyticus* rather than to *V. cholerae*. This implies that *V. vulnificus* is closer to *V. parahaemolyticus* than to *V. cholerae* from the evolutionary point of view. This result is also supported by the study of Lin et al. (2005), which determines that the block-interchange distance between *V. vulnificus* and *V. parahaemolyticus* is smaller than that between *V. vulnificus* and *V. cholerae*.

The goal of this experiment is to determine if the LRD aligner can achieve similar results to Chen et al. (2003), Lin et al. (2005), using the evaluation procedure for clustering an unknown organism presented in this chapter. Hence, the experiment consists of aligning simulated reads from the *V. vulnificus* chromosomes into *V. parahaemolyticus* and *V. cholerae*. It is important to note that three test cases were considered. In the first test case, simulated reads of chromosome VV1 are aligned into VP1 and VC1, respectively. In the second case, simulated reads of chromosome VV2 are aligned into VP2 and VC2, respectively. Finally, the simulated reads from both chromosomes of *V. vulnificus* are aligned into the two chromosomes of *V. parahaemolyticus* on one hand, and into the two chromosomes of *V. cholerae* on the other hand.

In this experiment, reads of 100 bases long were simulated using the default parameters of the *wgsim* tool (Li 2011). More precisely, the reads were generated using an error rate of 0.02, a mutation rate of 0.001, a fraction of indels of 0.15 (out of the total number of mutations), and a probability of extending an indel of 0.30. In this experiment, 30,000 simulated reads per chromosome are used, which corresponds to an average base coverage of 1. As in the previous experiment, half of the simulated reads from each genome are reverse complements. The LRD aligner is based on 3-mers with a maximum offset between paired 3-mers of 36. As in the previous experiments, the maximum distance threshold is set to 1000.

The scores of simulated reads from *V. vulnificus* chromosomes I and II aligned into *V. parahaemolyticus* and *V. cholerae* using the LRD aligner are shown in Table 7.13. The empirical results for all the three test cases are presented in this table. Each score is given by the average minimum Local Rank Distances of the aligned reads divided by the number of aligned reads on each genome. The results of the LRD aligner are similar to the results obtained by Chen et al. (2003), Lin et al. (2005). More precisely, the score between *V. vulnificus* and *V. parahaemolyticus* is lower than that between *V. vulnificus* and *V. cholerae* for both chromosomes of the three vibrio species. Even

**Table 7.13** The results of the rank-based aligner on vibrio species

| Reads source | Reference | LRDa score |
|---|---|---|
| VV1 | VP1 | 606.2 |
| VV1 | VC1 | 643.9 |
| VV2 | VP2 | 773.0 |
| VV2 | VC2 | 849.9 |
| VV1 + VV2 | VP1 + VP2 | 641.7 |
| VV1 + VV2 | VC1 + VC2 | 697.7 |

The LRD aligner is based 3-mers, a maximal offset of 36, and a Local Rank Distance threshold of 1000. The scores obtained by the LRD aligner for simulated reads of *V. vulnificus* chromosomes I and II aligned into *V. parahaemolyticus* and *V. cholerae* are presented in this table. The first column indicates the source chromosome of the simulated reads. The second column indicates the reference chromosome. The third and fourth columns show the scores of the two aligners computed with the evaluation tool provided in the software package

if chromosomes I and II are combined, *V. vulnificus* is found to be more similar to *V. parahaemolyticus*.

Some concern regarding the results obtained in this experiment might be that the results are influenced by the length difference between the reference genomes of *V. parahaemolyticus* and *V. cholerae*. First of all, the difference between the scores obtained by the LRD aligner is much higher than the difference between the lengths of the chromosomes VP1 and VC1. However, the study might be affected by the significant length difference between VP2 and VC2. While the number of simulated reads is fixed, the alignment tool excludes the reads that show a distance that is higher than the maximum threshold of 1000. The threshold should remove most of the reads that are aligned by chance, thus giving a score that is not influenced by the longer length of the VP2 chromosome.

## 7.8 Discussion

Designed to conform to more general principles, while being well adapted to DNA strings, LRD comes to improve several state-of-the-art methods for DNA sequence analysis. Phylogenetic experiments show that trees produced by LRD are better or at least comparable with those reported in the literature (Dinu and Ionescu 2012a; Dinu and Sgarro 2006; Li et al. 2004; Reyes et al. 2000). Furthermore, the empirical results showed that the fast LRD aligner can be considered as a viable alternative to standard alignment tools, since it can often be more accurate.

The results of the LRD aligners presented in this chapter are obtained using 3-mers and a maximum offset of 36. The maximum offset depends on the read length, more precisely it should be less or equal to the read length. The $k$-mers length should also be adjusted with regard to the read length. For reads of length 100, 3-mers are a reasonable choice since the chances of finding matching pairs of 3-mers between a read and the genome are very high. Nonetheless, 4-mers and 5-mers should also work well, especially if the reads and the reference genome belong to the same species. If longer reads are considered for alignment, even longer $k$-mers can be used for a better accuracy and speed. On the other hand, longer $k$-mers are likely to reduce the accuracy of the aligner when the mutation and error rates are high, since the longer is the $k$-mer the greater is the probability of containing a mutation or error. For instance, if a $k$-mer contains a point mutation, the $k$-mer will not be matched correctly when LRD is computed. Even if LRD is designed to handle such situations, a carefully chosen $k$-mer length can make the most of the aligner proposed in this work. Alternatively, the Hamming distance could be used to compare $k$-mers in the computation of LRD. This seems to be more appropriate from a biological point of view, in that it allows the pairing of $k$-mers with mutations. A faster version of LRD, that considers only the significant or the most frequent $k$-mers, is also of great interest for sequence alignment or related tasks. Significant $k$-mers could be those that encode genes, for example.

Overall, the LRD aligner gives the most accurate results and it seems to be very robust for reads that contain many errors or mutations. However, the accurate results of LRD come with a cost. The time that LRD takes to align the same number of reads is higher than the time of the state-of-the-art aligners evaluated in this chapter. Nevertheless, the empirical results presented in this work show that the LRD aligner can produce very accurate results in the same amount of time as the other alignment tools, simply by using a lower base coverage. There is still enough room to speed up the LRD algorithm. By implementing it on GPU, the LRD aligner will be comparable (in terms of time) with the other aligners that favor efficiency over correctness. The LRD aligner can be considered as a useful tool for sequence alignment, being highly accurate from a biological (or evolutionary) point of view.

The results presented in this chapter can be considered as a strong argument in favor of using Local Rank Distance for computational biology tasks, in order to obtain results that are often more accurate from a biological point of view. Local Rank Distance (Ionescu 2013) is related to the rearrangement distance (Amir et al. 2006). The rearrangement distance works with indexed $k$-mers and is based on a process of converting a string into another, in a similar fashion to the edit distance. Unlike

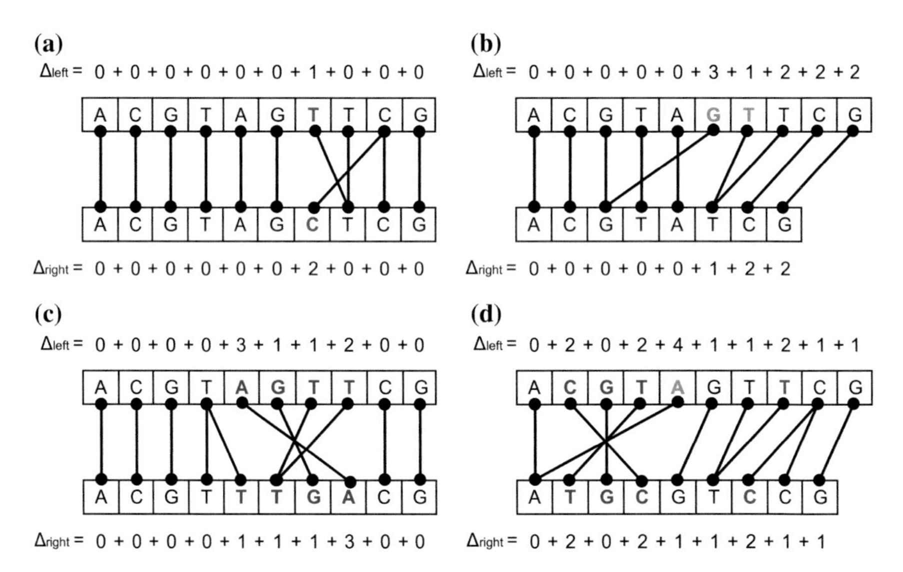


**Fig. 7.11** Local Rank Distance computed in the presence of different types of DNA changes such as point mutations, indels, and inversions. In the first three cases **a–c**, a single type of DNA polymorphism is included in the second (*bottom*) string. The last case **d** shows how LRD measures the differences between the two DNA strings when all the types of DNA changes occur in the second string. The nucleotides affected by changes are marked with bold. To compare the results for the different types of DNA changes, the first string is always the same in all the four cases. Note that in all the four examples, LRD is based on 1-mers. In each case, $\Delta_{LRD} = \Delta_{\text{left}} + \Delta_{\text{right}}$. **a** Measuring LRD with point mutations. The $T$ at index 7 is substituted with $C$. **b** Measuring LRD with indels. The substring $GT$ is deleted. **c** Measuring LRD with inversions. The substring $AGTT$ is inverted. **d** Measuring LRD with point mutations, indels, and invensions

the edit distance or the rearrangement distance, LRD does not impose such global constraints. Instead, LRD tries to capture only the local changes in DNA. This seems to be more natural from an evolutionary point of view, since changes in DNA, such as point mutations or indels, occur at the local level. Perhaps this is the key insight of why Local Rank Distance should be expected to give more accurate results than the other distance measures. For instance, the edit distance counts the minimum number of operations required to transform one string into the other. It is clear that the actual number of DNA changes that did occur may be higher than the minimum number of operations. The Hamming distance sides with Local Rank Distance regarding the local aspect. However, the Hamming distance is greatly affected by indels. A single character that is inserted (or deleted) into one of the two strings will damage the Hamming distance computation for the rest of string. On the other hand, Local Rank Distance is more robust to changes such as indels or duplications, since it sums up the positional offsets of identical $k$-mers. When two DNA sequences are identical, the sum of these positional offsets is zero. If the two DNA sequences are affected by various types of DNA changes, the positional offsets of identical $k$-mers increase mostly in the affected DNA regions. Consequently, the Local Rank Distance will be higher, since it finds displaced $k$-mers. When more point mutations, indels, reversals, or other kinds of errors occur in the DNA, LRD will indicate an even higher distance between the DNA sequences. Intuitively, Local Rank Distance reflects the total amount of local changes between two DNA sequences. This intuition can be better observed in Fig. 7.11, which shows how the Local Rank Distance between two DNA sequences changes when one of the two sequences is affected by different types of DNA polymorphisms. Another key insight of why the rank-based approach should work better is that Local Rank Distance can capture very fine differences between strings, unlike the more commonly used edit distance or Hamming distance. More results that support this statement are presented in the empirical study performed by Dinu and Ionescu (2012b), which compares rank distance with Hamming distance and edit distance, respectively.

## References


Altschul S, Gish W, Miller W, Myers E, Lipman D (1990) Basic local alignment search tool. J Mol Biol 215(3):403–410

Amir A, Aumann Y, Benson G, Levy A, Lipsky O, Porat E, Skiena S, Vishne Uzi (2006) Pattern matching with address errors: rearrangement distances. In: Proceedings of SODA, pp 1221–1229

Ander C, Schulz-Trieglaff O, Stoye J, Cox AJ (2013) metaBEETL: high-throughput analysis of heterogeneous microbial populations from shotgun DNA sequences. BMC Bioinform 14(S-5):S2

Barnes C, Goldman DB, Shechtman E, Finkelstein A (2011) The PatchMatch randomized matching algorithm for image manipulation. Commun ACM 54(11):103–110

Cao Y, Janke A, Waddell PJ, Westerman M, Takenaka O, Murata S, Okada N, Paabo S, Hasegawa M (1998) Conflict among individual mitochondrial proteins in resolving the phylogeny of Eutherian orders. J Mol Evol 47:307–322

Chen CY, Wu KM, Chang YC, Chang CH, Tsai HC, Liao TL, Liu YM, Chen HJ, Shen AB, Li JC, Su TL, Shao CP, Lee CT, Hor LI, Tsai SF (2003) Comparative genome analysis of Vibrio vulnificus, a marine pathogen. Genome Res 13(12):2577–2587

Chimani M, Woste M, Bocker S (2011) A closer look at the closest string and closest substring problem. In: Proceedings of ALENEX, pp 13–24

Cho TS, Avidan S, Freeman WT (2010) The patch transform. IEEE Trans Pattern Anal Mach Intell 32(8):1489–1501

Dinu LP, Ghetu F (2011) Circular rank distance: a new approach for genomic applications. In: DEXA Workshops, pp 397–401

Dinu LP, Ionescu RT, Popescu M (2012) Local Patch Dissimilarity for images. In: Proceedings of ICONIP 7663:117–126

Dinu LP, Ionescu RT (2012a) Clustering based on rank distance with applications on DNA. In: Proceedings of ICONIP 7667:722–729

Dinu LP, Ionescu RT (2012b) An efficient rank based approach for closest string and closest substring. PLoS ONE 7(6):e37576, 06

Dinu LP, Ionescu RT (2013) Clustering based on median and closest string via rank distance with applications on DNA. Neural Comput Appl 24(1):77–84

Dinu LP, Ionescu RT, Tomescu AI (2014) A rank-based sequence aligner with applications in phylogenetic analysis. PLoS ONE 9(8):e104006, 08. doi:10.1371/journal.pone.0104006

Dinu LP, Popescu M, Dinu A (2008) Authorship identification of romanian texts with controversial paternity. In: Proceedings of LREC

Dinu LP, Sgarro A (2006) A low-complexity distance for DNA strings. Fundam Informaticae 73(3):361–372

Dinu A, Dinu LP (2005) On the syllabic similarities of romance languages. In: Proceedings of CICLing 3406:785–788

Dinu LP, Manea F (2006) An efficient approach for the rank aggregation problem. Theoret Comput Sci 359(1–3):455–461

Heidelberg JF, Eisen JA, Nelson WC, Clayton RA, Gwinn ML, Dodson RJ, Haft DH, Hickey EK, Peterson JD, Umayam L, Gill SR, Nelson KE, Read TD, Tettelin H, Richardson D, Ermolaeva MD, Vamathevan J, Bass S, Qin H, Dragoi L, Sellers P, McDonald L, Utterback T, Fleishmann RD, Nierman WC, White O, Salzberg SL, Smith HO, Colwell RR, Mekalanos JJ, Venter JC, Fraser CM (2000) DNA sequence of both chromosomes of the cholera pathogen Vibrio cholerae. Nature 406(6795):477–483

Homer, N, Merriman B, Nelson SF (2009) BFAST: an alignment tool for large scale genome resequencing. PLoS ONE 4(11):e7767+

Huson DH, Auch AF, Qi J, Schuster SC (2007) MEGAN analysis of metagenomic data. Genome Res 17(3)

Ionescu RT, Popescu M, Cahill A (2014) Can characters reveal your native language? A language-independent approach to native language identification. In: Proceedings of EMNLP, pp 1363–1373

Ionescu RT (2013) Local Rank Distance. In: Proceedings of SYNASC, pp 221–228

Ionescu RT (2015) A fast algorithm for Local Rank Distance: Application to Arabic native language identification. In: Proceedings of ICONIP 9490:390–400

Joseph F (2004) Inferring phylogenies. Sinauer Associates, Sunderland

Langmead B, Trapnell C, Pop M, Salzberg SL (2009) Ultrafast and memory-efficient alignment of short DNA sequences to the human genome. Genome Biol 10(3):R25–10

Li H (2011) wgsim—read simulator for next generation sequencing. http://github.com/lh3/wgsim

Li H, Durbin R (2009) Fast and accurate short read alignment with Burrows-Wheeler transform. Bioinformatics 25(14):1754–1760

Li M, Chen X, Li X, Ma B, Vitanyi PMB (2004) The similarity metric. IEEE Trans Inform Theory 50(12):3250–3264

Lin YC, Lu CL, Chang H-Y, Tang CY (2005) An efficient algorithm for sorting by block-interchanges and its application to the evolution of vibrio species. J Comput Biol 12(1):102–112

Makino K, Oshima K, Kurokawa K, Yokoyama K, Uda T, Tagomori K, Iijima Y, Najima M, Nakano M, Yamashita A, Kubota Y, Kimura S, Yasunaga T, Honda T, Shinagawa H, Hattori M, Iida T (2003) Genome sequence of Vibrio parahaemolyticus: a pathogenic mechanism distinct from that of V cholerae. Lancet 361(9359):743–749

Meyer F, Paarmann D, D'Souza M, Olson R, Glass EM, Kubal M, Paczian T, Rodriguez A, Stevens R, Wilke A, Wilkening J, Edwards RA (2008) The metagenomics RAST server—a public resource for the automatic phylogenetic and functional analysis of metagenomes. BMC Bioinform 9(1):386

Melsted P, Pritchard J (2011) Efficient counting of k-mers in DNA sequences using a bloom filter. BMC Bioinform 12(1):333

Popescu, M, Grozea C (2012) Kernel methods and string kernels for authorship analysis. CLEF (Online Working Notes/Labs/Workshop)

Popescu M, Ionescu RT (2013) The story of the characters, the DNA and the native language. In: Proceedings of the Eighth Workshop on Innovative Use of NLP for Building Educational Applications, pp 270–278

Popov YV (2007) Multiple genome rearrangement by swaps and by element duplications. Theoret Comput Sci 385(1–3):115–126

Reyes A, Gissi C, Pesole G, Catzeflis FM, Saccone C (2000) Where do rodents fit? Evidence from the complete mitochondrial genome of sciurus vulgaris. Mol Biol Evol 17(6):979–983

Shapira D, Storer JA (2003) Large edit distance with multiple block operations. In: Proceedings of SPIRE 2857:369–377

Vezzi F, Fabbro CD, Tomescu AI, Policriti A (2012) rNA: a fast and accurate short reads numerical aligner. Bioinformatics 28(1):123–124

# Chapter 8
# Native Language Identification with String Kernels

## 8.1 Introduction

The most common approach in text mining classification tasks such as text categorization, authorship identification, or plagiarism detection is to rely on features like words, part-of-speech tags, stems, or some other high-level linguistic features. Perhaps surprisingly, recent results indicate that methods handling the text at the character level can also be very effective (Escalante et al. 2011; Grozea et al. 2009; Kate and Mooney 2006; Lodhi et al. 2002; Popescu 2011; Popescu and Dinu 2007; Popescu and Grozea 2012; Sanderson and Guenter 2006). By avoiding to explicitly consider features of natural language such as words, phrases, or meaning, an approach that works at the character level has an important advantage in that it is language independent and linguistic theory neutral. In this context, a state-of-the-art machine learning system for Native Language Identification (NLI) that works at the character level is presented in this chapter.

NLI is the task of identifying the native language of a writer, based on a text they have written in a language other than their mother tongue. This is an interesting subtask in forensic linguistic applications such as plagiarism detection and authorship identification, where the native language of an author is just one piece of the puzzle (Estival et al. 2007). NLI can also play a key role in second language acquisition (SLA) applications where NLI techniques are used to identify language transfer patterns that help teachers and students focus feedback and learning on particular areas of interest (Jarvis and Crossley 2012; Rozovskaya and Roth 2010).

The system based on string kernels described in this chapter was initially designed to participate in the 2013 NLI Shared Task (Tetreault et al. 2013) and obtained third place (Popescu and Ionescu 2013). The approach was further extended by Ionescu et al. (2014) to combine several string kernels via multiple kernel learning (MKL). The system revealed some interesting facts about the kinds of string kernels and kernel learning methods that worked well for NLI. For instance, one of the best performing kernels is the (histogram) intersection kernel, which is inspired from



R.T. Ionescu and M. Popescu, *Knowledge Transfer between Computer Vision and Text Mining*, Advances in Computer Vision and Pattern Recognition,
DOI 10.1007/978-3-319-30367-3_8

computer vision (Maji et al. 2008). Another kernel based on Local Rank Distance is inspired from biology (Dinu et al. 2014; Ionescu 2013). Two kernel classifiers are alternatively used for the learning task, namely Kernel Ridge Regression (KRR) and Kernel Discriminant Analysis (KDA). Interestingly, these kernel classifiers give better results than the widely used Support Vector Machines (SVM). Several experiments that demonstrate the state-of-the-art performance of the system based on string kernels are presented in the work of Ionescu et al. (2014). However, the system was only used for the task of identifying the native language of a person from text written in English. Furthermore, a study about the language transfer patterns captured by the system was not given. This chapter, which is essentially based on the work of Ionescu et al. (2016), aims to clarify these points and to give an extended presentation and evaluation of the string kernels approach.

The first goal of this chapter is to demonstrate that the system is indeed language independent, as claimed by Ionescu et al. (2014), and to show that it can achieve state-of-the-art performance on a consistent basis. To fulfill this goal, the system is evaluated on several corpora of text documents written in three different languages, namely Arabic, English, and Norwegian. Having a general character-based approach that works well across languages could be useful, for example, in a streamlined intelligence application that is required to work efficiently on a wide range of languages for which NLP tools or language experts might not be readily available, or where the user does not desire a complex customization for each language. On the other hand, kernels based on a different kind of information (for example, syntactic and semantic information (Moschitti et al. 2011)) can be combined with string kernels via MKL to improve accuracy in some specific situations. Indeed, others (Bykh and Meurers 2014; Malmasi and Dras 2014a; Tetreault et al. 2012) have obtained better NLI results by using ensemble models based on several features, such as Context-Free Grammar production rules, Stanford dependencies, function words, and part-of-speech $n$-grams, than using models based on each of these features alone. Combing string kernels with other kind of information is not covered in this book.

The second goal is to investigate which properties of the simple kernel-based approach are the main driver behind the higher performance than more complex approaches that take words, lemmas, syntactic information, or even semantics into account. The language transfer analysis provided in Sect. 8.6 shows that there are generalizations to the kinds of mistakes that certain non-native speakers make that can be captured by $p$-grams of different lengths. Interestingly, using a range of $p$-grams implicitly generates a very large number of features including (but not limited to) stop words, stems of content words, word suffixes, entire words, and even $p$-grams of short words. Rather than doing feature selection before the training step, which is the usual NLP approach, the kernel classifier, aided by regularization, selects the most relevant features during training. With enough training samples, the kernel classifier does a better job of selecting the right features from a very high feature space. This may be one reason for why the string kernel approach works so well. To gain additional insights into why this technique works so well, the features selected by the classifier as being most discriminating are analyzed in this work. The analysis of the discriminant features also provides some information (useful in the context of

SLA) about localized language transfer effects, namely word choice (lexical transfer) and morphological differences.

The chapter is organized as follows. Work related to native language identification and to methods that work at the character level is presented in Sect. 8.2. Section 8.3 presents several similarity measures for strings, namely string kernels. The learning methods used in the experiments are described in Sect. 8.4. Section 8.5 presents details about the experiments. Section 8.6 discusses the discriminant features of the string kernels approach and the observed language transfer effects. Finally, conclusions are drawn in Sect. 8.7.

## 8.2 Related Work

### *8.2.1 Native Language Identification*

As defined in the introduction, the goal of automatic native language identification (NLI) is to determine the native language of a language learner, based on a piece of writing in a foreign language. Most research has focused on identifying the native language of English language learners, though there have been some efforts recently to identify the native language of writing in other languages, such as Chinese (Malmasi and Dras 2014c) or Arabic (Malmasi and Dras 2014a).

The first work to study automated NLI was that of Tomokiyo and Jones (2001). In their study, a Naïve Bayes model is trained to distinguish speech transcripts produced by native versus non-native English speakers. A few years later, a second study on NLI appeared (Jarvis et al. 2004). In their work, Jarvis et al. (2004) tried to determine how well a Discriminant Analysis classifier could predict the L1 language of nearly 500 English learners from different backgrounds. To make the task more challenging, they included pairs of closely related L1 languages, such as Portuguese and Spanish. The seminal paper by Koppel et al. (2005) introduced some of the best-performing features for the NLI task: character, word, and part-of-speech $n$-grams along with features inspired by the work in the area of second language acquisition such as spelling and grammatical errors. In general, most approaches to NLI have used multi-way classification with SVM or similar models along with a range of linguistic features. The book of Jarvis and Crossley (2012) presents some of the state of the art approaches used up until 2012. Being the first book of its kind, it focuses on the automated detection of L2 language-use patterns that are specific to different L1 backgrounds, with the help of text classification methods. Additionally, the book presents methodological tools to empirically test language transfer hypotheses, with the aim of explaining how the languages that a person knows interact in the mind.

In 2013, the first shared task in the field was organized by Tetreault et al. (2013). This allowed researchers to compare approaches for the first time on a specifically designed NLI corpus that was much larger than previously available data sets. In the shared task, 29 teams submitted results for the test set, and one of the most successful aspects of the competition was that it drew submissions from teams working in a variety of research fields. The submitted systems utilized a wide range of machine learning approaches, combined with several innovative feature contributions. The best performing system in the closed task achieved an overall accuracy of 83.6 % on the 11-way classification of the test set, although there was no significant difference between the top teams. Since the 2013 NLI Shared Task, two more systems have reported results above the top scoring system of the NLI Shared Task. Bykh and Meurers (2014) present an ensemble classifier based on lexicalized and non-lexicalized local syntactic features. They explored the Context-Free Grammar (CFG) production rules as syntactic features for the task of NLI. Currently, the state-of-the-art NLI system for this data set is that of Ionescu et al. (2014). It is based on a wide range of character $p$-grams. A full presentation of this state-of-the-art system is given in this chapter.

Another interesting linguistic interpretation of native language identification data was only recently addressed, specifically the analysis of second language usage patterns caused by native language interference. Usually, language transfer is studied by Second Language Acquisition researchers using manual tools. Language transfer analysis based on automated native language identification methods has been the approach of Jarvis and Crossley (2012). The work of Swanson and Charniak (2014) also defines a computational methodology that produces a ranked list of syntactic patterns that are correlated with language transfer. Their methodology allows the detection of fairly obvious language transfer effects, without being able to detect underused patterns. The first work to address the automatic extraction of underused and overused features on a per native language basis is (Malmasi and Dras 2014b). Further similar studies are likely to appear in the years to come. The current work also addresses the automatic extraction of underused and overused features captured by character $p$-grams. In a similar manner to Malmasi and Dras (2014b), the discriminant features learned by a one-versus-all kernel classifier are analyzed.

### *8.2.2 Methods that Work at the Character Level*

In recent years, methods of handling text at the character level have demonstrated impressive performance levels in various text analysis tasks (Escalante et al. 2011; Grozea et al. 2009; Kate and Mooney 2006; Lodhi et al. 2002; Popescu 2011; Popescu and Dinu 2007; Popescu and Grozea 2012; Sanderson and Guenter 2006). String kernels are a common form of using information at the character level. They are a particular case of the more general convolution kernels (Haussler 1999), often used in various NLP tasks (Croce et al. 2011). Lodhi et al. (2002) proposed to use string kernels for document categorization, showing very good results. String kernels were also successfully used in authorship identification (Popescu and Dinu 2007; Popescu and Grozea 2012; Sanderson and Guenter 2006). For example, the system described by Popescu and Grozea (2012) ranked first in most problems and overall in the PAN 2012 Traditional Authorship Attribution tasks.

Character $p$-grams have been used by some of the systems developed for native language identification. In work where feature ablation results have been reported, the performance with only character $p$-gram features was modest compared to other types of features (Tetreault et al. 2012). Initially, most work limited the character features to unigrams, bigrams, and trigrams, perhaps because longer $p$-grams were considered too expensive to compute or unlikely to improve performance. However, some of the top systems in the 2013 NLI Shared Task were based on longer character $p$-grams, up to 9-grams (Jarvis et al. 2013; Popescu and Ionescu 2013). The results presented in this work are also obtained using a range of $p$-grams. Combining all $p$-grams in a range would generate millions of features, which are indeed expensive to compute and represent. The key of avoiding the computation of such a large number of features lies in using the dual representation provided by the string kernel. String kernel similarity matrices can be computed much faster and are extremely useful when the number of samples is much lower than the number of features.

## 8.3 Similarity Measures for Strings

### *8.3.1 String Kernels*

The kernel function gives kernel methods the power to naturally handle input data that is not in the form of numerical vectors, for example strings. The kernel function captures the intuitive notion of similarity between objects in a specific domain and can be any function defined on the respective domain that is symmetric and positive definite. For strings, many such kernel functions exist with various applications in computational biology and computational linguistics (Shawe-Taylor and Cristianini 2004).

Perhaps one of the most natural ways to measure the similarity of two strings is to count how many substrings of length $p$ the two strings have in common. This gives rise to the $p$-spectrum kernel. Formally, for two strings over an alphabet $\Sigma$, $s, t \in \Sigma^*$, the $p$-spectrum kernel is defined as:

$$k_p(s,t) = \sum_{v \in \Sigma^p} \text{num}_v(s) \cdot \text{num}_v(t),$$

where $\text{num}_v(s)$ is the number of occurrences of string $v$ as a substring in $s$.[1] The feature map defined by this kernel associates a vector of dimension $|\Sigma|^p$ containing the histogram of frequencies of all its substrings of length $p$ ($p$-grams) with each string. Example 11 shows how to compute the $p$-spectrum kernel between two strings using 2-grams.

[1] The notion of substring requires contiguity. Shawe-Taylor and Cristianini (2004) discuss the ambiguity between the terms *substring* and *subsequence* across different domains: biology, computer science.

*Example 11* Given $s =$ "pineapple" and $t =$ "apple pie" over an alphabet $\Sigma$, and the substring length $p = 2$, the set of 2-grams that appear in $s$ is denoted by $S$, while the set of 2-grams that appear in $t$ is denoted by $T$. The two sets $S$ and $T$ are given by:

$$S = \{\text{"}pi\text{"}, \text{"}in\text{"}, \text{"}ne\text{"}, \text{"}ea\text{"}, \text{"}ap\text{"}, \text{"}pp\text{"}, \text{"}pl\text{"}, \text{"}le\text{"}\},$$
$$T = \{\text{"}ap\text{"}, \text{"}pp\text{"}, \text{"}pl\text{"}, \text{"}le\text{"}, \text{"}e\_\text{"}, \text{"}\_p\text{"}, \text{"}pi\text{"}, \text{"}ie\text{"}\}.$$

The $p$-spectrum kernel between $s$ and $t$ can be computed as follows:

$$k_2(s,t) = \sum_{v \in \Sigma^2} \text{num}_v(s) \cdot \text{num}_v(t),$$

where $\Sigma^2 = S \cup T$. If a 2-gram does not appear in one of the strings, the corresponding term in the above sum is zero. As the frequency of each 2-gram in $s$ or $t$ is not greater than one, the $p$-spectrum kernel is given by $|S \cap T|$, namely the number of common 2-grams among the two strings. Thus, $k_2(s,t) = 5$.

A variant of this kernel can be obtained if the embedding feature map is modified to associate a vector of dimension $|\Sigma|^p$ containing the presence bits (instead of frequencies) of all its substrings of length $p$ with each string. Thus, the character $p$-grams presence bits kernel is obtained:

$$k_p^{0/1}(s,t) = \sum_{v \in \Sigma^p} \text{in}_v(s) \cdot \text{in}_v(t),$$

where $\text{in}_v(s)$ is 1 if string $v$ occurs as a substring in $s$, and 0 otherwise.

In computer vision, the (histogram) intersection kernel has successfully been used for object class recognition from images (Maji et al. 2008; Vedaldi and Zisserman 2010). In the work of Ionescu et al. (2014), the intersection kernel is used for the first time as a kernel for strings. The intersection string kernel is defined as follows:

$$k_p^{\cap}(s,t) = \sum_{v \in \Sigma^p} \min\{\text{num}_v(s), \text{num}_v(t)\},$$

where $\text{num}_v(s)$ is the number of occurrences of string $v$ as a substring in $s$.

For the $p$-spectrum kernel, the frequency of a $p$-gram has a very significant contribution to the kernel, since it considers the product of such frequencies. On the other hand, the frequency of a $p$-gram is completely disregarded in the $p$-grams presence bits kernel. The intersection kernel lies somewhere in the middle between the $p$-grams presence bits kernel and $p$-spectrum kernel, in the sense that the frequency of a $p$-gram has a moderate contribution to the intersection kernel. More precisely, the following inequality that describes the relation between the three kernels holds:

$$k_p^{0/1}(s,t) \le k_p^{\cap}(s,t) \le k_p(s,t).$$

What is actually more interesting is that the intersection kernel assigns a high score to a $p$-gram if it has a high frequency in both strings, since it considers the minimum of the two frequencies. The $p$-spectrum kernel assigns a high score even when the $p$-gram has a high frequency in only one of the two strings. Thus, the intersection kernel captures something more about the correlation between the $p$-gram frequencies in the two strings.

Depending on the task, the kernels can yield different results, since one of the kernels can be more suitable than the other for a certain task. For example, the $p$-spectrum kernel seems to be appropriate for tasks such as authorship identification or text categorization by topic, where the frequency of a $p$-gram plays an important role. Indeed, the author's style can be easily detected if a certain writing pattern appears several times. On the other hand, a single occurrence of a certain type of mistake could sometimes be enough to make a strong assumption about the native language of a person. From this perspective, the $p$-gram frequencies could actually contain more noise than useful information. Therefore, using the $p$-grams presence bits kernel could be more suitable for the NLI task. It is also possible that some mistakes are specific to more than one native language, but the frequency of those mistakes can be different for each L1. In this case, the intersection kernel can capture such language-specific frequency correlations. Various hypotheses in favor of one or the other kernel can be put forward, but instead of choosing a single kernel function, a better approach could be to combine the kernels and use them together. Nevertheless, empirical evidence is necessary for each of the above assumptions.

Data normalization helps to improve machine learning performance for various applications. Since the range of values of raw data can have large variation, classifier objective functions will not work properly without normalization. Features are usually normalized through a process called *standardization*, which makes the values of each feature in the data have zero-mean and unit-variance. By normalization, each feature has an approximately equal contribution to the similarity between two samples. Kernel normalization is discussed in Chap. 2.

To ensure a fair comparison of strings of different lengths, normalized versions of the $p$-spectrum kernel, the $p$-grams presence bits kernel, and the intersection kernel are being used:

$$\hat{k}_p(s,t) = \frac{k_p(s,t)}{\sqrt{k_p(s,s) \cdot k_p(t,t)}},$$

$$\hat{k}_p^{0/1}(s,t) = \frac{k_p^{0/1}(s,t)}{\sqrt{k_p^{0/1}(s,s) \cdot k_p^{0/1}(t,t)}},$$

$$\hat{k}_p^{\cap}(s,t) = \frac{k_p^{\cap}(s,t)}{\sqrt{k_p^{\cap}(s,s) \cdot k_p^{\cap}(t,t)}}.$$

Taking into account $p$-grams of different length and summing up the corresponding kernels, new kernels, termed *blended spectrum kernels*, can be obtained. It is worth mentioning that open source Java implementations of the blended string kernels presented in this chapter are provided at http://string-kernels.herokuapp.com.

The string kernel implicitly embeds the texts in a high-dimensional feature space. Then, a kernel-based learning algorithm implicitly assigns a weight to each feature, thus selecting the features that are important for the discrimination task. For example, in the case of text categorization the learning algorithm enhances the features representing stems of content words (Lodhi et al. 2002), while in the case of authorship identification, the same learning algorithm enhances the features representing function words (Popescu and Dinu 2007).

### 8.3.2 *Kernel based on Local Rank Distance*

The string distance measure presented in Chap. 7 is transformed into a kernel and used for the NLI task, in a learning context. It is important to note that an upper bound for LRD can be computed as the product between the maximum offset $m$ and the number of pairs of compared $p$-grams. Thus, LRD can be normalized to a value in the [0, 1] interval. By normalizing, LRD becomes a dissimilarity measure. However, LRD needs to be used as a kernel in the experiments. The classical way to transform a distance or dissimilarity measure into a similarity measure is by using the RBF kernel (Shawe-Taylor and Cristianini 2004):

$$\hat{k}_p^{LRD}(s, t) = \exp\left(-\frac{\Delta_{LRD}(s, t)}{2\sigma^2}\right),$$

where $s$ and $t$ are two strings and $p$ is the $p$-grams length. The parameter $\sigma$ is usually chosen so that values of $\hat{k}(s, t)$ are well scaled. In the above equation, $\Delta_{LRD}$ is already normalized to a value in the [0, 1] interval to ensure a fair comparison of strings of different length.

## 8.4 Learning Methods

The approach used for the NLI task is based on kernel methods. An overview of kernel methods is given in Chap. 2. Some specific topics on kernel methods that are used in the context of NLI, which are not fully addressed in Chap. 2, are discussed in more detail in this section.

As stated in Chap. 2, kernel-based learning algorithms implicitly embed the data into a Hilbert feature space by specifying the inner product between each pair of points. More precisely, a kernel matrix that contains the pairwise similarities between every pair of training samples is used in the learning stage to assign a vector of weights to the training samples. Let $\alpha$ denote this weight vector. In the test stage, the pairwise similarities between a test sample $x$ and all the training samples are computed. Then, the following binary classification function assigns a positive or a negative label to the test sample:

$$g(x) = \sum_{i=1}^{n} \alpha_i \cdot k(x, x_i), \tag{8.1}$$

where $x$ is the test sample, $n$ is the number of training samples, $X = \{x_1, x_2, \ldots, x_n\}$ is the set of training samples, $k$ is a kernel function, and $\alpha_i$ is the weight assigned to the training sample $x_i$. In the primal form, the same binary classification function can be expressed as:

$$g(x) = \langle w, x \rangle, \tag{8.2}$$

where $\langle \cdot, \cdot \rangle$ denotes the scalar product, $x \in \mathbb{R}^m$ is the test sample represented as a vector of features, and $w \in \mathbb{R}^m$ is a vector of feature weights that can be computed as follows:

$$w = \sum_{i=1}^{n} \alpha_i \cdot x_i, \tag{8.3}$$

given that the kernel function $k$ can be expressed as a scalar product between samples, as follows:

$$k(s, t) = \langle \phi(s), \phi(t) \rangle,$$

where $\phi$ is the embedding feature map associated to the kernel $k$. An important remark is that not all kernels have a corresponding finite feature space induced by $\phi$. It can be trivially observed that the $p$-spectrum kernel or the $p$-grams presence bits kernel both have a finite feature space. On the other hand, it is well known (Shawe-Taylor and Cristianini 2004) that the embedding feature map of the RBF kernel induces an infinite feature space. Therefore, the RBF kernel based on LRD can only be used in the dual form. Furthermore, using some kernel functions in the primal form is inherently prohibitive. With enough effort, a finite feature representation for such a kernel can be determined in order to be used in the primal form, but trying to interpret the weights associated with the primal features becomes impossible or at least very difficult. The intersection kernel falls in this category. The language transfer analysis presented in Sect. 8.6 is based only on the $p$-grams presence bits kernel, which is extensively used throughout the experiments, since the weights $w$ corresponding to the primal representation of the $p$-grams presence bits kernel can be obtained with Eq. (8.3).

The advantage of using the dual representation induced by the kernel function becomes clear if the dimension of the feature space $m$ is taken into consideration. Since string kernels are based on character $p$-grams, the feature space is indeed very high. For instance, using 5-grams based only on the 26 letters of the English alphabet will result in a feature space of $26^5 = 11{,}881{,}376$ features. By considering only the 5-grams that occur in a given corpus, there will be less features since some $p$-grams will never appear in the corpus. However, in the experiments conducted on the TOEFL11 corpus the feature space includes 5-grams along with 6-grams, 7-grams 8-grams, and 9-grams. The actual number of features generated from this corpus using 5–9 $p$-grams is precisely $m = 4{,}662{,}520$. As long as the number of samples $n$ is much lower than the number of features $m$, it can be more efficient to use the dual representation given by the kernel matrix. This fact is also known as the *kernel trick* (Shawe-Taylor and Cristianini 2004).

For the NLI experiments, two binary kernel classifiers are used, namely the SVM and the KRR. In the case of binary classification problems, Support Vector Machines try to find the vector of weights that defines the hyperplane that maximally separates the images in the Hilbert space of the training examples belonging to the two classes. Kernel Ridge Regression selects the vector of weights that simultaneously has small empirical error and small norm in the Reproducing Kernel Hilbert Space generated by the kernel function. More details about SVM and KRR are provided in Chap. 2. The important fact is that the above optimization problems are solved in such a way that the coordinates of the embedded points are not needed, only their pairwise inner products which in turn are given by the kernel function.

SVM and KRR produce binary classifiers, but native language identification is usually a multi-class classification problem. There are many approaches for combining binary classifiers to solve multi-class problems. Typically, the multi-class problem is broken down into multiple binary classification problems using common decomposing schemes such as: one-versus-all and one-versus-one. There are also kernel methods that take the multi-class nature of the problem directly into account, for instance Kernel Discriminant Analysis. The KDA classifier is sometimes able to improve accuracy by avoiding the masking problem (Hastie and Tibshirani 2003). In the case of multi-class native language identification, the masking problem may appear, for instance, when non-native English speakers have acquired, as the second language, a different language to English. Another case of masking is when a native language $A$ is somehow related to two other native languages $B$ and $C$, in which case the samples that belong to class $A$ can sit in the middle between the samples of classes $B$ and $C$, as illustrated in Fig. 8.1. In this case, the class in the middle is masked by the other two classes, as it never dominates. KDA can solve such unwanted situations automatically, without having to identify what languages are related by any means, such as geographical position, quantitative linguistic analysis, and so on.

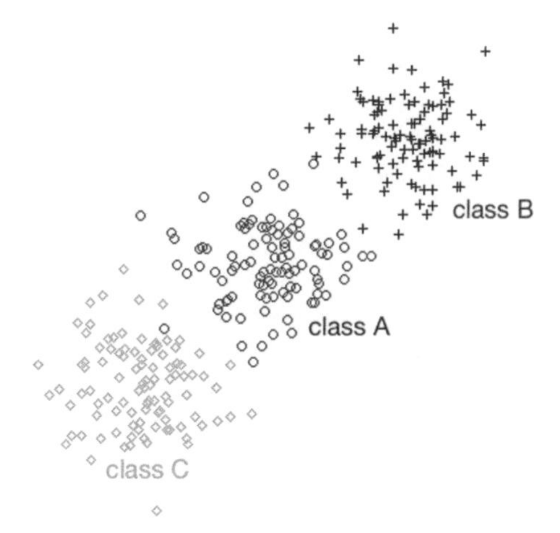


**Fig. 8.1** An example with three classes that illustrates the masking problem. Class *A* is masked by classes *B* and *C*

## 8.5 Experiments

The main purpose of the experiments is to demonstrate that techniques that work at the character level and are inspired by advances in computer vision can obtain state-of-the-art performance in NLI and are truly language independent. The secondary purpose is to show what kind of kernels and which kernel method works best in the context of NLI.

### *8.5.1 Data Sets Description*

In this chapter, experiments are carried out on five data sets: a modified version of the ICLEv2 corpus (Granger et al. 2009), the ETS Corpus of Non-Native Written English, or TOEFL11 (Blanchard et al. 2013), the TOEFL11-Big corpus as used by Tetreault et al. (2012), a subset of the second version of the Arabic Learner Corpus (ALC) (Alfaifi et al. 2014), and a subset of the ASK Corpus (Tenfjord et al. 2006). A summary of the corpora is given in Table 8.1.

**Table 8.1** Summary of corpora used in the experiments

| Corpus | L2 | L1 languages | Documents |
|---|---|---|---|
| ICLEv2 | English | 7 | 770 |
| TOEFL11 | English | 11 | 12,100 |
| TOEFL11-Big | English | 11 | 87,502 |
| ALC subset | Arabic | 7 | 329 |
| ASK corpus subset | Norwegian | 7 | 1400 |

The ICLEv2 is a corpus of essays written by highly proficient non-native college-level students of English. A modified version of the corpus that has been normalized as much as possible for topic and character encoding (Tetreault et al. 2012) is used in the experiments. This version of the corpus contains 110 essays each for 7 native languages: Bulgarian, Chinese, Czech, French, Japanese, Russian, and Spanish.

The ETS Corpus of Non-Native Written English (TOEFL11) was first introduced by Tetreault et al. (2012) and extended for the 2013 Native Language Identification Shared Task (Tetreault et al. 2013). The TOEFL11 corpus contains a balanced distribution of essays per prompt (topic) per native language. The corpus contains essays written by speakers of the following 11 languages: Arabic, Chinese, French, German, Hindi, Italian, Japanese, Korean, Spanish, Telugu, and Turkish. For the shared task, the 12,100 essays were split into 9,900 for training, 1,100 for development and 1,100 for testing.

Tetreault et al. (2012) present a corpus, TOEFL11-Big, to investigate the performance of their NLI system on a very large data set. This data set contains the same languages as TOEFL11, but with no overlap in content. It contains a total of over 87 thousand essays written to a total of 76 different prompts. The distribution of L1 per prompt is not as even as for TOEFL11, though all topics are represented for all L1s.

The second version of the Arabic Learner Corpus was recently introduced by Alfaifi et al. (2014). The corpus includes essays by Arabic learners studying in Saudi Arabia. There are 66 different mother tongue (L1) representations in the corpus, but the majority of these have less than 10 essays. To compare the string kernels approach with the state-of-the-art method (Malmasi and Dras 2014a), the subset of ALC used by Malmasi and Dras (2014a) is also used in this chapter. The subset is obtained by considering only the top 7 native languages by number of essays. The ALC subset used in the experiments contains 329 essays in total, which are not evenly distributed per native language, as shown in Table 8.2.

The ASK Corpus is a language learner corpus of Norwegian as a second language developed by Tenfjord et al. (2006). It contains 1,938 essays collected from language tests on two different proficiency levels. There are 9 native languages in the corpus, but only 7 of them are equally represented by 200 essays each. To avoid any bias

**Table 8.2** Distribution of the documents per native language in the ALC subset

| Language | Documents |
|---|---|
| Chinese | 76 |
| English | 35 |
| French | 44 |
| Fulani | 36 |
| Malay | 46 |
| Urdu | 64 |
| Yoruba | 28 |
| Total | 329 |

from the uneven distribution of documents per language, the experiments presented in this work include only the 7 languages that contain 200 essays each: English, Dutch, German, Polish, Russian, Serbo-Croatian, and Spanish. The corpus has been tokenized, and no raw version is available.

### 8.5.2 Parameter Tuning and Implementation Choices

In the string kernels approach presented in this work, documents or essays from the corpora are treated as strings. Therefore, the notions of *string* and *document* are used interchangeably. Because the approach works at the character level, there is no need to split the texts into words, or to do any NLP-specific preprocessing. The only editing done to the texts was the replacing of sequences of consecutive whitespace characters (space, tab, new line, and so on) with a single space character. This normalization is needed in order to prevent an incorrect measure of similarity between texts, simply as a result of different spacing. All uppercase letters were converted to the corresponding lowercase ones. The letter case normalization has no effect on the documents of the ALC subset, since there are no distinct upper and lower case letter forms in the Arabic alphabet.

A series of preliminary experiments were conducted in order to select the best-performing learning method. In these experiments the string kernel was fixed to the $p$-spectrum normalized kernel of length 5 ($\hat{k}_5$), because the goal was to select the best learning method, and not to find the best kernel. The following learning methods were evaluated: one-versus-one SVM, one-versus-all SVM, one-versus-one KRR, one-versus-all KRR, and KDA. A 10-fold cross-validation procedure was carried out on the TOEFL11 training set to evaluate the classifiers. The preliminary results indicate that the one-versus-all KRR and the KDA classifiers produce the best results. Therefore, they are selected for the remaining experiments.

Another set of preliminary experiments were performed to determine the range of $p$-grams that gives the most accurate results. Each language was separately considered to determine the optimal range. This is motivated by how different the Arabic writing system is from English and Norwegian. For example, short vowels can be omitted from the Arabic writing, and consequently produce shorter words. There could also be differences between English and Norwegian, since Norwegian has a productive compound noun system that may occasionally result in very long words, although of course the two languages are more closely related to each other than to Arabic, because they belong to the branch of Germanic languages.

First, a 10-fold cross-validation procedure was carried out on the TOEFL11 training set. All the $p$-grams in the range 2–10 were evaluated. Furthermore, experiments with different blended kernels were conducted to see whether combining $p$-grams of different lengths could improve the accuracy. The best results were obtained when all the $p$-grams with length in the range 5–8 were used. Other authors (Bykh and Meurers 2012; Popescu and Ionescu 2013) have also reported better results by using $p$-grams with the length in a range, rather than using $p$-grams of fixed length. Consequently,

**Table 8.3** Average word length and optimal $p$-gram range for the TOEFL11 corpus (English L2), the ALC subset (Arabic L2), and the ASK corpus (Norwegian L2)

| Corpus | Average word length | Range of $n$-grams |
|---|---|---|
| TOEFL11 | 4.46 | 5–8 |
| ALC subset | 4.36 | 3–5 |
| ASK corpus subset | 4.34 | 5–8 |

Spaces and punctuation were not taken into consideration when the average word lengths were computed

the results reported on the English corpora are based on blended string kernels based on 5–8 $p$-grams. Using a similar procedure on the ASK corpus, the range of $p$-grams found to work best was again 5–8. A 10-fold cross-validation procedure was also conducted on the ALC subset of 329 documents in order to evaluate the $p$-grams in the range 2–7. The best performance was obtained using 4-grams. In general, a blended spectrum of $p$-grams produces better results, and therefore the combination of 3–5 $p$-grams (centered around 4-grams) was chosen for the rest of the experiments on the ALC corpus. Preliminary results on the Arabic corpus indicate that using the range of 3–5 $p$-grams gives a roughly 2 % improvement over using $p$-grams of length 4 alone. Other ranges of $p$-grams were not evaluated to prevent any overfitting on the ALC corpus. An interesting note is that the only difference between the systems evaluated on the English corpora and those evaluated on the Arabic corpus is that they are based on different $p$-gram lengths. On the other hand, the range of $p$-grams is identical for the English and the Norwegian corpora, probably because the two languages are more closely related to each other than to Arabic. Table 8.3 shows the average word length for the three corpora on which the range of $p$-grams was independently tuned. Remarkably, the average word lengths computed in the current study are consistent with other statistics reported in literature. Indeed, according to a study conducted by Intellaren,[2] the average word length in the Quran (Arabic) is 4.25 letters, and, according to Manning et al. (2008), the average English word length is about 4.5 letters. In general, it seems that the range of $p$-grams is related to the average word length in a specific language. More precisely, the $p$-grams having the same length as the average word or the ones that are slightly longer produce better accuracy rates. This is an encouraging result because it confirms that language-specific knowledge (e.g., average word length) is not required (although it can be helpful), but can be empirically established as described above, if necessary.

Some preliminary experiments were also performed to establish the type of kernel to be used, namely the blended $p$-spectrum kernel ($\hat{k}_{5-8}$), the blended $p$-grams presence bits kernel ($\hat{k}^{0/1}_{5-8}$), the blended $p$-grams intersection kernel ($\hat{k}^{\cap}_{5-8}$), or the kernel based on LRD ($\hat{k}^{LRD}_{5-8}$). These different kernel representations are obtained from the same data. The idea of combining all these kernels is natural when one wants to improve the performance of a classifier. When multiple kernels are combined, the features are actually embedded in a higher dimensional space. As a consequence,

[2]http://www.intellaren.com/articles/en/qss.

the search space of linear patterns grows, which helps the classifier to select a better discriminant function. The most natural way of combining two kernels is to sum them up. Summing up kernels or kernel matrices is equivalent to feature vector concatenation. Another option is to combine kernels by kernel alignment (Cristianini et al. 2001). Instead of simply summing kernels, kernel alignment assigns weights for each of the two kernels based on how well they are aligned with the ideal kernel $YY'$ obtained from training labels. The kernels were evaluated alone and in various combinations. The best kernels are the blended $p$-grams presence bits kernel and the blended $p$-grams intersection kernel. The best kernel combinations include the blended $p$-grams presence bits kernel, the blended $p$-grams intersection kernel, and the kernel based on LRD. As the kernel based on LRD is slightly slower than the other string kernels, the kernel combinations that include it were only evaluated on the TOEFL11 corpus, the ICLE corpus, the ALC subset, and the ASK corpus.

### 8.5.3 Experiment on TOEFL11 Corpus

This section describes the results on the TOEFL11 corpus and, for the sake of completion, it also includes results for the 2013 *Closed* NLI Shared Task. In the closed shared task the goal is to predict the native language of testing examples, restricted to learning only from the training and the development data. The additional information from *prompts* or the English language proficiency level were not used in the approach presented here.

The regularization parameters were tuned on the development set. In this case, the systems were trained on the entire training set. A 10-fold cross-validation (CV) procedure was done on the training and the development sets. The folds were provided along with the TOEFL11 corpus. Finally, the results of the proposed systems are also reported on the NLI Shared Task test set. For testing, the systems were trained on both the training set and the development set. The results are summarized in Table 8.4.

The results presented in Table 8.4 show that string kernels can reach state-of-the-art accuracy levels for this task. Overall, it seems that KDA is able to obtain better results than KRR. The intersection kernel alone is able to obtain slightly better results than the presence bits kernel. The kernel based on LRD gives significantly lower accuracy rates, but it is able to improve the performance when it is combined with the blended $p$-grams presence bits kernel. In fact, most of the kernel combinations give better results than each of their components. The best kernel combination is that of the presence bits kernel and the intersection kernel. Results are quite similar when they are combined either by summing them up or by kernel alignment. The best performance on the test set (85.3 %) is obtained by the system that combines these two kernels via kernel alignment and learns using KDA. This system is 1.7 % better than the state-of-the-art system of Jarvis et al. (2013) based on SVM and word features, this being the top scoring system in the 2013 NLI Shared Task. It is also 2.6 % better than the state-of-the-art system based on string kernels of Popescu and Ionescu (2013). On the cross-validation procedure, there are three systems that

**Table 8.4** Accuracy rates on TOEFL11 corpus (English L2) of various classification systems based on string kernels compared with other state-of-the-art approaches

| Method | Dev (%) | 10-fold CV (%) | Test (%) |
|---|---|---|---|
| Ensemble model (Tetreault et al. 2012) | – | 80.9 | – |
| KRR and string kernels (Popescu and Ionescu 2013) | – | 82.6 | 82.7 |
| SVM and word features (Jarvis et al. 2013) | – | **84.5** | 83.6 |
| Ensemble model and CFG (Bykh and Meurers 2014) | – | – | 84.8 |
| KRR and $\hat{k}^{0/1}_{5-8}$ | 85.4 | 82.5 | 82.0 |
| KRR and $\hat{k}^{\cap}_{5-8}$ | 84.9 | 82.2 | 82.6 |
| KRR and $\hat{k}^{LRD}_{5-8}$ | 78.7 | 77.1 | 77.5 |
| KRR and $\hat{k}^{0/1}_{5-8} + \hat{k}^{LRD}_{5-8}$ | 85.7 | 82.6 | 82.7 |
| KRR and $\hat{k}^{\cap}_{5-8} + \hat{k}^{LRD}_{5-8}$ | 84.9 | 82.2 | 82.0 |
| KRR and $\hat{k}^{0/1}_{5-8} + \hat{k}^{\cap}_{5-8}$ | 85.5 | 82.6 | 82.5 |
| KRR and $a_1\hat{k}^{0/1}_{5-8} + a_2\hat{k}^{\cap}_{5-8}$ | 85.5 | 82.6 | 82.5 |
| KDA and $\hat{k}^{0/1}_{5-8}$ | 86.2 | 83.6 | 83.6 |
| KDA and $\hat{k}^{\cap}_{5-8}$ | 85.2 | 83.5 | 84.6 |
| KDA and $\hat{k}^{LRD}_{5-8}$ | 79.7 | 78.5 | 79.2 |
| KDA and $\hat{k}^{0/1}_{5-8} + \hat{k}^{LRD}_{5-8}$ | **87.1** | 84.0 | 84.7 |
| KDA and $\hat{k}^{\cap}_{5-8} + \hat{k}^{LRD}_{5-8}$ | 85.8 | 83.4 | 83.9 |
| KDA and $\hat{k}^{0/1}_{5-8} + \hat{k}^{\cap}_{5-8}$ | 86.4 | 84.1 | 85.0 |
| KDA and $a_1\hat{k}^{0/1}_{5-8} + a_2\hat{k}^{\cap}_{5-8}$ | 86.5 | 84.1 | **85.3** |
| KDA and $\hat{k}^{0/1}_{5-8} + \hat{k}^{\cap}_{5-8} + \hat{k}^{LRD}_{5-8}$ | 87.0 | 84.1 | 84.8 |

The best accuracy rates on each set of experiments are highlighted in bold. The weights $a_1$ and $a_2$ from the weighted sums of kernels are computed by kernel alignment

reach an accuracy of 84.1 %. All of them are based on KDA and various kernel combinations. The greatest accuracy rate of 84.1 % reported for the cross-validation procedure is 3.2 % above the state-of-the-art system of Tetreault et al. (2012) and 0.4 % below the top scoring system of Jarvis et al. (2013). The empirical results obtained in this experiment demonstrate that the approach proposed in this chapter can reach state-of-the-art accuracy levels.

The results presented in Table 8.4 are obtained on the data released for the 2013 NLI Shared Task. The documents provided for the task were already tokenized. Access to the raw text documents was not provided during the task. As the TOEFL11 corpus was later released through LDC,[3] the raw text documents also became available for use. It is particularly interesting to evaluate the various string kernel systems on the raw text documents, in order to determine whether the string kernels approach is indeed neutral to any linguistic theory. The process of breaking a stream of text up into tokens is not an easy task for some languages and it is often based on linguistic theories. Even if tokenization is fairly straightforward in languages that use spaces

[3]https://catalog.ldc.upenn.edu/LDC2014T06.

**Table 8.5** Accuracy rates on the raw text documents of the TOEFL11 corpus (English L2) of various classification systems based on string kernels

| Method | Dev (%) | 10-fold CV (%) | Test (%) |
|---|---|---|---|
| KRR and $\hat{k}^{0/1}_{5-9}$ | 86.3 | 82.6 | 82.8 |
| KRR and $\hat{k}^{\cap}_{5-9}$ | 84.5 | 82.5 | 82.6 |
| KRR and $\hat{k}^{LRD}_{5-9}$ | 79.9 | 77.9 | 77.6 |
| KRR and $\hat{k}^{0/1}_{5-9} + \hat{k}^{LRD}_{5-9}$ | 86.1 | 82.6 | 83.1 |
| KRR and $\hat{k}^{\cap}_{5-9} + \hat{k}^{LRD}_{5-9}$ | 84.9 | 82.4 | 82.1 |
| KRR and $\hat{k}^{0/1}_{5-9} + \hat{k}^{\cap}_{5-9}$ | 85.6 | 82.9 | 82.6 |
| KRR and $a_1\hat{k}^{0/1}_{5-9} + a_2\hat{k}^{\cap}_{5-9}$ | 85.6 | 82.9 | 82.6 |
| KDA and $\hat{k}^{0/1}_{5-9}$ | 86.2 | 83.7 | 84.6 |
| KDA and $\hat{k}^{\cap}_{5-9}$ | 85.2 | 83.6 | 84.3 |
| KDA and $\hat{k}^{LRD}_{5-9}$ | 81.2 | 79.1 | 79.6 |
| KDA and $\hat{k}^{0/1}_{5-9} + \hat{k}^{LRD}_{5-9}$ | 86.6 | 83.9 | 84.7 |
| KDA and $\hat{k}^{\cap}_{5-9} + \hat{k}^{LRD}_{5-9}$ | 85.9 | 83.5 | 83.9 |
| KDA and $\hat{k}^{0/1}_{5-9} + \hat{k}^{\cap}_{5-9}$ | 85.9 | **84.1** | 85.1 |
| KDA and $a_1\hat{k}^{0/1}_{5-9} + a_2\hat{k}^{\cap}_{5-9}$ | 86.1 | **84.1** | **85.3** |
| KDA and $\hat{k}^{0/1}_{5-9} + \hat{k}^{\cap}_{5-9} + \hat{k}^{LRD}_{5-9}$ | **86.7** | 84.0 | 84.9 |

The best accuracy rates on each set of experiments are highlighted in bold. The weights $a_1$ and $a_2$ from the weighted sums of kernels are computed by kernel alignment

to separate words, there are many edge cases including contractions and hyphenated words that require linguistic knowledge. Therefore, analyzing the results of the string kernels approach when tokenization is not being used at all will reveal if the linguistic knowledge helps to improve the performance or not. Table 8.5 contains the results on the raw text documents of various strings kernels combined either with KRR and KDA. An important remark is that the tokenization process can influence the average word length in the corpus. For instance, “don’t” is divided into two tokens, namely “do” and “n’t”. Therefore, the range of $p$-grams used for computing the blended string kernels was adjusted based on the development set for the raw text documents. All $p$-grams in the range 3–9 were evaluated and the best results were obtained when all the $p$-grams with length in the range 5–9 were used. Not surprisingly, the range of $p$-grams for the raw text documents is fairly close to the range of $p$-grams that worked best for the tokenized text.

The empirical results shown in Table 8.5 indicate that the presence bits kernel and the kernel based on LRD obtain better results on the raw text documents. On the other hand, the results of the intersection kernel are roughly the same, or only slightly better compared to the results presented in Table 8.4. When the various kernels are mixed together, the results are nearly identical to the results obtained on tokenized text documents. Overall, the results are better when the string kernels are computed on the raw text, although the improvements are not considerably different, since the accuracy difference is less than 1 % in all cases. However, the improvement could be the result of removing tokenization or the results of using a different range of $p$-grams.

To find which of these two hypotheses is responsible for the performance gain, the presence bits kernel was computed on the raw text documents using a range of 5–8 $p$-grams. The accuracy obtained by KRR and $\hat{k}^{0/1}_{5-9}$ on raw text was 82.6 %, which is still better than the accuracy of the very same system when tokenization was included (82.0 %). Therefore, it can be concluded that both hypotheses are responsible for the improved results. According to some linguistic theories, it is recommended to separate the text into tokens before any other processing is done. However, this is not necessary in the case of string kernels. Indeed, the empirical results confirmed that the approach based on character $p$-grams does not rely on any linguistic theory to reach state-of-the-art performance levels.

It is worth mentioning that a significance test performed by the organizers of the 2013 NLI Shared Task showed that the top systems that participated in the competition are not essentially different. Their conclusion also implies that adding or removing text tokenization in the string kernel approach does not have a significant influence on the results, even if they appear to be slightly better when tokenization is not used on the TOEFL11 corpus. Nevertheless, the rest of the experiments on English corpora are conducted on tokenized text documents, to ensure a fair comparison with the other state-of-the-art approaches that use text tokenization as a preprocessing step. The second reason for applying the string kernel approach on tokenized texts is that the range of $p$-grams is smaller (5–8 instead of 5–9) which translates into a faster technique that requires less time to compute the blended spectrum kernels. Note, though, that since the texts in the ALC corpus are not tokenized, the string kernel approach is applied directly on the raw text documents. In other words, the string kernel approach is evaluated on tokenized text documents as long as these documents are included in the corpus. On the other hand, the language transfer analysis given in Sect. 8.6 includes only the features computed on the raw text documents of the TOEFL11 corpus, in order to avoid any confusion between the language transfer patterns and the patterns induced by tokenization. For instance, adding spaces before punctuation is part of the tokenization process, but it could be mistaken for a language transfer pattern as well.

### 8.5.4 Experiment on ICLE Corpus

The results on the ICLEv2 corpus using a 5-fold cross-validation procedure are summarized in Table 8.6. To adequately compare the results with a state-of-the-art system, the same 5-fold cross-validation procedure used by Tetreault et al. (2012) was also used in this experiment. Table 8.6 shows that the results obtained by the presence bits kernel and by the intersection kernel are systematically better than the state-of-the-art system of Tetreault et al. (2012). While both KRR and KDA produce accuracy rates that are better than the state-of-the-art accuracy rate, it seems that KRR is slightly better in this experiment. Again, the idea of combining kernels seems to produce more robust systems. The best systems are based on combining the presence bits kernel either with the kernel based on LRD or the intersection kernel. Overall,

Table 8.6 Accuracy rates on ICLEv2 corpus (English L2) of various classification systems based on string kernels compared with a state-of-the-art approach

| Method | 5-fold CV (%) |
|---|---|
| Ensemble model (Tetreault et al. 2012) | 90.1 |
| KRR and $\hat{k}^{0/1}_{5-8}$ | 91.2 |
| KRR and $\hat{k}^{\cap}_{5-8}$ | 90.5 |
| KRR and $\hat{k}^{LRD}_{5-8}$ | 81.8 |
| KRR and $\hat{k}^{0/1}_{5-8} + \hat{k}^{LRD}_{5-8}$ | **91.3** |
| KRR and $\hat{k}^{\cap}_{5-8} + \hat{k}^{LRD}_{5-8}$ | 90.1 |
| KRR and $\hat{k}^{0/1}_{5-8} + \hat{k}^{\cap}_{5-8}$ | 90.9 |
| KRR and $\hat{k}^{0/1}_{5-8} + \hat{k}^{\cap}_{5-8} + \hat{k}^{LRD}_{5-8}$ | 90.6 |
| KDA and $\hat{k}^{0/1}_{5-8}$ | 90.5 |
| KDA and $\hat{k}^{\cap}_{5-8}$ | 90.5 |
| KDA and $\hat{k}^{LRD}_{5-8}$ | 82.3 |
| KDA and $\hat{k}^{0/1}_{5-8} + \hat{k}^{LRD}_{5-8}$ | 90.8 |
| KDA and $\hat{k}^{\cap}_{5-8} + \hat{k}^{LRD}_{5-8}$ | 90.4 |
| KDA and $\hat{k}^{0/1}_{5-8} + \hat{k}^{\cap}_{5-8}$ | 91.0 |
| KDA and $\hat{k}^{0/1}_{5-8} + \hat{k}^{\cap}_{5-8} + \hat{k}^{LRD}_{5-8}$ | 90.8 |

The accuracy rates are reported using a 5-fold cross-validation (CV) procedure. The best accuracy rate is highlighted in bold

the reported accuracy rates are higher than the state-of-the-art accuracy rate. The best performance (91.3 %) is achieved by the KRR classifier based on combining the presence bits kernel with the kernel based on LRD. This represents a 1.2 % improvement over the state-of-the-art accuracy rate of Tetreault et al. (2012). Two more systems are able to obtain accuracy rates greater than 91.0 %. These are the KRR classifier based on the presence bits kernel (91.2 %) and the KDA classifier based on the sum of the presence bits kernel and the intersection kernel (91.0 %). The overall results on the ICLEv2 corpus show that the string kernels approach can reach state-of-the-art accuracy levels, although a paired-sample Student's t-test indicated that the results are not significantly different. It is worth mentioning the purpose of this experiment was to use the same approach determined to work well in the TOEFL11 corpus. To serve this purpose, the range of $p$-grams was not tuned on this data set. Furthermore, other classifiers were not tested in this experiment. Nevertheless, better results can probably be obtained by adding these aspects into the equation.

### 8.5.5 *Experiment on TOEFL11-Big Corpus*

The presence bits kernel and the intersection kernel are evaluated on the TOEFL11-Big corpus using the 10-fold CV procedure. The folds are chosen according to Tetreault et al. (2012), in order to fairly compare the string kernels with their state-of-the-art ensemble model. The kernel based on LRD is not evaluated in this experiment on purpose, since it obtained lower accuracy rates in the TOEFL11 and ICLE experiments. The second reason for not including LRD is that it is more computationally expensive than the presence bits kernel or the intersection kernel.

As in the previous experiments, both KRR and KDA were tested on TOEFL11-Big, but KDA was not trained using all the training samples because it runs out of memory. The machine used for this experiment has 256 GB of RAM, and it seems that training KDA on 78 thousand samples requires slightly more memory. However, a KDA classifier can successfully be trained on 75 thousand samples on the same machine. Therefore, the results of KDA presented in this section are obtained by training it on a random sample of 75 thousand samples selected during each fold. The results of KDA might be slightly different (but probably less than 1 %) if the entire available training data would be used. The regularization parameters of KRR and KDA were tuned such that an accuracy of roughly 98 % is obtained on the training set. In other words, the classifiers are allowed to disregard around 2 % of the training documents to prevent overfitting. From this perspective, the disregarded training documents are not considered helpful for the NLI task. Preliminary results showed that trying to fit more of the training data leads to lower accuracy rates (not more than 1–2 % though) in the CV procedure.

The results obtained by the presence bits kernel and by the intersection kernel are presented in Table 8.7. The two kernels are also combined together. When considering the individual kernels, KDA seems to work much better than KRR, even if it is not using the entire available training data. For instance, the KDA classifier based on the intersection kernel attains the same accuracy as the state-of-the-art ensemble model of Tetreault et al. (2012), while KRR based on the same kernel reaches an accuracy of only 82.3 %, which is 2.3 % lower than the state-of-the-art (84.6 %). On the other hand, KRR overturns the results when the two kernels are combined through MKL. Indeed, KRR based on the sum of $\hat{k}^{0/1}_{5-8}$ and $\hat{k}^{\cap}_{5-8}$ reaches an accuracy of 84.8 %, while KDA based on the kernel combination obtains 84.7 %. Both of them are better

**Table 8.7** Accuracy rates on TOEFL11-Big (English L2) corpus of various classification systems based on string kernels compared with a state-of-the-art approach

| Method | Test (%) |
|---|---|
| Ensemble model (Tetreault et al. 2012) | 84.6 |
| KRR and $\hat{k}^{0/1}_{5-8}$ | 82.6 |
| KRR and $\hat{k}^{\cap}_{5-8}$ | 82.3 |
| KRR and $\hat{k}^{0/1}_{5-8} + \hat{k}^{\cap}_{5-8}$ | **84.8** |
| KDA and $\hat{k}^{0/1}_{5-8}$ | 84.5 |
| KDA and $\hat{k}^{\cap}_{5-8}$ | 84.6 |
| KDA and $\hat{k}^{0/1}_{5-8} + \hat{k}^{\cap}_{5-8}$ | 84.7 |

The accuracy rates are reported using a 10-fold cross-validation (CV) procedure. The best accuracy rate is highlighted in bold

than the state-of-the-art model. Although an improvement of 0.2 % or 0.1 % over the state-of-the-art does not look like much, the number of samples in the TOEFL11-Big data set can tell a different story, since 0.1 % means roughly 90 samples. From this point of view, KRR based on the kernel combination assigns the correct labels for more than 170 extra samples compared to the state-of-the-art system. To draw a conclusion on the empirical results, the idea of combining kernels through MKL proves to be extremely helpful for boosting the performance of the classification system. KDA seems to be more stable with respect to the type of kernel, while KRR gives the top accuracy only when the kernels are combined.

### 8.5.6 Cross-Corpus Experiment

In this experiment, various systems based on KRR or KDA are trained on the TOEFL11 corpus and tested on the TOEFL11-Big corpus. The state-of-the-art ensemble model (Tetreault et al. 2012) was trained and tested in exactly the same manner. The goal of this cross-corpus evaluation is to show that the string kernel approach is not performing well simply because of potential topic bias in the corpora. Again, the kernel based on LRD was not included in this experiment since it is more computationally expensive. Therefore, only the presence bits kernel and the intersection kernel were evaluated on the TOEFL11-Big corpus. The results are summarized in Table 8.8. The same regularization parameters determined to work well on the TOEFL11 development set were used.

The most interesting fact is that all the proposed systems are at least 30 % better than the state-of-the-art system. Considering that the TOEFL11-Big corpus contains 87 thousand samples, the 30 % improvement is significant without any doubt. Diving

**Table 8.8** Accuracy rates on TOEFL11-Big corpus (English L2) of various classification systems based on string kernels compared with a state-of-the-art approach

| Method | Test (%) |
|---|---|
| Ensemble model (Tetreault et al. 2012) | 35.4 |
| KRR and $\hat{k}^{0/1}_{5-8}$ | 66.7 |
| KRR and $\hat{k}^{\cap}_{5-8}$ | 67.2 |
| KRR and $\hat{k}^{0/1}_{5-8} + \hat{k}^{\cap}_{5-8}$ | **67.7** |
| KRR and $a_1\hat{k}^{0/1}_{5-8} + a_2\hat{k}^{\cap}_{5-8}$ | **67.7** |
| KDA and $\hat{k}^{0/1}_{5-8}$ | 65.6 |
| KDA and $\hat{k}^{\cap}_{5-8}$ | 65.7 |
| KDA and $\hat{k}^{0/1}_{5-8} + \hat{k}^{\cap}_{5-8}$ | 66.2 |
| KDA and $a_1\hat{k}^{0/1}_{5-8} + a_2\hat{k}^{\cap}_{5-8}$ | 66.2 |

The systems are trained on the TOEFL11 corpus and tested on the TOEFL11-Big corpus. The best accuracy rate is highlighted in bold. The weights $a_1$ and $a_2$ from the weighted sums of kernels are computed by kernel alignment

into details, it can be observed that the results obtained by KRR are higher than those obtained by KDA. However, both methods perform very well compared to the state of the art. Again, kernel combinations are better than each of their individual kernels alone.

It is important to mention that the significant performance increase is not due to the learning method (KRR or KDA), but rather due to the string kernels that work at the character level. It is not only the case that string kernels are language independent, but for the same reasons they can also be topic independent. The topics (prompts) from TOEFL11 are different from the topics from TOEFL11-Big, and it becomes clear that a method that uses words as features is strongly affected by topic variations, since the distribution of words per topic can be completely different. Nonetheless, mistakes that reveal the native language can be captured by character $p$-grams that can appear more often even in different topics. The results indicate that this is also the case of the approach based on string kernels, which seems to be more robust to such topic variations of the data set. The best system has an accuracy rate that is 32.3 % better than the state-of-the-art system of Tetreault et al. (2012). Overall, the empirical results indicate that the string kernels approach can achieve significantly better results than other state-of-the-art approaches.

### 8.5.7 Experiment on ALC Subset Corpus

The string kernels approach is evaluated on Arabic data in order to demonstrate that the approach is language independent by showing that it can obtain state-of-the-art results. The work of Malmasi and Dras (2014a) is the first (and only) to show NLI results on Arabic data. In their evaluation, the 10-fold CV procedure was used to evaluate an SVM model based on several types of features including CFG rules, function words, and part-of-speech $n$-grams. To directly compare the string kernel systems with the approach of Malmasi and Dras (2014a), the same folds should have been used. As the folds are not publicly available, two alternative solutions are adopted to fairly compare the two NLI approaches. First of all, each classification system based on string kernels is evaluated by repeating the 10-fold CV procedure for 20 times and averaging the resulted accuracy rates. The folds are randomly selected at each trial. This helps to reduce the amount of accuracy variation introduced by using a different partition of the data set for the 10-fold CV procedure. To give an idea of the amount of variation in each trial, the standard deviations for the computed average accuracy rates are also reported. Second of all, the leave-one-out (LOO) cross-validation procedure is also adopted because it involves a predefined partitioning of the data set. Furthermore, the LOO procedure can easily be performed on a small data set, such as the subset of 329 samples of the ALC. Thus, the LOO procedure is more suitable for this NLI experiment, since it is straight forward to compare newly developed systems with the previous state of art systems. Malmasi and Dras (2014a) were kindly asked to re-evaluate their system using the LOO CV procedure and they readily provided additional results

**Table 8.9** Accuracy rates on ALC subset (Arabic L2) of various classification systems based on string kernels compared with a state-of-the-art approach

| Method | 10-fold CV (%) | LOO CV (%) | Bootstrap (%) |
|---|---|---|---|
| SVM and combined features (Malmasi and Dras 2014a) | 41.0 | 41.6 | – |
| KRR and $\hat{k}^{0/1}_{3-5}$ | $55.5 \pm 1.3$ | 58.7 | 48.83–49.91 |
| KRR and $\hat{k}^{\cap}_{3-5}$ | $55.2 \pm 1.4$ | 56.8 | 48.56–49.29 |
| KRR and $\hat{k}^{LRD}_{3-5}$ | $50.1 \pm 1.2$ | 51.4 | 47.67–48.75 |
| KRR and $\hat{k}^{0/1}_{3-5} + \hat{k}^{LRD}_{3-5}$ | $54.9 \pm 1.2$ | 56.8 | 48.86–49.97 |
| KRR and $\hat{k}^{\cap}_{3-5} + \hat{k}^{LRD}_{3-5}$ | $54.7 \pm 1.4$ | 57.2 | 48.44–49.86 |
| KRR and $\hat{k}^{0/1}_{3-5} + \hat{k}^{\cap}_{3-5}$ | **$56.8 \pm 1.3$** | **59.3** | **48.93–50.14** |
| KRR and $\hat{k}^{0/1}_{3-5} + \hat{k}^{\cap}_{3-5} + \hat{k}^{LRD}_{3-5}$ | $56.5 \pm 1.6$ | 59.0 | 48.60–49.89 |
| KDA and $\hat{k}^{0/1}_{3-5}$ | $55.3 \pm 1.1$ | 57.4 | 47.34–48.57 |
| KDA and $\hat{k}^{\cap}_{3-5}$ | $51.2 \pm 1.5$ | 53.2 | 46.02–47.20 |
| KDA and $\hat{k}^{LRD}_{3-5}$ | $45.9 \pm 1.1$ | 47.1 | 42.59–43.58 |
| KDA and $\hat{k}^{0/1}_{3-5} + \hat{k}^{LRD}_{3-5}$ | $52.7 \pm 1.3$ | 54.4 | 46.00–47.15 |
| KDA and $\hat{k}^{\cap}_{3-5} + \hat{k}^{LRD}_{3-5}$ | $50.2 \pm 1.3$ | 51.1 | 45.66–47.07 |
| KDA and $\hat{k}^{0/1}_{3-5} + \hat{k}^{\cap}_{3-5}$ | $53.4 \pm 1.2$ | 55.3 | 46.55–47.93 |
| KDA and $\hat{k}^{0/1}_{3-5} + \hat{k}^{\cap}_{3-5} + \hat{k}^{LRD}_{3-5}$ | $53.8 \pm 1.1$ | 54.7 | 46.01–47.26 |

The accuracy rates are reported using a studentized bootstrap procedure and two cross-validation (CV) procedures, one based on 10 folds and one based on leave-one-out (LOO). The 10-fold CV procedure was repeated for 20 times and the results were averaged to reduce the accuracy variation introduced by randomly selecting the folds. The standard deviations for the computed average accuracy rates are also given. The bootstrap procedure is based on 200 iterations, and the reported confidence intervals are based on a 95 % confidence level. The best accuracy rate is highlighted in bold

which are included in Table 8.9. Lastly, a studentized bootstrap procedure based on 200 iterations was employed to provide confidence intervals with a confidence level of 95 %. The bootstrap procedure is suitable for small data sets such as the ALC subset.

Table 8.9 presents the results of the classification systems based on string kernels in contrast to the results of the SVM model based on several combined features. The results clearly indicate the advantage of using an approach that works at the character level. When the 10-fold CV procedure is used, the average accuracy rates of the systems based on string kernels range between 45.9 and 56.8 %. The lowest scores are obtained by LRD, while the presence bits kernel and the intersection kernel obtain better results. Even so, the kernel based on LRD is far better than the model of Malmasi and Dras (2014a). Combining the kernels through MKL, again proves to be a good idea. Certainly, the top scoring system in the 10-fold CV experiment is the KRR based on sum of the presence bits kernel and the intersection kernel. Its accuracy (56.8 %) is 15.8 % above the accuracy of the SVM based on combined features (41.0 %). An important remark is that the standard deviations computed over the 20 trials are between 1.1 and 1.6 % for all systems, which means the amount of accuracy variation is too small to have an influence on the overall conclusion. Furthermore, all the results obtained using the 10-fold CV procedure are consistent with the results obtained using the leave-one-out CV procedure or the bootstrap procedure. The accuracy rates are generally lower when bootstrap is employed for evaluation, because the bootstrap procedure is a rather more pessimistic way of estimating the performance compared to the CV procedures. With no exception, the models reach better accuracy rates when the LOO procedure is used, most likely because there are more samples available for training. When the LOO CV procedure is used, the systems based on string kernels attain accuracy rates that range between 47.1 and 59.3 %. Better accuracy rates are obtained when the kernels are combined. When $\hat{k}^{0/1}_{3-5}$ and $\hat{k}^{\cap}_{3-5}$ are summed together and the training is performed by KRR, the best accuracy rate (59.3 %) is obtained, which brings an improvement of 17.7 % over accuracy of the SVM model (41.6 %). Regarding the classification systems, KRR obtains better results than KDA for every kernel type, even if the differences are not that high. Still, both KRR and KDA give results that are much better than the SVM of Malmasi and Dras (2014a). In conclusion, each and every classification systems based on string kernels is significantly better that the SVM model of Malmasi and Dras (2014a) on the Arabic data set.

The topic distribution per native language is unknown for the ALC subset, so Malmasi and Dras (2014a) removed any lexical information from their model to prevent any topic bias. However, the authors of the paper presenting the second version of the Arabic Learner Corpus (Alfaifi et al. 2014) state that: "We decided to choose two common genres in the learner corpora, narrative and discussion. For the written part, learners had to write narrative or discussion essays under two specific titles which were likely to suit both native and non-native learners of Arabic, entitled *a vacation trip* for the narrative and *my study interest* for the discussion type.", which means that, at a coarse-grained level, there are two main topics (prompts) in the corpus. However, the topic distribution can also be discussed at a fine-grained level. Naturally, each of the two main topics can be further divided into a broad range of subtopics. Since there is a large number of potential subtopics, it is unlikely that their distribution per native language is biased in such a way that it will have an impact on the results. Considering the coarse-grained topic distribution, there is no chance that the accuracy rates obtained by the various systems based on string kernels are due any kind of topic bias, since there are only two topics and seven native languages in the ALC subset used in this experiment. Even in the worst case scenario, if the documents for each L1 were to belong to a single topic, a method that relies on topic distribution to predict the native language could obtain at most 42.6 %, assuming the most frequent L1 languages in the corpus (Chinese and Urdu) belong to different topics. More precisely, 42.6 % can be obtained only when the texts written by Chinese speakers belong to one topic, the texts written by Urdu speakers to the other topic, and the classifier based on topic distribution can predict the right topic with 100 % accuracy for each of the two native languages. It becomes obvious that the best result reported in this section using string kernels (59.3 %) cannot be obtained with a classifier that relies on the topics distribution per native language. This observation is consistent with the results obtained in the cross-corpus experiment (Sect. 8.5.6), which demonstrate that the string kernel approach naturally avoids topic bias. In light of these comments, it is worth mentioning that the SVM model of Malmasi and Dras (2014a) could benefit from lexical information without the risk of introducing topic bias. Indeed, their model extended to include character 3-grams reaches an accuracy of 51.2 % using the LOO procedure. Still, this result is not as good as using character $p$-grams alone, as the results of string kernels presented in Table 8.9 indicate.

### 8.5.8 Experiment on ASK Corpus

The work of Pepper (2012) presents an extensive set of experiments on the ASK corpus. The 10-fold CV procedure and the leave-one-out CV procedure were alternatively used to evaluate various LDA models based on linguistic features. Pepper (2012) used only 100 essays per native language and formed several subsets each of five languages. All the subsets included four languages in common, namely English, German, Polish, and Russian. The fifth language was different in each subset. The three subsets that included Spanish, Dutch, and Serbo-Croatian as the fifth language have also been used in the present work in order to compare the results of string kernels with the LDA model of Pepper (2012). For a direct comparison, the same 100 essays per native language should have been used, but this information is not available. For this reason, the string kernels are evaluated by repeating the LOO CV procedure 20 times, and in each trial 100 essays per native language are randomly selected from the 200 essays available for each language. Table 8.10 reports the average accuracy rates along with their standard deviations which give some indication about the amount of variation in each trial. The results obtained on the subset that includes Spanish (SP) along with English, German, Polish, and Russian (EGPR) are given on the second column. In a similar way, the results obtained on the subset that includes Dutch as the fifth language are listed on the third column, while the results obtained on the subset that includes Serbo-Croatian as the fifth language are given on the last column. The best results obtained by Pepper (2012) using the LOO CV procedure are listed on the first row. An important remark is that Pepper (2012) reported much lower accuracy rates when the 10-fold CV procedure was used. Because of this, the comparison is carried out only by using the LOO CV procedure, which also helps to reduce the amount of variation in the 20 trials. However, it must be mentioned the results of string kernels are only 2 % lower when the 10-fold CV procedure is being used, but any further details of that experiment are not reported here.

The empirical results presented in Table 8.10 are consistent with the results on the ALC corpus. Indeed, all the systems based on string kernels are significantly better than the state-of-the-art approach, in all three experiments. KDA and KRR seem to produce fairly similar results, but there is an interesting difference between the two classifiers that stands out. This difference refers to the fact that KRR produces better results when the sum of $\hat{k}^{0/1}_{5-8}$ and $\hat{k}^{\cap}_{5-8}$ is being used, while KDA produces better results when the kernel based on LRD is also added into the sum. When

**Table 8.10** Accuracy rates on three subsets of five languages of the ASK corpus (Norwegian L2) of various classification systems based on string kernels compared with a state-of-the-art approach

| Method | EGPR+SP (%) | EGPR+DU (%) | EGPR+SC (%) |
|---|---|---|---|
| LDA (Pepper 2012) | 51.2 | 55.0 | 55.0 |
| KRR and $\hat{k}^{0/1}_{5-8}$ | $67.4 \pm 2.1$ | $68.2 \pm 1.8$ | $68.1 \pm 2.0$ |
| KRR and $\hat{k}^{\cap}_{5-8}$ | $67.5 \pm 1.7$ | $68.6 \pm 2.0$ | $68.2 \pm 2.0$ |
| KRR and $\hat{k}^{LRD}_{5-8}$ | $62.2 \pm 1.9$ | $62.9 \pm 1.9$ | $61.5 \pm 2.1$ |
| KRR and $\hat{k}^{0/1}_{5-8} + \hat{k}^{LRD}_{5-8}$ | $66.6 \pm 2.1$ | $67.5 \pm 1.6$ | $65.6 \pm 1.4$ |
| KRR and $\hat{k}^{\cap}_{5-8} + \hat{k}^{LRD}_{5-8}$ | $66.2 \pm 2.1$ | $67.7 \pm 1.9$ | $65.7 \pm 2.0$ |
| KRR and $\hat{k}^{0/1}_{5-8} + \hat{k}^{\cap}_{5-8}$ | $\mathbf{67.8 \pm 1.9}$ | $69.0 \pm 1.4$ | $\mathbf{68.5 \pm 1.8}$ |
| KRR and $\hat{k}^{0/1}_{5-8} + \hat{k}^{\cap}_{5-8} + \hat{k}^{LRD}_{5-8}$ | $67.5 \pm 1.9$ | $68.4 \pm 2.0$ | $67.1 \pm 1.9$ |
| KDA and $\hat{k}^{0/1}_{5-8}$ | $67.2 \pm 1.7$ | $68.7 \pm 1.4$ | $67.9 \pm 1.5$ |
| KDA and $\hat{k}^{\cap}_{5-8}$ | $67.3 \pm 2.0$ | $68.8 \pm 1.4$ | $67.3 \pm 2.2$ |
| KDA and $\hat{k}^{LRD}_{5-8}$ | $61.9 \pm 1.8$ | $63.4 \pm 1.7$ | $61.0 \pm 2.0$ |
| KDA and $\hat{k}^{0/1}_{5-8} + \hat{k}^{LRD}_{5-8}$ | $66.6 \pm 2.1$ | $67.4 \pm 1.5$ | $65.7 \pm 1.4$ |
| KDA and $\hat{k}^{\cap}_{5-8} + \hat{k}^{LRD}_{5-8}$ | $65.9 \pm 1.8$ | $67.2 \pm 1.2$ | $65.9 \pm 1.8$ |
| KDA and $\hat{k}^{0/1}_{5-8} + \hat{k}^{\cap}_{5-8}$ | $67.4 \pm 1.8$ | $68.5 \pm 1.9$ | $67.2 \pm 1.6$ |
| KDA and $\hat{k}^{0/1}_{5-8} + \hat{k}^{\cap}_{5-8} + \hat{k}^{LRD}_{5-8}$ | $67.7 \pm 1.4$ | $\mathbf{69.1 \pm 1.9}$ | $68.0 \pm 1.3$ |

All subsets include samples of English, German, Polish, and Russian (EGPR). The fifth language is different in each subset. The first subset includes Spanish (SP), the second one includes Dutch (DU), and the last subset includes Serbo-Croatian (SC) as the fifth language. The accuracy rates on each subset are reported using the leave-one-out cross-validation procedure, which was repeated for 20 times and the results were averaged to reduce the accuracy variation introduced by randomly selecting the documents (100 per language). The standard deviations for the computed average accuracy rates are also given. The best accuracy rate for each subset of five languages is highlighted in bold

the kernels are used independently, it can be observed that LRD produces accuracy rates that are nearly 5 % below the accuracy rates of the intersection kernel and the presence bits kernel, respectively. Nevertheless, the kernel based on LRD is still significantly better than the LDA model used by Pepper (2012). For instance, when Spanish is used as the fifth language, all the string kernels, including the one based on LRD, attain accuracy rates that are at least 10 % better than LDA. In all three evaluation sets, the best results are obtained when kernels are combined together. The best accuracy on the first subset (67.8 %) is given by the KRR based on the kernel combination of $\hat{k}^{0/1}_{5-8}$ and $\hat{k}^{\cap}_{5-8}$. Compared to the LDA model, the accuracy improvement is 16.6 %. Likewise, the best accuracy on the third subset (68.5 %) is again obtained by the KRR based on the kernel combination of $\hat{k}^{0/1}_{5-8}$ and $\hat{k}^{\cap}_{5-8}$. This time, the accuracy improvement over the state-of-the-art LDA model is 13.5 %. The best accuracy on the second subset of five languages (69.1 %) is obtained by the KDA based on the combination of all three string kernels. The accuracy improvement over the LDA model is 14.1 %, which is consistent with the improvement demonstrated on the other two subsets that include Spanish and Serbo-Croatian, respectively. Even if the string kernels are significantly better than the LDA model, the two distinct approaches seem to agree on the difficulty of each subset. Indeed, both approaches produces better results when Dutch is used as the fifth language and worse results when Spanish is the fifth language. This could indicate that Dutch native speakers writing in Norwegian can be more easily distinguished than Spanish speakers writing in Norwegian. While one might expect Dutch, German and English speakers to be more similar (and therefore potentially more difficult to distinguish) than say English or Spanish speakers writing in Norwegian, this does not appear to be the case, given that the overall performance is better when Dutch is added to the mix of English, German, Polish, and Russian. What it can be said for sure is that the approach based on string kernels attains a significant performance improvement over the LDA based on linguistic features. The standard deviations computed over the 20 trials are less than 2.2 % for all systems, which indicates that the amount of accuracy variation is small enough to support the conclusion that the string kernel approach works considerably better.

The string kernels are further evaluated on all the seven languages that are equally represented in the ASK corpus. The purpose of this evaluation is to provide an easy way of comparing other newly developed systems (in the future) with the systems based on string kernels. Two standard cross-validation procedures are used. First, each classification systems based on string kernels is evaluated by repeating the 10-fold CV procedure for 20 times and averaging the resulted accuracy rates. The folds are randomly selected at each trial. Second, the leave-one-out CV procedure was also adopted because it involves a predefined partitioning of the data set. The results for the seven-way classification task are given in Table 8.11.

The results presented in Table 8.11 are similar to those presented in Table 8.10. The extra number of documents per native language (200 instead of 100) seems to compensate for having to discriminate between seven languages instead of five. When the 10-fold CV procedure is used, the average accuracy rates of the systems based on string kernels range between 60.5 and 68.2 %. The lowest scores are obtained by LRD, while the presence bits kernel and the intersection kernel obtain better results. Combining the kernels through MKL, proves to work again as a robust approach that usually improves performance. It must be noted that combining the LRD kernel with either one of the other kernels ($\hat{k}^{0/1}_{5-8}$ or $\hat{k}^{\cap}_{5-8}$) gives slightly lower results than using the respective kernels independently. On the other hand, the top scoring system in the 10-fold CV experiment is the KDA based on sum of the presence bits kernel and the intersection kernel. All the results obtained using the LOO CV procedure are consistently higher than the results obtained using the 10-fold CV procedure, because there are more samples available for training. When the LOO CV procedure is used, the systems based on string kernels reach accuracy rates that range between 61.9 and 69.6 %. The best system proves to be the KDA based on the sum of the presence bits kernel, the intersection kernel, and the kernel based on LRD. Overall, KDA seems to produce better results than KRR in the seven-way classification task, but none of the two classifiers seems to be better than the other in the five-way classification tasks. Nevertheless, all the results produced by string kernels are very good, especially

**Table 8.11** Accuracy rates on the ASK corpus (Norwegian L2) of various classification systems based on string kernels

| Method | 10-fold CV (%) | LOO CV (%) |
|---|---|---|
| KRR and $\hat{k}^{0/1}_{5-8}$ | $67.1 \pm 0.6$ | 68.2 |
| KRR and $\hat{k}^{\cap}_{5-8}$ | $67.4 \pm 0.6$ | 68.1 |
| KRR and $\hat{k}^{LRD}_{5-8}$ | $60.5 \pm 0.7$ | 62.0 |
| KRR and $\hat{k}^{0/1}_{5-8} + \hat{k}^{LRD}_{5-8}$ | $66.7 \pm 0.5$ | 67.5 |
| KRR and $\hat{k}^{\cap}_{5-8} + \hat{k}^{LRD}_{5-8}$ | $66.9 \pm 0.6$ | 68.7 |
| KRR and $\hat{k}^{0/1}_{5-8} + \hat{k}^{\cap}_{5-8}$ | $67.6 \pm 0.5$ | 68.1 |
| KRR and $\hat{k}^{0/1}_{5-8} + \hat{k}^{\cap}_{5-8} + \hat{k}^{LRD}_{5-8}$ | $67.9 \pm 0.6$ | 69.0 |
| KDA and $\hat{k}^{0/1}_{5-8}$ | $67.6 \pm 0.7$ | 69.2 |
| KDA and $\hat{k}^{\cap}_{5-8}$ | $67.7 \pm 0.7$ | 69.1 |
| KDA and $\hat{k}^{LRD}_{5-8}$ | $60.8 \pm 0.5$ | 61.9 |
| KDA and $\hat{k}^{0/1}_{5-8} + \hat{k}^{LRD}_{5-8}$ | $66.7 \pm 0.8$ | 68.0 |
| KDA and $\hat{k}^{\cap}_{5-8} + \hat{k}^{LRD}_{5-8}$ | $67.0 \pm 0.5$ | 68.3 |
| KDA and $\hat{k}^{0/1}_{5-8} + \hat{k}^{\cap}_{5-8}$ | $\mathbf{68.2 \pm 0.7}$ | 69.2 |
| KDA and $\hat{k}^{0/1}_{5-8} + \hat{k}^{\cap}_{5-8} + \hat{k}^{LRD}_{5-8}$ | $67.9 \pm 0.5$ | **69.6** |

The accuracy rates are reported using two cross-validation (CV) procedures, one based on 10 folds and one based on leave-one-out (LOO). The 10-fold CV procedure was repeated for 20 times and the results were averaged to reduce the accuracy variation introduced by randomly selecting the folds. The standard deviations for the computed average accuracy rates are also given. The best accuracy rate is highlighted in bold

when compared to the LDA based on linguistic features (Pepper 2012). The results on the Norwegian corpus along with the results on the Arabic corpus demonstrate that the string kernel approach is language independent.

## 8.6 Language Transfer Analysis

In order to better understand why these simple character $p$-gram models work so well, a focused manual analysis of the most discriminant features for each L1 is carried out next. The features that are about to be analyzed are learned from the TOEFL11 training and development sets, which contain 11,000 documents together. Following the analysis of Malmasi and Dras (2014b), character sequences that are overused or underused by a particular group of L1 speakers are considered to be discriminating.

As previously mentioned in Sect. 8.4, the features considered for the language transfer analysis were generated by the blended spectrum presence bits kernel of 5–9 $p$-grams, which was computed on the raw text documents. The dual weights $\alpha$ used in the analysis are learned by KRR using the one-versus-all scheme. According to the one-versus-all scheme, a binary classifier was built to discriminate each L1 from the others, resulting in a set of 11 dual weight vectors. As the string kernels are based on the dual representation, the first step in order to determine which features are most discriminant, is to obtain the weights $w$ corresponding to the primal representation. The high dimensionality of the explicit feature space can pose interesting challenges in terms of computational complexity and depending of the type of the kernel, elaborated and efficient solutions have been proposed in literature (Pighin and Moschitti 2009, 2010). Fortunately, in the case of the blended spectrum presence bits kernel of 5–9 $p$-grams, the dimensionality of the primal feature space (4,662,520) is manageable. This enables the direct reconstruction of the feature weights $w$ from the dual weights $\alpha$ and the examples (represented in the primal space) using Eq. (8.3). The primal weights of each binary classifier indicate both the overused and the underused patterns that help to discriminate a certain L1 language from the others. More precisely, the features that are associated with positive weights are those that are overused in the respective L1 language, while those that are associated with negative weights correspond to underused patterns. There are 4,662,520 features in total generated from the blended spectrum presence bits kernel, and each feature corresponds to a particular $p$-gram. It becomes clear that performing an analysis on this tremendous amount of features per native language is out of the question. Instead, the analysis presented in this work is focused only on the top (positive and negative) features, that indicate which character sequences are most likely to help discriminate a particular L1 from the others. The features that are not helpful for the classification task have weights close to zero and they are excluded from the analysis.

Table 8.12 lists interesting character sequences among the most discriminating overuse sequences for each L1 in the TOEFL11 corpus. Some interesting patterns can be observed here. First, although the corpus is not topic-biased, some of the most discriminant character sequences for many languages contain strings relating to country, language or city names. This is evident from strings such as "german", "franc", "chin", "japan", and so on. Noticeably, for some L1s these strings do not appear in the top discriminant features. For example, Arabic and Spanish which are spoken in many different countries tend not to have country or city names as the top features. This may be because these writers talk about a wide range of different countries and cities and so the weights of any individual country feature might get demoted. Several teams in the 2013 NLI Shared Task also observed this result (Abu-Jbara et al. 2013; Gebre et al. 2013; Henderson et al. 2013; Malmasi et al. 2013; Mizumoto et al. 2013; Tsvetkov et al. 2013). Abu-Jbara et al. (2013) carried out a simple experiment where they only used features corresponding to country names and achieved an accuracy of 21.3 % on the development set. The work of Gebre et al. (2013) presents a related experiment where the authors remove all country names from the set of features. They show that the accuracy of their system only drops 1.2 % when these features are removed.

Second, there are some spelling errors particular to some sets of L1s that appear in the top features. For Spanish L1s, the spelling of "different" with only one 'f' seems to be a common misspelling. Moreover, the missing space in "a lot" for Arabic L1s is quite discriminant, as is the misspelling of "their" as "thier". The misspelling of words such as "additionally", "personally" that duplicates the 'n' incorrectly seems to be quite a typical error for French native speakers. This is similar to some of the findings of Malmasi and Dras (2014b), who discover several misspellings as being typical for a particular group of L1 speakers. This is also a well-established phenomenon observed in work on second language acquisition and language transfer (Jarvis and Crossley 2012).

Other discriminant features indicating sentence-initial preferences can also be observed. For example, Chinese speakers are more likely to write a sentence that starts with "take" over all other L1s. Korean speakers tend to prefer starting a sentence with "as" or "also", while Japanese speakers more often start their sentences with "if". There are also discriminating use of adverbs. Similar to Malmasi and Dras (2014b), the overused patterns presented in Table 8.12 show that Hindi speakers tend to use "hence" more often than other speakers. It can also be observed that Chinese speakers tend to use "just" or "still" more often than other L1s, while German speakers are inclined to prefer the use of "often" or "special".

The discriminant features confirm the hypothesis that the model based on character *p*-grams is capturing information both about lexical items but also about morphology and grammatical preferences. For example, stems such as "german" which appears in context referring to Germany, Germans, German, Germanic are common in the list of discriminant features. On the other hand, a preference for a definite article before "life" can be noticed in Italian. Hints of grammatical usage tendencies can also be easily observed. For example, one of the most discriminant patterns for German speakers is that they include a comma before "that". This reflects the grammatical requirement of the equivalent "dass" clause in German. Hindi speakers most often use the phrase "as compared". French speakers like to use the phrase "To conclude" rather than "In conclusion" or something similar, and Turkish speakers prefer to have the complete sentence "I agree." without any complement clause.

**Table 8.12** Examples of discriminant overused character sequences with their ranks (left) according to the KRR model based on blended spectrum presence bits kernel extracted from the TOEFL11 corpus (English L2)

| German | | French | | Arabic | | Hindi | | Spanish | | Chinese | |
|---|---|---|---|---|---|---|---|---|---|---|---|
| 1 | , that | 1 | indeed | 1 | alot | 2 | as compa | 1 | , is | 2 | t most |
| 6 | german | 19 | onnal | 9 | any | 9 | hence | 2 | difer | 4 | chin |
| 11 | . but | 21 | is to | 13 | them | 16 | then | 13 | , but | 7 | just |
| 13 | often | 26 | franc | 16 | thier | 17 | indi | 15 | , etc | 8 | still |
| 207 | special | 28 | to concl | 19 | his | 21 | towards | 17 | cesar | 14 | . take |
| **Italian** | | **Japanese** | | **Korean** | | **Telugu** | | **Turkish** | | | |
| 1 | ital | 1 | japan | 1 | korea | 1 | i concl | 1 | i agree. | | |
| 3 | o beca | 15 | . if | 24 | e that | 6 | days | 11 | turk | | |
| 4 | fact | 19 | i disa | 27 | . as | 7 | .the | 21 | . becau | | |
| 9 | , for | 27 | . the | 30 | soci | 11 | where as | 32 | s about | | |
| 24 | the life | 38 | . it | 36 | . also | 13 | e above | 37 | being | | |

**Table 8.13** Examples of discriminant underused character sequences (ranks omitted for readability) according to the KRR model based on blended spectrum presence bits kernel extracted from the TOEFL11 corpus (English L2)

| German | French | Arabic | Hindi | Spanish | Chinese |
|---|---|---|---|---|---|
| . for ex | cause | a lot | s will | after | reasons |
| true | . secon | .for | main | nd so | an a |
| s, and | , becaus | too | time. | especial | have to |
| keep | table | .they | deep | but now | if the |
| using | subjects | conclu | great | have dif | . even |
| **Italian** | **Japanese** | **Korean** | **Telugu** | **Turkish** | |
| s. so | gree tha | in japan | . also | most ad | |
| . and i | in my | talk | somet | thier | |
| into | give | them. | maybe | . first, | |
| . it is | eem much | full | how t | grow | |
| . that | it can | both | focus | well | |

In Table 8.13, some of the most negatively weighted character sequences for each language are listed. The assumption is that if a character sequence receives a high negative weight in the model, then it indicates that that character sequence is under used by speakers of that language. Simply looking at the most negatively weighted features for a specific L1, only tells which character sequences are helpful in predicting that an essay does not belong to the respective L1. Many of the negatively weighted character sequences are common sequences shared by speakers of almost all languages, and they may even correspond to overused patterns in some other L1 language. Therefore, the analysis of underused patterns is restricted to the top 1000 negatively weighted features, and for each L1, the character sequences that only appear in the top 1000 for that one language are chosen. In this way, the character sequences that do not offer interesting insights are removed from the analysis.

Some interesting patterns also emerge for the underused character sequences for each L1, including, for example, some pairs of character sequences. In Arabic the misspelling "alot" is a highly weighted feature, and correspondingly the correct spelling "a lot" receives a large negative weight in the model. The misspelling "thier" was overused by Arabic speakers, but underused by Turkish speakers. Also, the phrase "in japan" is most likely to be written by Japanese speakers, but noticeably very rarely by Korean speakers. There are also patterns of syntactic construction underuse, particularly in how different L1 speakers begin sentences. Germans tend to refrain from beginning a sentence with "for example", Turkish speakers seem to refrain from beginning sentences with "first" and Italian speakers refrain from starting sentences with "and I", "it is" or "that". The table also includes examples of disfavored lexical choices, including Chinese speakers underuse of "reasons", Telugu speaker underuse of "maybe", and Japanese speaker underuse of "give".

The features presented in Table 8.12 can be directly compared to the overuse category in the paper of Malmasi and Dras (2014b) where lexical features are listed for two languages, Arabic and Hindi. The second highest ranked feature for Arabic discovered by Malmasi and Dras (2014b), "anderstand", roughly appears at rank 691 in the KRR and $\hat{k}^{0/1}_{5-9}$ model, while "mony" (rank 4 in their model) appears at rank 141. Interestingly, the models are finding similar features, however the ranking is different. Similarly, the features presented in Table 8.13 can be directly compared to the underuse category presented by Malmasi and Dras (2014b) (though it is unclear how they treated features with a high negative weight across several languages). The underuse of the determiner "an" in Chinese observed by Malmasi and Dras (2014a) can also be observed in Table 8.13.

## 8.7 Discussion

A comprehensive overview and evaluation of a state-of-the-art approach to native language identification was presented in this chapter. The system works at the character level, making the approach completely language independent, topic independent, and linguistic theory neutral. The state-of-the-art accuracy rates have been surpassed in all the experiments presented in this work, sometimes by a very large margin. The idea of combining the string kernels, either by kernel sum or by kernel alignment, proved to be extremely useful, always producing the best NLI performance. Certainly, the best system presented in this work is based on combining the intersection and the presence string kernels through MKL and on deciding the class label either with KDA or KRR. The best string kernels approach is 1.7 % above the top scoring system of the 2013 NLI Shared Task. Furthermore, it has an impressive generalization capacity, achieving results that are 30 % higher than the state-of-the-art method in the cross-corpus experiment. The best string kernels system also shows an improvement of 17.7 % over the previous best accuracy on Arabic data (Malmasi and Dras 2014a), and an improvement of 14–16 % over the results reported by Pepper (2012) on Norwegian data, proving that string kernels are indeed language independent.

All in all, the experiments on the five corpora presented in this chapter show that the approach based on string kernels generalizes well across corpora, topics and languages. However, it should also be pointed out that the corpora available for these kinds of studies are typically academic in nature (i.e., from language learners). Applying these models to other genres of text (tweets for example) is a challenge that has not been explored in this work. While many of the language transfer patterns and characteristics that the string kernels approach is learning (e.g., morphological irregularities, lexical choice, among others) would also apply across genre, it is unclear whether such signals would be strong enough in shorter texts (such as a single tweet). On the other hand, these models should be able to perform similarly well on other kinds of extended writing where there would be sufficient language transfer signals for the models to pick up on. These explorations can be addressed in future work, if such corpora become available.

Language transfer effects confirmed by analyzing the features selected by the string kernel classifier as being more discriminating were also discussed in this work. This analysis offered some clues as to why the string kernel approach is able to work so well and surpass state-of-the-art accuracy levels. In future work, string kernels can be used to identify the native language for different L2 languages, as long as public data sets become available. Furthermore, string kernels can be used as a tool to analyze language transfer effects on the respective L2 languages. There are, however, some limitations of this approach for the task of analyzing language transfer, since a limited number of local syntactic effects can ever be captured.

## References


Abu-Jbara A, Jha R, Morley E, Radev D (2013) Experimental results on the native language identification shared task. In: Proceedings of the Eighth Workshop on Innovative Use of NLP for Building Educational Applications, pp 82–88

Alfaifi A, Atwell E, Hedaya I (2014) Arabic Learner Corpus (ALC) v2: a new written and spoken corpus of arabic learners. In: Proceedings of the learner corpus studies in Asia and the world

Bykh S, Meurers D (2012) Native language identification using recurring $n$-grams—investigating abstraction and domain dependence. In: Proceedings of COLING, pp 425–440

Bykh S, Meurers D (2014) Exploring syntactic features for native language identification: a variationist perspective on feature encoding and ensemble optimization. In: Proceedings of COLING, pp 1962–1973

Blanchard D, Tetreault J, Higgins D, Cahill A, Chodorow M (2013) TOEFL11: A Corpus of Non-Native English. Technical report, Educational Testing Service

Cristianini N, Shawe-Taylor J, Elisseeff A, Kandola JS (2001) On kernel-target alignment. In: Proceedings of NIPS, pp 367–373

Croce D, Moschitti A, Basili R (2011) Structured lexical similarity via convolution kernels on dependency trees. In: Proceedings of EMNLP, pp 1034–1046

Dinu LP, Ionescu RT, Tomescu AI (2014) A rank-based sequence aligner with applications in phylogenetic analysis. PLoS ONE 9(8):e104006 08. doi:10.1371/journal.pone.0104006

Escalante HJ, Solorio T, Montes-y-Gómez M (2011) Local histograms of character n-grams for authorship attribution. In: Proceedings of ACL: HLT 1:288–298

Estival D, Gaustad T, Pham S-B, Radford W, Hutchinson B (2007) Author profiling for English emails. In: Proceedings of PACLING, pp 263–272

Gebre BG, Zampieri M, Wittenburg P, Heskes T (2013) Improving native language identification with tf-idf weighting. In: Proceedings of the Eighth Workshop on Innovative Use of NLP for Building Educational Applications, pp 216–223

Granger S, Dagneaux E, Meunier F (2009) The international corpus of learner english: handbook and CD-ROM, version 2. Presses Universitaires de Louvain, Louvain-la-Neuve

Grozea C, Gehl C, Popescu M (2009) ENCOPLOT: pairwise sequence matching in linear time applied to plagiarism detection. In: 3rd PAN workshop. Uncovering plagiarism, authorship and social software misuse, pp 10

Hastie T, Tibshirani R (2003) The elements of statistical learning. Springer, corrected edition. ISBN 0387952845

Haussler D (1999) Convolution kernels on discrete structures. Technical report UCS-CRL-99-10, University of California at Santa Cruz, Santa Cruz

Henderson J, Zarrella G, Pfeifer C, Burger JD (2013) Discriminating non-native english with 350 words. In: Proceedings of the Eighth Workshop on Innovative Use of NLP for Building Educational Applications, pp 101–110

Ionescu RT (2013) Local Rank Distance. In: Proceedings of SYNASC, pp 221–228
Ionescu RT, Popescu M, Cahill A (2014) Can characters reveal your native language? A language-independent approach to native language identification. In: Proceedings of EMNLP, pp. 1363–1373
Ionescu RT, Popescu M, Cahill A (2016) String kernels for native language identification: insights from behind the curtains. Comput Linguist
Jarvis S, Crossley S (eds) (2012) Approaching language transfer through text classification: explorations in the detection-based approach, vol 64. Multilingual Matters Limited, Bristol
Jarvis S, Castañeda JG, Nielsen R (2004) Investigating L1 lexical transfer through learners wordprints. Second Language Research Forum (SLRF)
Jarvis S, Bestgen Y, Pepper S (2013) Maximizing classification accuracy in native language identification. In: Proceedings of the Eighth Workshop on Innovative Use of NLP for Building Educational Applications, pp 111–118
Kate RJ, Mooney RJ (2006) Using string-kernels for learning semantic parsers. In: Proceedings of ACL, pp 913–920
Koppel M, Schler J, Zigdon K (2005) Automatically determining an anonymous author's native language. In: Proceedings of ISI, pp 209–217
Lodhi H, Saunders C, Shawe-Taylor J, Cristianini N, Watkins CJCH (2002) Text classification using string kernels. J Mach Learn Res 2:419–444
Maji S, Berg AC, Malik J (2008) Classification using intersection kernel support vector machines is efficient. In: Proceedings of CVPR
Malmasi S, Dras M (2014a) Arabic Native Language Identification. In: Proceedings of the EMNLP 2014 workshop on arabic natural language processing (ANLP), pp 180–186
Malmasi S, Dras M (2014b) Language transfer hypotheses with linear SVM weights. In: Proceedings of EMNLP, pp 1385–1390
Malmasi S, Dras M (2014c) Chinesenative language identification. In: Proceedings of EACL 2:95–99
Malmasi S, Wong S-MJ, Dras M (2013) NLI shared task 2013: Mq submission. In: Proceedings of the Eighth Workshop on Innovative Use of NLP for Building Educational Applications, pp 124–133
Manning CD, Raghavan P, Schütze H (2008) Introduction to information retrieval. Cambridge University Press, New York
Mizumoto T, Hayashibe Y, Sakaguchi K, Komachi M, Matsumoto Y (2013) Naist at the NLI 2013 shared task. In: Proceedings of the Eighth Workshop on Innovative Use of NLP for Building Educational Applications, pp 134–139
Moschitti A, Chu-Carroll J, Patwardhan S, Fan J, Riccardi G (2011) Using syntactic and semantic structural kernels for classifying definition questions in jeopardy! In: Proceedings of EMNLP, pp 712–724
Pepper S (2012) Lexical transfer in Norwegian interlanguage: a detection-based approach. Master's thesis, University of Oslo, Olso, Norway
Pighin D, Moschitti A (2009) Reverse engineering of tree kernel feature spaces. In: Proceedings of EMNLP, pp 111–120
Pighin D, Moschitti A (2010) On reverse feature engineering of syntactic tree kernels. In: Proceedings of CoNLL, pp. 223–233
Popescu M (2011) Studying translationese at the character level. In: Proceedings of RANLP, pp 634–639
Popescu M, Dinu LP (2007) Kernel methods and string kernels for authorship identification: the federalist papers case. In: Proceedings of RANLP
Popescu M, Grozea C (2012) Kernel methods and string kernels for authorship analysis. CLEF (Online Working Notes/Labs/Workshop)
Popescu M, Ionescu RT (2013) The story of the characters, the DNA and the native language. In: Proceedings of the Eighth Workshop on Innovative Use of NLP for Building Educational Applications, pp 270–278

Rozovskaya A, Roth D (2010) Generating confusion sets for context-sensitive error correction. In: Proceedings of EMNLP, pp 961–970

Sanderson C, Guenter S (2006) Short text authorship attribution via sequence kernels, Markov chains and author unmasking: an investigation. In: Proceedings EMNLP, pp 482–491

Shawe-Taylor J, Cristianini N (2004) Kernel methods for pattern analysis. Cambridge University Press

Swanson B, Charniak E (2014) Data driven language transfer hypotheses. In: Proceedings of EACL, pp 169–173

Tenfjord K, Meurer P, Hofland K (2006) The ASK Corpus—a language learner corpus of norwegian as a second language. In: Proceedings of LREC, pp 1821–1824

Tetreault J, Blanchard D, Cahill A (2013) A report on the first native language identification shared task. In: Proceedings of the Eighth Workshop on Innovative Use of NLP for Building Educational Applications, pp 48–57

Tetreault J, Blanchard D, Cahill A, Chodorow M (2012) Native tongues, lost and found: resources and empirical evaluations in native language identification. In: Proceedings of COLING, pp 2585–2602

Tomokiyo LM, Jones R (2001) You're not from 'round here, are you? Naive Bayes detection of non-native utterances. In: Proceedings of NAACL

Tsvetkov, Y., Twitto N, Schneider N, Ordan N, Faruqui M, Chahuneau V, Wintner S, Dyer C (2013) Identifying the l1 of non-native writers: the CMU-Haifa system. In: Proceedings of the Eighth Workshop on Innovative Use of NLP for Building Educational Applications, pp 279–287

Vedaldi A, Zisserman A (2010) Efficient additive kernels via explicit feature maps. In: Proceedings of CVPR, pp 3539–3546

# Chapter 9
# Spatial Information in Text Categorization

## 9.1 Introduction

The *bag of words* (BOW) model is one of the most popular representations used in various text classification tasks, from text categorization by topic (Joachims 1998; Sebastiani 2002) to sentiment analysis (Pang et al. 2002) and others. Although it disregards the spatial relations among words (a document is treated simply as a collection of words), it achieves very good performance for text categorization by topic (Joachims 1998) and similar text mining tasks (Manning and Schütze 1999; Manning et al. 2008).

Computer vision researchers have demonstrated that the performance of the bag of visual words model can be improved by including the spatial information in the model. One such example is the Spatial Non-Alignment Kernel (SNAK) presented in Chap. 5, among many others presented in Chap. 3. Given the significant improvements in object recognition from images provided by the use of spatial information (Ionescu and Popescu 2015; Koniusz and Mikolajczyk 2011; Krapac et al. 2011; Lazebnik et al. 2006; Sánchez et al. 2012; Uijlings et al. 2009), an interesting question arises, namely whether spatial information can also be useful in text classification tasks. Intuitively, the position of a word in a text document can disclose some information about the respective word. For instance, if the word appears near the beginning or near the end of a text, it probably belongs to the introduction or the conclusion. A word that appears in these rather more important sections of the document is typically more important. Function words are excluded from this example, as they can be treated independently. Nonetheless, the general intuition that spatial information is useful in text classification is supported by the fact that spatial information has already been used in some way or another in text analysis tasks (Johnson and Zhang 2015; Pu et al. 2007; Tan et al. 2002; Xue and Zhou 2009). One of the most popular approaches to capture the spatial relation between words is to use word $n$-grams. However, this approach can only be used to recover the spatial relations of words that appear very close to each other in text. While word $n$-grams are useful



R.T. Ionescu and M. Popescu, *Knowledge Transfer between Computer Vision and Text Mining*, Advances in Computer Vision and Pattern Recognition,
DOI 10.1007/978-3-319-30367-3_9

in certain situations (Tan et al. 2002), a general approach should also provide the means to encode the spatial relations of distantly situated words. However, trying to increase the length of $n$-grams will generate an exponential expansion of the feature vector space, leading to a greater number of parameters which are difficult to estimate. An attempt to avoid the dimensionality inflation of the feature vector space given by word $n$-grams is the Local Word Bag (LWB) model of Pu et al. (2007). As Pu et al. (2007) have found in their work, longer $n$-grams tend to occur fewer times in a collection of documents, and they become worthless in practice. Hence, the LWB model is based on 5-grams in the experiments conducted by Pu et al. (2007), which can only provide information about the words located in the immediate vicinity of other words.

In this chapter, two general frameworks of encoding information in images are adapted to work on text documents. Contrary to the approaches based on word $n$-grams (Pu et al. 2007; Tan et al. 2002), the proposed approaches are aimed at encoding the location of words at a coarse level. More specifically, they roughly indicate the region of text in which a word appears, without telling its position in a local context. The first approach is to build a spatial pyramid representation for text by dividing the text into increasingly fine parts (chunks of text) and by representing each part as a bag of words. Similar to its computer vision counterpart (Lazebnik et al. 2006), the final spatial pyramid representation is obtained by concatenating the representations of each chunk of text. The second approach is based on adapting the SNAK framework (Ionescu and Popescu 2015) to the BOW model used for text analysis. For each word, the average position and the standard deviation is computed based on all the occurrences of the word in the text. The pairwise similarity of two text documents is then computed by taking into account the difference between the average positions and the difference between the standard deviations of each word in the two documents. The main difference from the image version of SNAK is that the words in text have only one coordinate instead of two, so the representation induced by SNAK is only three times larger than the standard BOW representation.

While spatial information plays an important role in object recognition form images, this chapter shows evidence that the spatial information is also useful in text categorization by topic. More precisely, text categorization experiments are conducted to compare the standard BOW model with two enhanced BOW models that leverage the use of spatial information. The empirical results show that spatial information can improve the performance by more than 1 %, which indicates that spatial information is indeed useful. Furthermore, the proposed approaches are fairly simple to implement and use, having the right ingredients to become very popular in the field of text mining, similar to the spatial pyramid for images that enjoys a great notoriety in computer vision (Lazebnik et al. 2006; Szeliski 2010).

The chapter is organized as follows. Section 9.2 gives an overview of the existing methods for using the spatial information in text analysis. The proposed approaches to encode spatial information inspired from computer vision are described in Sect. 9.3. The text categorization experiments are presented in Sect. 9.4. Finally, a discussion is given in Sect. 9.5.

## 9.2 Related Work

While the bag of words achieves very high performance in many practical situations, a clear disadvantage of this approach is that the spatial relations between words in a text document are lost. Researchers have sought to recover this information and use it to obtain better performance (Johnson and Zhang 2015; Pu et al. 2007; Tan et al. 2002; Xue and Zhou 2009). Perhaps one of the most popular approaches is to use word $n$-grams (Tan et al. 2002). This approach is able to recover the relation of words located in the vicinity of other words. However, if two words are located more than $n$ words apart from each other, their relation will be lost when using word $n$-grams.

Inspired by the pyramid match kernel proposed by Grauman and Darrell (2005), Pu et al. (2007) propose the Local Word Bag model. In their model, a document is represented as a set of several overlapping bags of words. The LWB model incorporates the word co-occurrence patterns at sentence or phrase level, just like $n$-grams. The advantage of the LWB model is that longer $n$-grams can be used without worrying about the dimension of the feature space. This is achieved by employing a modified version of the pyramid match kernel of Grauman and Darrell (2005). In the end, the LWB model suffers for the same problem as a representation based on word $n$-grams, namely that the spatial relations between words that are far apart are lost. Similar to the approach of Grauman and Darrell (2005), the LWB model is a rather complex approach, which is a shortcoming that constrains its utility in practical applications. Starting from the idea of Grauman and Darrell (2005), Lazebnik et al. (2006) present a simple and computationally efficient method for recognizing scene categories in images based on approximate global geometric correspondence. Their technique works by partitioning the image into increasingly fine subregions and computing histograms of local features found inside each subregion. In this chapter, the simple yet effective approach of Lazebnik et al. (2006) is adapted for text documents as described in Sect. 9.3.1. The spatial pyramid for text holds information about the region of text in which a word appears, without being able to pinpoint its exact position in text. The SNAK framework presented in Sect. 9.3.2 is based on the same principle.

Xue and Zhou (2009) design some distributional features to measure the characteristics of a word's distribution in a document. They consider the compactness of the occurrences of a word, which indicates whether a word appears in a specific part of a document or spreads over the entire document. They also consider the position of the first occurrence of a word, based on the intuition that the author naturally mentions the important contents in the earlier parts of a document. The SNAK framework uses about the same information as the distributional features of Xue and Zhou (2009). Instead of the position of first occurrence of a word, the SNAK framework uses the average position of a word in a document. Different from the approach of Xue and Zhou (2009), SNAK includes the spatial information into a kernel function which measures the difference between the average positions and the difference between the standard deviations of each word in the two documents. As shown in Sect. 9.4, another advantage of the SNAK framework is that it yields better results than the approach of Xue and Zhou (2009).

Johnson and Zhang (2015) apply convolutional neural networks (CNN) to text categorization, in order to make use of the one dimensional structure of text data so that each unit in the convolution layer responds to a small region of a text document, namely a sequence of words. Remarkably, the approach of Johnson and Zhang (2015) is inspired by work in computer vision (Krizhevsky et al. 2012; Simonyan and Zisserman 2014; Szegedy et al. 2015), in a similar way to the frameworks presented in Sect. 9.3, which are inspired by Lazebnik et al. (2006) and Ionescu and Popescu (2015), respectively. In computer vision, a convolution layer consists of several computation units, and each unit responds to a small region of an input image. In each layer of the CNN model, the information from the previous layers is used to represent more complex patterns, that cover increasingly larger regions of the input data. In other words, convolutional neural networks build a hierarchy of features, which inherently embeds spatial information of the input data.

## 9.3 Methods to Encode Spatial Information

The frameworks to encode spatial information proposed in this chapter are based on an underlying bag of words model particularly adapted to text categorization by topic. The following steps are required to obtain a bag of words representation suited for the text categorization task. First of all, the text is broken down into tokens. After applying the tokenization process, the next step is to eliminate the stop words,[1] as they do not provide useful information in the context of text categorization by topic. The remaining words are stemmed using the Porter stemmer algorithm (Porter 1980).[2] This algorithm removes the commoner morphological and inflexional endings from words in English. The resulted terms from the entire collection of documents are collected into a vocabulary. The frequency of each term is then computed on a per document basis. Let $f_{t,d}$ denote the raw frequency of a term $t$ in a document $d$, namely the number of times $t$ occurs in $d$. The bag of words representation used in this chapter is obtained by computing the log normalized *term frequency* as follows:

$$tf(t,d) = \begin{cases} 1 + log\ f_{t,d}, \text{ if } f_{t,d} > 0 \\ 0, \text{ if } f_{t,d} = 0 \end{cases}. \tag{9.1}$$

To make things completely clear, an example is given next. Indeed, Example 12 shows how to compute the term frequency in a particular case.

[1] Stop words are the most common words in a language, usually function words, such as *what*, *is*, *this*.

[2] Stemming is the process that reduces a word to its root form.

*Example 12* Given a document $d$ = "He lives in a big house with a big garage in a big city." and a term $t$ = "big", the number of occurrences of $t$ in $d$ is $f_{t,d} = 3$. Hence, the log normalized term frequency is:

$$tf(t, d) = 1 + log\ f_{t,d} = 1 + log3 \approx 1 + 0.4771 \approx 1.4771.$$

### 9.3.1 *Spatial Pyramid for Text*

The work of Lazebnik et al. (2006) presents a method for recognizing scene categories in images based on aggregating statistics of local features (visual words) over fixed subregions (bins). Their technique works by partitioning the image into bins and computing histograms of visual words found inside each bin. This process is repeated at multiple levels, and the resulted histograms are concatenated into a single representation known as the *spatial pyramid*. For example, if the spatial pyramid is based on three levels, the convention followed by Lazebnik et al. (2006) is to divide the image into $1 \times 1$, $2 \times 2$, and $4 \times 4$ bins. In other words, the number of bins at level $l$ is $2^{l-1} \times 2^{l-1}$. At the first level there is a single bin that coincides with the entire image to be analyzed.

The spatial pyramid representation is a simple and efficient extension of the bag of visual words representation, that contains information about the visual words that appear in a predefined region of the input image. In a similar fashion, a spatial pyramid for text can be developed. More precisely, at each level $l$ of the spatial pyramid, the text is divided into $l$ parts of equal length, and a bag of words representation is computed for each part. The bag of words representation contains log normalized term frequencies of the terms that appear in a given part of the input text documents. Thus, it can also be described as a word histogram.

It is worth noting that the convention to use $2^{l-1}$ bins at level $l$ is not kept in the spatial pyramid for text. This is primarily motivated by the fact that a text can naturally be structured into a number of parts that is not necessarily a power of two. For example, an essay can be divided into an introduction, a body, and a conclusion. The narrative structure of a novel can also be divided into three sections known as the setup, the conflict, and the resolution. Although the spatial pyramid approach is based on a naive approach that does not involve dividing the text into meaningful parts, it has a greater chance of approximating these meaningful parts if the text is divided into $l$ parts at a level $l$.

Let $L$ be the number of levels chosen for the spatial pyramid. The total number of word histograms $T$ is

$$T = 1 + 2 + \cdots + L = \frac{L(L + 1)}{2}.$$

The word histograms are concatenated into a single feature vector that represents the entire text document. The final feature vector is termed spatial pyramid for text.

Given a vocabulary of terms $V$, the number of features in the spatial pyramid is $|V| \cdot T$, where $|V|$ is the size of the vocabulary. For example, using a spatial pyramid based on three levels, will generate a representation that is six times larger than the standard BOW representation ($T = 1 + 2 + 3 = 6$).

After the spatial pyramid representation for text is obtained, the final implementation issue that needs to be settled is the normalization. The same approach as (Lazebnik et al. 2006) is adopted here. More precisely, all histograms are normalized by the total weight of all words in the text, which gives maximum computational efficiency. This kind of normalization is enough to deal with the effects of variable text lengths. Last but not least, it is worth mentioning the spatial pyramid is a fairly simple approach, being very easy to implement and use in many practical applications.

### 9.3.2 Spatial Non-Alignment Kernel for Text

The Spatial Non-Alignment Kernel (SNAK) is a framework presented in Chap. 5 that includes spatial information into the bag of visual words model. Along with Local Patch Dissimilarity and Local Rank Distance, it stems from the idea of measuring the spatial non-alignment among two objects. In computer vision, the SNAK framework can roughly determine the spatial non-alignment between two images by measuring how the spatial distribution of each visual word varies in the two images. In text mining, the SNAK framework can be adapted in a straightforward manner to measure the spatial non-alignment between two text documents by considering words instead of visual words.

As in computer vision, additional information for each word needs to be stored in the feature representation of a text document. Indeed, the average position and the standard deviation of all the occurrences of a word in the text document need to be computed. While in the case of image data these statistics are computed for each of the two image coordinates, this is no longer necessary for text data. The SNAK feature vector of a text document includes the average position and the standard deviation of a term together with the log normalized frequency of the respective term, resulting in a feature space that is three times greater than the original feature space corresponding to the standard bag of words. The size of the feature space is identical to a spatial pyramid for text based on two levels, but it is two times smaller than a spatial pyramid based on three levels.

Let $U$ represent the SNAK feature vector of a text document. For each term at an index $i$ in a vocabulary, $U$ will contain triplets as defined below:

$$u(i) = \left(tf^u(i), m^u(i), s^u(i)\right),$$

where the first component of $u(i)$ represents the log normalized term frequency as defined in Eq. (9.1), $m(i)$ represents the mean (or average) position of the $i$-th term, and $s(i)$ represents the standard deviation of the $i$-th term. It is important to note that the last two components of $u(i)$ are normalized with respect to the length of the text document, to reduce the effects of variable text lengths in a collection of documents. If the visual word $i$ does not appear in the text document ($tf^u(i) = 0$), the last two components are undefined. In fact, $m(i)$ and $s(i)$ are not being used at all, if $tf^u(i) = 0$. Example 13 shows how to compute a triplet for a certain term that appears in a text document.

*Example 13* Given a document $d =$ "He lives in a big house with a big garage in a big city." and a term $t =$ "big", it can be easily observed that $t$ appears precisely three times in $d$ at positions 5, 9 and 13, respectively. The length of $d$ is 14. Let $U$ represent the SNAK feature vector of $d$ and let $i$ denote the index of $t$ in the vocabulary of terms. The components in the triplet $u(i)$ are computed as follows:

$$tf^u(i) = 1 + log3 \approx 1 + 0.4771 \approx 1.4771,$$

$$m^u(i) = \frac{5+9+13}{3} \cdot \frac{1}{14} = \frac{9}{14} \approx 0.6429,$$

$$s^u(i) = \sqrt{\frac{1}{3-1}\left((5-9)^2 + (9-9)^2 + (13-9)^2\right)} \cdot \frac{1}{14} = \sqrt{\frac{32}{2}} \cdot \frac{1}{14} \approx 0.2857.$$

Finally, the triplet corresponding to the term $t$ is:

$$u(i) \approx (1.4771, 0.6429, 0.2857).$$

As in the case of visual words, the SNAK kernel between two feature vectors $U$ and $V$ can be defined as follows:

$$k_{\text{SNAK}}(U, V) = \sum_{i=1}^{n} \exp\left(-c_1 \cdot \Delta_{\text{mean}}(u(i), v(i))\right) \cdot \exp\left(-c_2 \cdot \Delta_{std}(u(i), v(i))\right), \tag{9.2}$$

where $n$ is the number of terms in the vocabulary, $c_1$ and $c_2$ are two parameters with positive values, $u(i)$ is the triplet in $U$ corresponding to the $i$-th term in the vocabulary, $v(i)$ is the triplet in $V$ corresponding to the $i$-th term in the vocabulary, and $\Delta_{\text{mean}}$ and $\Delta_{std}$ are defined as follows:

$$\Delta_{\text{mean}}(u, v) = \begin{cases} (m^u - m^v)^2, & \text{if } tf^u, tf^v > 0 \\ \infty, & \text{otherwise} \end{cases}$$

$$\Delta_{std}(u, v) = \begin{cases} (s^u - s^v)^2, & \text{if } tf^u, tf^v > 0 \\ \infty, & \text{otherwise} \end{cases}$$

where $m$ and $s$ are components of the triplets $u$ and $v$. If a term does not appear in at least one of the two compared text documents, its contribution to $k_{\text{SNAK}}$ is zero, as $\Delta_{\text{mean}}$ and $\Delta_{std}$ are infinite. Since the definition of SNAK for text data is essentially the same as in computer vision, it remains a kernel function. As in computer vision, the SNAK framework is a fairly simple approach, that can be easily generalized and combined with many other kernel functions. The following equation shows how to combine SNAK with the another kernel $k^*$ that takes into account the log normalized term frequency:

$$k_{\text{SNAK}}(U, V) = \sum_{i=1}^{n} k^*(tf^u(i), tf^v(i)) \cdot \\ \cdot \exp(-c_1 \cdot \Delta_{\text{mean}}(u(i), v(i))) \cdot \exp(-c_2 \cdot \Delta_{std}(u(i), v(i))). \quad (9.3)$$

In the experiments, the linear kernel is used as $k^*$ in Eq. (9.3).

## 9.4 Experiments

### *9.4.1 Data Set Description*

The Reuters-21578 corpus (Lewis 1997) is one of the most widely used test collections for text categorization research. It contains 21,578 articles collected from Reuters newswire. Following the procedure of Joachims (1998) and that of Yang and Liu (1999), the categories that have at least one document in the training set and one in the test set are selected. This leads to a total of 90 categories. Two evaluation modes are then used. In the first mode, unlabeled documents are eliminated. After removing the unlabeled documents, there are 10,787 documents left that belong to 90 categories. Each document belongs to one or more categories and the average number of categories per document is 1.235. The collection is split into 7,768 documents in the training set and 3,019 documents in test set. In the second evaluation mode, the unlabeled documents are kept in the collection. Hence, the second evaluation mode is a little more difficult. The second evaluation mode leads to a corpus of 9,598 training documents and 3,299 test documents. Using two slightly different evaluation modes is motivated by the fact that previous works use one of the two modes, but the results are not directly comparable. For instance, the first evaluation mode is used in the work of Xue and Zhou (2009), while the second evaluation mode is used by Joachims (1998). The Reuters-21578 corpus is available at http://www.daviddlewis.com/resources/testcollections/reuters21578/.

### *9.4.2 Implementation Choices*

The bag of words representation is obtained by eliminating the stop words and by stemming the rest of the words. The standard bag of words is used as a baseline model in the text categorization experiments. In computer vision, the spatial pyramid is usually based on two or three levels in practical situations. It has been observed that adding more pyramid levels (Lazebnik et al. 2006) does not necessarily increase performance, and it becomes hard to compensate for the fact that it requires more space. Thus, the spatial pyramids evaluated in the following experiments are based on two and three levels, respectively. The SNAK framework takes both the average position and the standard deviation of each term into account. In object recognition from images, empirical results demonstrated that they have an almost equal contribution to the proposed framework. Hence, the two constants $c_1$ and $c_2$ from Eq. (9.3) are set to the same value in the text categorization experiments. Since these statistics are normalized with respect to the document length, a good choice for $c_1$ and $c_2$ is 0.5. Although no tuning is performed in the case of SNAK, it is worth mentioning that tuning the parameters $c_1$ and $c_2$ is likely to improve performance. As the other two evaluated methods (standard BOW and spatial pyramid) do not require tuning, it is perhaps better to refrain from tuning $c_1$ and $c_2$ for a fair comparative study. In all the experiments, Kernel Ridge Regression is the method of choice for the learning stage. It was chosen in favor of Support Vector Machines, because it was found to work slightly better in a series of preliminary experiments.

### 9.4.3 Evaluation Procedure

To evaluate and compare the text categorization approaches, the precision and the recall are first computed based on the confusion matrix presented in Table 9.1. The *precision* is given by the number of true positive documents ($TP$) divided by the number of documents predicted as positive by the classifier ($TP + FP$), while the *recall* is given by the number of true positive documents ($TP$) divided by the total number of documents marked as positive by a trusted expert judge ($TP + FN$). To capture precision and recall into a single representative number, the $F_1$ measure can be employed. The $F_1$ measure can be interpreted as a weighted average of the precision and recall given by:

$$F_1 = 2 \cdot \frac{precision \cdot recall}{precision + recall}.$$

For each category, a binary classifier is trained to predict the positive and negative labels for the test documents. However, the performance of the classifier needs to

**Table 9.1** Confusion matrix (also known as contingency table) of a binary classifier with labels +1 or −1

| | Labels | Expert judgments | |
|---|---|---|---|
| | | +1 | −1 |
| Classifier | +1 | True positive ($TP$) | False positive ($FP$) |
| Predictions | −1 | False negative ($FN$) | True negative ($TN$) |

be evaluated at the global level (over all categories). Two approaches are used in literature to aggregate the $F_1$ measure over multiple categories. One is based on computing a confusion matrix for each category, which can be used to subsequently calculate the $F_1$ measure for each category. Finally, the global $F_1$ measure is obtained by averaging all the $F_1$ measures. This first measure is known as macro-averaged $F_1$ ($macroF_1$). The other approach is based on computing a global confusion matrix for all the categories by summing the documents that fall in each of the four conditioned sets, namely true positives, true negatives, false positives, and false negatives. The global $F_1$ measure is immediately computed with the values provided by the global confusion matrix. This second measure is known as micro-averaged $F_1$ ($microF_1$). As noted by Xue and Zhou (2009), the classifier's performance on rare categories has more impact on the macro-averaged $F_1$ measure, while the performance on common categories has more impact on the micro-averaged $F_1$ measure. Thus, it makes sense to report both these measures in the following experiments.

### *9.4.4 Experiment on Reuters-21578 Corpus*

The text categorization results on the Reuters-21578 corpus are presented in Table 9.2. The macro-averaged and the micro-averaged $F_1$ measures are reported for two evaluation modes, one that includes unlabeled documents and one that excludes them. The standard word histogram representation is compared with two approaches that incorporate spatial information into the bag of words model. In both evaluation modes, the spatial pyramids and the SNAK framework improve performance over the standard bag of words.

When the unlabeled documents are included in the experiment, the spatial pyramid based on three levels works slightly better than the spatial pyramid based on two levels. For example, the macro-averaged $F_1$ measure grows by nearly 1 % (from 0.511 to 0.520) when using the spatial pyramid based on two levels, and by nearly 2 % (from 0.511 to 0.529) when using the spatial pyramid based on three levels.

**Table 9.2** Empirical results on the Reuters-21578 corpus obtained by the standard bag of words versus two methods that encode spatial information, namely spatial pyramids and SNAK

| Representation | Unlabeled documents included | | Unlabeled documents excluded | |
|---|---|---|---|---|
| | $microF_1$ | $macroF_1$ | $microF_1$ | $macroF_1$ |
| Histogram | 0.865 | 0.511 | 0.875 | 0.523 |
| Spatial pyramid (2 levels) | 0.870 | 0.520 | 0.882 | 0.537 |
| Spatial pyramid (3 levels) | 0.872 | 0.529 | 0.881 | 0.538 |
| SNAK | **0.877** | **0.549** | **0.886** | **0.561** |

The macro-averaged and the micro-averaged $F_1$ measures are reported for two evaluation modes, one that includes unlabeled documents and one that excludes unlabeled documents. The learning is always done by KRR. The best scores are highlighted in bold

The improvements in terms of the micro-averaged $F_1$ measure are not equally high. Interestingly, the SNAK framework attains better performance than both spatial pyramids. Its results are even better than the spatial pyramid based on three levels, which requires twice the space. The macro-averaged $F_1$ score given by SNAK is almost 4 % better than the standard bag of words and 2 % better than the spatial pyramid based on three levels. The micro-averaged $F_1$ score given by SNAK is also roughly 1 % better than the standard representation and 0.5 % better than the spatial pyramid based on three levels.

The relative improvements provided by spatial pyramids and by SNAK are about the same when the unlabeled documents are excluded from the experiment. The only difference is that the spatial pyramid based on two levels gives almost identical results to the spatial pyramid based on three levels. Even so, they are both able to improve performance. More precisely, they attain macro-averaged $F_1$ scores that are nearly 1.5 % over the word histogram, and micro-averaged $F_1$ scores that are nearly 0.7 % over the word histogram. It is important to note that all methods give better results when the unlabeled documents are excluded, since the task becomes a little more easy (the classifier has a lower chance of making a mistake). As in the previous evaluation mode, the SNAK framework yields better performance than the spatial pyramids. Moreover, the improvements of the SNAK approach over the other representations are consistent in both evaluation modes.

Overall, the best results are obtained by the SNAK framework. Remarkably, the SNAK framework was also found to work better than the spatial pyramid in computer vision (Ionescu and Popescu 2015). Thus, an interesting pattern seems to take shape. More precisely, it appears that the SNAK framework gives better results regardless of the data type, image, or text. Given that it is more compact than a spatial pyramid based on three levels, it should probably be preferred in favor of the spatial pyramid.

## 9.5 Discussion

Two methods for including spatial information into the widely used bag of words model have been described in this chapter. Both of them are inspired by research in computer vision. The spatial pyramid is perhaps the best known method for including spatial information in the bag of visual words. The SNAK framework is a recent development of Ionescu and Popescu (2015) that exhibits better performance in object recognition from images than the popular spatial pyramid, while being more compact in terms of space. In this chapter, the spatial pyramid and the SNAK framework have been adapted to text data. Moreover, these two frameworks have been used for the first time in a text mining task, namely text categorization by topic. The empirical results presented in this chapter indicate that the spatial pyramid and the SNAK framework can help to improve performance in text categorization by topic. However, it is important to mention that spatial information should not be expected to improve performance in every text mining task. For example, computer vision researchers have found that spatial pyramids are not useful in texture classification from images. The spatial pyramid recovers some information about the location of objects in images, but this information is useless in texture analysis, since the patterns that form a certain type of texture are uniform across the entire area covered by the respective texture. In a similar way, spatial information may be found to be worthless in some specific text mining tasks. Hence, the proposed approaches should be evaluated on each individual task before drawing a general conclusion.

## References

Grauman K, Darrell T (2005) The pyramid match kernel: discriminative classification with sets of image features. In: Proceedings of ICCV 2005(2):1458–1465

Ionescu RT, Popescu M (2015) Have a SNAK. Encoding spatial information with the Spatial Non-Alignment Kernel. In: Proceedings of ICIAP 9279:97–108

Joachims T (1998) Text categorization with Suport Vector Machines: learning with many relevant features. In: Proceedings of ECML 137–142

Johnson R, Zhang T (2015) Effective use of word order for text categorization with convolutional neural networks. In: Proceedings of NAACL 103–112

Koniusz P, Mikolajczyk K (2011) Spatial coordinate coding to reduce histogram representations, dominant angle and colour pyramid match. In: Proceedings of ICIP 661–664

Krapac J, Verbeek J, Jurie F (2011). Modeling spatial layout with Fisher vectors for image categorization. In: Proceedings of ICCV 1487–1494

Krizhevsky A, Sutskever I, Hinton GE (2012) ImageNet classification with deep convolutional neural networks. In: Proceedings of NIPS 1106–1114

Lazebnik S, Schmid C, Ponce J (2006) Beyond bags of features: spatial pyramid matching for recognizing natural scene categories. In: Proceedings of CVPR 2:2169–2178

Lewis D (1997) The Reuters-21578 text categorization test collection. http://www.daviddlewis.com/resources/testcollections/reuters21578/

Manning CD, Schütze H (1999) Foundations of statistical natural language processing. MIT Press, Cambridge

Manning CD, Raghavan P, Schütze H (2008) Introduction to information retrieval. Cambridge University Press, New York

Pang B, Lee L, Vaithyanathan S (2002) Thumbs up? sentiment classification using machine learning techniques. In: Proceedings of EMNLP 10:79–86

Porter MF (1980) An algorithm for suffix stripping. Program 14(3):130–137

Pu W, Liu N, Yan S, Yan J, Xie K, Chen Z (2007) Local word bag model for text categorization. In: Proceedings of ICDM 625–630

Sánchez J, Perronnin F, de Campos T (2012) Modeling the spatial layout of images beyond spatial pyramids. Pattern Recognit Lett 33(16):2216–2223. ISSN 0167–8655

Sebastiani F (2002) Machine learning in automated text categorization. ACM Comput Surv 34(1):1–47

Simonyan K, Zisserman A (2014) Very deep convolutional networks for large-scale image recognition. CoRR arXiv:abs/1409.1556

Szegedy C, Liu W, Jia Y, Sermanet P, Reed S, Anguelov D, Erhan D, Vanhoucke V, Rabinovich A (2015) Going deeper with convolutions. In: Proceedings of CVPR

Szeliski R (2010) Computer vision: algorithms and applications, 1st edn. Springer-Verlag New York Inc., New York

Tan C-M, Wang Y-F, Lee C-D (2002) The use of bigrams to enhance text categorization. Inf Process Manag 38(4):529–546

Uijlings JRR, Smeulders AWM, Scha RJH (2009) What is the spatial extent of an object? In: Proceedings of CVPR 770–777
Xue X-B, Zhou Z-H (2009) Distributional features for text categorization. IEEE Trans Knowl Data Eng 21(3):428–442
Yang Y, Liu X (1999) A re-examination of text categorization methods. In: Proceedings of SIGIR 42–49

# Chapter 10
# Conclusions

## 10.1 Discussion and Conclusions

Machine learning is currently a vast area of research with applications in a variety of fields, such as computer vision (Forsyth and Ponce 2002; Krizhevsky et al. 2012; Szeliski 2010; Zhang et al. 2007), computational biology (Dinu and Ionescu 2013; Inza et al. 2010; Leslie et al. 2002), information retrieval (Chifu and Ionescu 2012; Ionescu et al. 2015b; Manning et al. 2008), natural language processing (Lodhi et al. 2002; Popescu and Grozea 2012; Sebastiani 2002), data mining (Han et al. 2011), and many others (Ionescu et al. 2015a). This book has proposed and presented several machine learning methods that are designed for specific tasks that belong to computer vision, computational biology or text mining. The studied tasks range from handwritten digit recognition, texture classification, object recognition, and facial expression recognition, to phylogenetic analysis, sequence alignment, native language identification and text classification by topic. For this broad range of applications several similarity-based learning methods presented in Chap. 2 have been employed. More specifically, this book has studied approaches such as nearest neighbor models, local learning, kernel methods, and clustering methods. The studied methods exhibit state-of-the-art performance levels in the approached tasks. To support this claim, it is important to mention the improved bag of visual words model (Ionescu et al. 2013) that has obtained the fourth place at the Facial Expression Recognition (FER) Challenge of the ICML 2013 Workshop in Challenges in Representation Learning (WREPL), and the system based on string kernels (Popescu and Ionescu 2013) that has ranked third in the closed Native Language Identification Shared Task of the BEA-8 Workshop of NAACL 2013. With the improvements presented in Chap. 8, the string kernels approach currently represents the state-of-the-art method in native language identification.

The applications approached in this book can be divided into two areas that are traditionally considered as different research fields, namely computer vision on one hand and string processing on the other. While computer vision deals with image data,



R.T. Ionescu and M. Popescu, *Knowledge Transfer between Computer Vision and Text Mining*, Advances in Computer Vision and Pattern Recognition,
DOI 10.1007/978-3-319-30367-3_10

string processing refers to the analysis of string data in the form of text documents, DNA strings, and so on. Although at first sight computer vision and string processing seem to be unrelated fields of study, recent results, such as the ones presented in this book, suggest that image and string analysis can be approached in similar ways. Indeed, the concept of treating image and text in a similar fashion has proven to be very fertile for particular applications in computer vision (Duygulu et al. 2002; Farhadi et al. 2010; Leung and Malik 2001; Sadeghi and Farhadi 2011; Sivic et al. 2005) and text mining (Barnard and Johnson 2005; Barnard et al. 2003; Johnson and Zhang 2015; Pu et al. 2007). The concept of treating image and text in a similar manner represents the cornerstone concept of this book. Hence, several methods that are based on this underlying concept have been thoroughly presented in individual chapters. First, a dissimilarity measure for images has been presented in Chap. 4. The dissimilarity measure is inspired from the rank distance measure (Dinu and Manea 2006) for strings. The main concern was to redesign rank distance in order to adapt it from one-dimensional input (strings) to two-dimensional input (digital images). While rank distance is a highly accurate measure for strings, the experiments presented in Chap. 4 suggest that the proposed extension of rank distance to images is very accurate for handwritten digit recognition and texture analysis. Second, some improvements to the popular bag of visual words model have been proposed in Chap. 5. This model is inspired by the bag of words model from text mining and information retrieval. Third, a new distance measure has been introduced in Chap. 7. It was inspired from the image dissimilarity measure presented in Chap. 4. Designed to conform to more general principles and adapted to DNA strings, Local Rank Distance has demonstrated that it can achieve better results than several state-of-the-art methods for DNA sequence analysis. Furthermore, another application of this novel distance measure for strings has been presented in Chap. 8. More precisely, a kernel based on this distance measure has been used for native language identification. Moreover, the intersection kernel, which is widely used in computer vision, has been applied on native language identification in Chap. 8. Lastly, two approaches for including spatial information in the bag of visual words, namely the spatial pyramid and the Spatial Non-Alignment Kernel, have been applied on text data in Chap. 9, improving the results of the bag of words model in the context of text categorization by topic. To summarize, all the approaches presented in this book come to support the concept of treating image and text in a similar manner. However, it must be pointed out that most approaches have to be redesigned or adapted in order to work on different data types. Usually, the amount of work required to transfer a specific concept or method from one domain to the other depends on the place of the respective concept or method in the processing pipeline. The closer the concept is to the raw data type, the harder it is to adapt it for a new data type. This is the case of the distance or similarity measures that work directly on image or text data, such as Local Patch Dissimilarity or Local Rank Distance. The concepts that sit somewhere in the middle of the processing pipeline can be adapted more easily. This is the case of the PQ kernel or the SNAK framework. Finally, it becomes almost trivial to transfer the concepts that are applied at the end of the processing pipeline. For instance, classification methods such as Support Vector Machines or Kernel Ridge

Regression can be directly applied on various classification tasks from computer vision or text mining, without requiring any other changes than parameter tuning. It must be pointed out that, in this book, the trivial cases have not been considered as proper examples of knowledge transfer, although such efforts are always appreciated in literature (Joachims 1998).

Although a significant amount of research has been conducted using the idea of borrowing and adapting concepts from text processing to computer vision, or from computer vision to text processing, the concept of treating image and text in a similar fashion is far from saturated. The methods presented in this book barely scratch the surface on this topic. Being the first book on knowledge transfer between computer vision and text mining, it comes to lay the ground for future exploration. Surely, there are still many concepts and methods studied and applied in a single field, waiting to be discovered and used by researchers in other fields of study. Nonetheless, it should be pointed out that it does not always make sense to consider knowledge transfer. Indeed, there are many approaches and tasks that are very specific to one domain or the other. One such example is the objectness measure (Alexe et al. 2010, 2012), which quantifies how likely it is for an image window to contain an object. In computer vision, the task of identifying image regions that contain objects is not trivial and it requires elaborate methods such as the objectness measure. Trying to develop an elaborate method to determine if a text window (the equivalent of an image window) contains a meaningful concept is really not necessary. This can be achieved simply by eliminating the stop words in the text window.

To conclude, this book represents a strong argument in favor of treating image and text in a similar fashion, a concept that is very promising and truly fertile for some specific applications in computer vision and text mining.

## References

Alexe B, Deselaers T, Ferrari V (2010) What is an object? In: Proceedings of CVPR, pp 73–80, June 2010

Alexe B, Deselaers T, Ferrari V (2012) Measuring the objectness of image windows. IEEE Trans Pattern Anal Mach Intell 34(11):2189–2202

Barnard K, Johnson M (2005) Word sense disambiguation with pictures. Artif Intell 167(1–2):13–30

Barnard K, Duygulu P, Forsyth D, de Freitas N, Blei DM, Jordan MI (2003) Matching words and pictures. J Mach Learn Res 3:1107–1135

Chifu A-G, Ionescu RT (2012) Word sense disambiguation to improve precision for ambiguous queries. Central Eur J Comput Sci 2(4):398–411

Dinu LP, Ionescu RT (2013) Clustering based on median and closest string via rank distance with applications on DNA. Neural Comput Appl 24(1):77–84

Dinu LP, Manea F (2006) An efficient approach for the rank aggregation problem. Theoret Comput Sci 359(1–3):455–461

Duygulu P, Barnard K, de Freitas JFG, Forsyth DA (2002) Object recognition as machine translation: learning a lexicon for a fixed image vocabulary. In: Proceedings of ECCV, pp 97–112

Farhadi A, Hejrati M, Sadeghi MA, Young P, Rashtchian C, Hockenmaier J, Forsyth D (2010) Every picture tells a story: generating sentences from images. In: Proceedings of ECCV, pp 15–29

Forsyth DA, Ponce J (2002) Computer vision: a modern approach. Prentice Hall Professional Technical Reference
Han J, Kamber M, Pei J (2011) Data mining: concepts and techniques, 3rd edn. Morgan Kaufmann Publishers Inc., San Francisco
Inza I, Calvo B, Armañanzas R, Bengoetxea E, Larrañaga P, Lozano JA (2010) Machine learning: an indispensable tool in bioinformatics. Methods Mol Biol (Clifton, N.J.) 593:25–48
Ionescu RT, Popescu M, Grozea C (2013) Local learning to improve bag of visual words model for facial expression recognition. In: Workshop on challenges in representation learning, ICML
Ionescu RT, Popescu AL, Popescu M, Popescu D (2015a) BiomassID: a biomass type identification system for mobile devices. Comput Electron Agric 113:244–253
Ionescu RT, Chifu A-G, Mothe J (2015b) DeShaTo: describing the shape of cumulative topic distributions to rank retrieval systems without relevance judgments. In: Proceedings of SPIRE 9309:75–82
Joachims T (1998) Text categorization with Suport Vector Machines: learning with many relevant features. In: Proceedings of ECML, pp 137–142
Johnson R, Zhang T (2015) Effective use of word order for text categorization with convolutional neural networks. In: Proceedings of NAACL, pp 103–112
Krizhevsky A, Sutskever I, Hinton GE (2012) ImageNet classification with deep convolutional neural networks. In: Proceedings of NIPS, pp 1106–1114
Leslie CS, Eskin E, Noble WS (2002) The spectrum kernel: a string kernel for SVM protein classification. In: Proceedings of Pacific symposium on biocomputing, pp 566–575
Leung T, Malik J (2001) Representing and recognizing the visual appearance of materials using three-dimensional textons. Int J Comput Vis 43(1):29–44
Lodhi H, Saunders C, Shawe-Taylor J, Cristianini N, Watkins CJCH (2002) Text classification using string kernels. J Mach Learn Res 2:419–444
Manning CD, Raghavan P, Schütze H (2008) Introduction to information retrieval. Cambridge University Press, New York
Popescu M, Grozea C (2012) Kernel methods and string kernels for authorship analysis. In: CLEF (Online Working Notes/Labs/Workshop), Sept 2012
Popescu M, Ionescu RT (2013) The story of the characters, the DNA and the native language. In: Proceedings of the Eighth Workshop on Innovative Use of NLP for Building Educational Applications, pp 270–278, June 2013
Pu W, Liu N, Yan S, Yan J, Xie K, Chen Z (2007) Local Word bag model for text categorization. In: Proceedings of ICDM, pp 625–630
Sadeghi MA, Farhadi A (2011) Recognition using visual phrases. In: Proceedings of CVPR, pp 1745–1752
Sebastiani F (2002) Machine learning in automated text categorization. ACM Comput Surv 34(1): 1–47
Sivic J, Russell BC, Efros AA, Zisserman A, Freeman WT (2005) Discovering objects and their localization in images. In: Proceedings of ICCV, pp 370–377
Szeliski R (2010) Computer vision: algorithms and applications, 1st edn. Springer-Verlag New York Inc., New York
Zhang J, Marszalek M, Lazebnik S, Schmid C (2007) Local features and kernels for classification of texture and object categories: a comprehensive study. Int J Comput Vis 73(2):213–238

# Index

**A**
Arabic Learner Corpus, 203–207, 210, 214–217
Artificial intelligence, 1, 3
ASK corpus, 203, 204, 206, 207, 217–220

**B**
Bag of visual words, 3, 4, 6–10, 42, 47, 48, 50, 90, 93, 99–103, 107, 113, 115, 118, 120–125, 128–130, 229, 233, 234, 239, 243, 244
Bag of words, 4, 6, 7, 9, 47, 48, 114, 136, 142, 229–234, 236–239, 244
Bhattacharyya coefficient, 27, 42, 79–81, 85, 89, 91, 95, 103
Bioinformatics, 2, 139, 153, 165
Biomass Texture data set, 86, 87, 89, 95, 96
Birds data set, 62–64, 79, 80, 100, 113–115, 118, 119, 121, 122
Blended spectrum kernel, 199, 210, 220–223
Brodatz data set, 54, 85–87, 89–94

**C**
Class masking problem, 26, 202, 203
Cluster analysis, 1, 8, 16, 30, 32, 140
Clustering, 1, 2, 8, 9, 16, 30, 31, 33, 42, 46, 130, 135, 140, 149, 163, 170, 171, 180, 182–184, 186, 243
Computational biology, 2, 8, 9, 15, 16, 22, 54, 135, 137, 139, 149, 151, 188, 197, 243
Computer vision, 2–8, 15, 19, 30, 41, 42, 45, 47, 49, 53, 99, 101, 103, 136, 153, 193, 198, 203, 229, 230, 232, 234–236, 239, 243–245
Cross-validation, 64, 69, 73, 74, 80, 91, 94, 96, 118, 119, 180, 205–208, 210–212, 214–220

**D**
Deep learning, 3, 6, 8, 42, 49
Dissimilarity measure, 6–9, 19, 53, 60, 62, 63, 74, 75, 81, 85, 88, 90, 92, 95, 96, 160, 200, 244
Distance measure, 7–9, 15–18, 20, 21, 32, 33, 42, 44, 53, 68, 79, 135, 136, 140, 141, 149, 151, 153, 154, 181, 189, 200, 244
Dual representation, 114, 197, 202, 221

**E**
Edit distance, 16, 54, 136, 137, 140, 149, 172, 174, 176, 177, 181, 184, 188, 189
Euclidean distance, 18, 32, 43, 58, 63, 64, 67, 71–75, 77–81, 159



R.T. Ionescu and M. Popescu, *Knowledge Transfer between Computer Vision and Text Mining*, Advances in Computer Vision and Pattern Recognition,
DOI 10.1007/978-3-319-30367-3

## F

Facial expression recognition, 2, 8, 9, 21, 42, 100, 101, 103, 122–125, 127, 129, 243
Feature map, 9, 23, 24, 28, 100, 104–106, 129, 197, 198, 201
Feature space, 15–18, 23, 24, 26, 31, 105, 108, 123, 127, 143, 194, 200–202, 221, 231, 234

## H

Hamming distance, 16, 44, 54, 106, 136, 140, 149, 160, 172, 187, 189
Handwritten digit recognition, 6, 8, 17–19, 26, 61, 79, 243, 244
Hellinger's kernel, 27, 28, 99, 103, 109, 114–116, 118, 119, 121
Hierarchical clustering, 31, 33, 34, 140, 167–171
Hilbert space, 22–24, 200, 202

## I

ICLE corpus, 203, 204, 207, 210–212
Image classification, 8, 41, 46, 47, 75, 81, 83, 96, 99, 101, 129
Image descriptors, 4, 8, 33, 41, 42, 45–48, 81, 101, 108
Image patches, 8, 45–48, 53, 54, 56, 58, 60, 61, 63–72, 74, 78–83, 85, 91, 92, 94–96, 108, 149, 152, 153
Inner product, 15, 22–24, 26, 200, 202
Intersection kernel, 7, 27, 99, 103, 109, 114–116, 118, 120–122, 193, 198, 199, 201, 206, 207, 209–213, 215, 218, 219, 244

## J

Jensen–Shannon kernel, 27, 28, 99, 103, 114–116, 118, 119

## K

K-means, 4, 31–33, 47, 81, 102, 123
Kendall's tau, 16, 99, 104, 106, 140, 149
Kernel alignment, 22, 30, 207–209, 213, 224
Kernel Discriminant Analysis, 22, 25, 26, 63, 80, 88–93, 95, 101, 123, 127, 128, 194, 202, 205, 207–214, 216–219, 224
Kernel function, 9, 22, 23, 26, 28, 29, 49, 100, 105, 107, 109, 130, 197, 199, 201, 202, 231, 235, 236
Kernel matrix, 23, 28–30, 78, 79, 116, 117, 127, 200, 202
Kernel methods, 2, 8, 9, 15, 16, 22, 23, 25, 26, 28, 29, 63, 64, 72, 78–80, 85, 88, 89, 92, 99, 101, 103, 109, 123, 127, 128, 135, 197, 200, 202, 203, 243
Kernel normalization, 28, 105, 114, 199
Kernel Partial Least Squares, 22, 26, 88, 89, 91, 95
Kernel Ridge Regression, 22, 25, 26, 63, 72–75, 78–80, 88, 89, 91, 95, 101, 123, 127, 128, 194, 202, 205, 207, 209–220, 222–224, 237, 238, 245
Kernel trick, 30, 106, 114, 127, 129, 202
Knowledge transfer, 2, 3, 6–10, 47, 49, 245

## L

Language transfer, 10, 193–196, 201, 210, 220, 222, 224, 225
Linear kernel, 26, 28, 73, 99, 103, 109, 114–116, 120, 121, 236
Linkage, 33, 34, 167, 171
Local learning, 7, 8, 16, 20, 21, 53, 54, 63, 75, 78, 79, 96, 101, 123, 125, 127, 129, 130, 243
Local Patch Dissimilarity, 6–8, 53, 54, 56–58, 60–81, 85, 88, 91, 96, 108, 135, 149, 151–153, 158, 234, 244
Local Rank Distance, 9, 10, 34, 135, 140, 141, 149–155, 157, 158, 160, 161, 163–189, 194, 200, 201, 206, 207, 209–213, 215, 217–219, 234, 244
Local Texton Dissimilarity, 8, 53, 54, 81–83, 85, 86, 88–96, 135

## M

Machine learning, 1, 2, 4, 7, 8, 10, 15, 17, 19, 24, 28, 30, 41, 53, 54, 64, 86, 96, 112, 135, 136, 142, 193, 196, 199, 243
MNIST data set, 18, 20, 62–75, 77–79, 96
Multiple kernel learning, 9, 22, 29, 30, 90, 91, 93, 96, 101, 109, 129, 130, 193, 194, 212, 213, 215, 219, 224

## N

Native language identification, 2, 10, 16, 26, 30, 135, 136, 142, 143, 149, 193–197,

199, 200, 202–204, 207, 208, 210, 212, 214, 221, 224, 243, 244
Natural language processing, 2–4, 6, 15, 49, 135–137, 141, 142, 144, 151, 194, 243
Nearest neighbor, 2, 7, 15–21, 45, 46, 62–72, 74–80, 88–92, 95, 96, 243
Neural networks, 1, 6, 41, 49, 61, 232

**O**
Object recognition, 2–5, 7–10, 15, 26, 41, 45, 49, 81, 99–103, 110, 113, 116, 119, 122, 129, 130, 198, 229, 230, 237, 239, 243
Objectness, 9, 49, 100, 110, 111, 114, 130, 245

**P**
P-grams, 10, 16, 49, 142, 144, 153–161, 194–202, 205–207, 209–211, 214, 217, 220–222, 229–231
P-spectrum kernel, 197–199, 201, 205, 206
Parameter tuning, 64, 78, 119, 127, 129, 205, 245
Pascal VOC data set, 4, 100, 112, 114–117, 119–121
Phylogenetic analysis, 2, 8, 9, 135, 140, 141, 149–151, 153, 165–167, 243
Phylogenetic tree, 9, 15, 135, 140, 149, 150, 167–171, 180
Positive semi-definite, 22, 24, 28
PQ kernel, 6, 9, 15, 27, 90, 93, 99–101, 103–107, 110, 114–116, 118, 119, 129, 130, 244
Presence bits kernel, 198, 199, 201, 206, 207, 209–213, 215, 218–223
Pyramid match kernel, 15, 22, 231

**R**
Radial Basis Function kernel, 20, 27, 63, 88, 109, 200, 201
Rank distance, 6–9, 16, 53–56, 58, 60, 100, 108, 135–138, 140, 149, 151–153, 158, 172, 189, 244
Reproducing Kernel Hilbert Space, 22, 24, 25, 202
Reuters-21578 corpus, 236, 238

**S**
Sequence alignment, 2, 139, 140, 149, 151, 153, 161, 167, 172, 174, 187, 188, 243
SIFT descriptors, 4, 41, 46–48, 102, 113, 123–125, 127
Similarity measure, 18, 19, 42, 63, 78, 88, 99, 160, 195, 197, 200, 244
Similarity-based learning, 2, 15, 16, 22, 42, 88, 243
Spatial information, 7, 9, 10, 30, 48, 49, 100, 101, 103, 107, 109, 114, 120–124, 128, 130, 136, 142, 229–232, 234, 238–240, 244
Spatial Non-Alignment Kernel, 7, 9, 10, 27, 49, 100, 101, 103, 107–111, 114, 119–122, 130, 135, 229–232, 234–239, 244
Spatial pyramid, 7, 9, 10, 48, 99, 100, 103, 108, 114–124, 127, 128, 130, 230, 231, 233, 234, 236–240, 244
String kernel, 9, 10, 16, 30, 135, 136, 142, 143, 193–198, 200, 202, 204–220, 224, 225, 243
String processing, 2, 3, 7–9, 135, 243, 244
Supervised learning, 1, 8, 30
Support Vector Machines, 1, 20–22, 25, 26, 41, 63, 72–75, 78–80, 88, 89, 91, 95, 101, 114–120, 123, 127–129, 143, 194, 202, 205, 207, 214–217, 237, 244

**T**
Tangent distance, 15, 19, 42, 44, 45, 75, 77, 78
Term frequency, 16, 232–234, 236
Text categorization, 2, 7, 10, 16, 26, 136, 142, 143, 199, 200, 229, 230, 232, 236–239, 244
Text mining, 2, 6–9, 16, 30, 47, 135, 141, 142, 193, 229, 230, 234, 239, 240, 243–245
Textons, 8, 47, 48, 53, 54, 80–83, 85, 90, 92, 94, 96
Texture classification, 2, 6, 8, 26, 47, 48, 54, 81–83, 85, 86, 88, 90, 93, 95, 96, 239, 243, 244
Time complexity, 8, 19, 53, 54, 58, 60, 63, 65, 79, 96, 106, 107, 155, 157, 164, 165
TOEFL11 corpus, 10, 202–214, 220–223
TOEFL11-Big corpus, 10, 203, 204, 211–214

**U**

UIUCTex data set, 54, 85, 86, 88, 91, 93–95
Unsupervised learning, 1, 30

**V**

Visual word histogram, 7–9, 47, 99, 101, 103, 105, 107, 114, 123, 124, 127, 130, 233